Hyperkahler Manifolds

Hyperkahler Manifolds

by
Misha Verbitsky and Dmitri Kaledin

International Press
www.intlpress.com

Hyperkahler Manifolds
by Misha Verbitsky and Dmitri Kaledin

ISBN 978-1-57146-209-1

Paperback reissue 2010.
Previously published in 1999 under ISBN 1-57146-071-3 (clothbound).

Printed in the United States of America.

Preface

Hyperkähler manifolds have appeared at first within the framework of differential geometry as an example of Riemannian manifolds with holonomy of a special restricted type. However, they have soon exhibited such diverse and unexpected links with various branches of mathematics that now hyperkähler geometry by itself forms a separate research subject. Among the traditional areas fused within this new subject are differential and algebraic geometry of complex manifolds, holomorphic symplectic geometry, geometric representation theory, Hodge theory and many others. The most recent addition to the list is the link between hyperkähler geometry and theoretical physics: it turns out that hyperkähler manifolds play a critical part in the modern version of the string theory, which is in itself the basis for the future unified field theory and quantum gravity.

Perhaps because the structure of a hyperkähler manifold is so rich, such manifolds are quite rare and hard to construct. Thus every new example or a class of examples of hyperkähler manifolds is of considerable interest. The main goal of this book is to describe two recent developments in this area, one dealing with compact hyperkähler manifolds, the other - with a rather general class of non-compact ones. In order to make the presentation as self-contained as possible, we have included much preliminary material and gave an exposition of most of the basic facts of the theory. We believe that this makes it possible to read the book without any prior knowledge of hyperkähler geometry and to use it as an introduction to the subject. On the other hand, it is our hope that the new examples of hyperkähler manifolds constructed here would be of interest to a specialist in the field.

For the detailed description of the new results proved in the book the reader is referred to the introductions to the individual chapters. In this general introduction we restrict ourselves to giving a brief historical overview of the theory of hyperkähler manifolds and indicating the place of our results in the general framework of hyperkähler geometry.

Historical overview

Recall that one can define a Kähler manifold as a Riemannian manifold M equipped with an almost complex structure parallel with respect to the Levi-Civita connection. It is well-known that such an almost complex structure is automatically integrable, thus every Kähler M is a complex manifold. Moreover, the Riemannian metric and the complex structure together define a non-degenerate closed 2-form ω on M, thus making M a symplectic manifold.

The notion of a hyperkähler manifold is obtained from this definition by replacing the field of complex numbers with the algebra of quaternions. A hyperkähler manifold is by definition a Riemannian manifold M equipped with two anticommuting almost complex structures parallel with respect to the Levi-Civita connection. These two almost complex structures generate an action of the quaternion algebra in the tangent bundle to M, which is also parallel. Every quaternion h with $h^2 = -1$ defines an almost complex struture on M. This almost complex structure is parallel, hence integrable. Thus every hyperkähler manifold is canonically complex, and in many different ways.

It is convenient to fix once and for all a quaternion I with $I^2 = -1$ and to consider a hyperkähler manifold M as complex by means of the corresponding complex structure. It is canonically Kähler. Moreover, one can combine the other complex structures on M with the Riemannian metric and obtain, apart from the Kähler 2-form ω, a canonical closed non-degenerate holomorphic 2-form Ω on M. Thus every hyperkähler manifold carries canonical Kähler and holomorphically symplectic structures.

A Riemannian manifold of dimension $4n$ is hyperkähler if and only if its holonomy group is contained in the symplectic group $Sp(n)$. As such, hyperkähler manifolds first appeared in the classification of all possible holonomy groups given by M. Berger [Ber]. The term "hyperkähler manifold" was introduced by E. Calabi in his paper [C], where he also constructed several non-trivial examples of hyperkähler metrics. All of Calabi's examples were non-compact. In fact, all these manifolds were total spaces of cotangent bundles to Kähler manifolds.

At the time of the original paper of Calabi's, it seemed that hyperkähler manifolds are a rather unusual phenomenon, not unlike, for example, sporadic simple groups. However, starting with the beginning of the eighties, there has been a wave of discoveries in the area, and we now know a lot of examples of hyperkähler metrics which occur "in the nature". These examples split naturally into two groups, depending on whether the underlying complex manifold is compact.

A powerful tool for constructing compact hyperkähler manifolds is the famous Calabi-Yau Theorem, which provides a Ricci-flat Kähler metric on every compact manifold of Kähler type with trivial canonial bundle. Its usefulness for the hyperkähler geometry lies in the fact that every hyperkähler manifold is canonically holomorphically symplectic. The converse statement is far from being true: a holomorphically symplectic Kähler manifold does not have to be hyperkähler. However, the converse is true if we require in addition that the hololomorphic symplectic form is parallel with respect to the Levi-Civita connection. It is easy to see that every Kähler manifold equipped with a parallel holomorphic symplectic form is hyperkähler.

In general it is very hard to check whether a holomorphic symplectic form on a compact Kähler manifold is parallel. However, there exists a theorem of S. Bochner's [Boch] which claims that this is always the case when the Kähler metric is Ricci-flat. Since the canonical bundle of a holomorphically symplectic manifold is trivial, indeed, trivialized by the power of the symplectic from, the Calabi-Yau Theorem shows that every compact holomorphically symplectic manifold of Kähler type carries a Ricci-flat Kähler metric. This metric must be hyperkähler by the Bochner Theorem. Thus every compact holomorphically symplectic manifold of Kähler type is hyperkähler.

Well-known examples of compact holomorphically symplectic manifolds of dimension 2 are abelian complex surfaces and K3 surfaces. In higher dimensions non-trivial examples of such manifolds have been given by A. Beauville [Beau], extending earlier results of A. Fujiki [F]. Beauville's examples are the Hilbert schemes of points on an abelian or a K3 surface. All these manifolds are of Kähler type, hence hyperkähler.

Given a compact hyperkähler manifold M, one can consider moduli spaces $\mathcal{M}$ of stable holomorphic vector bundles on M with fixed Chern classes. When M is 4-dimensional, hence either an abelian surface or a K3 surface, the moduli space $\mathcal{M}$ is known to be smooth and hyperkähler (see [Kob] for an excellent exposition of these results). When the Chern classes are such that $\mathcal{M}$ is compact, we obtain in this way a new compact hyperkähler manifold. This situation was studied in detail by S. Mukai in [M1], [M2]. It is now known that some of the compact moduli spaces of stable bundles on a K3 surface are deformationally equivalent to hyperkähler manifolds of the type constructed by Beauville. Conjecturally all these moduli spaces lie in the Beauville's deformation class. We refer the reader to [Huy] for an overview of this subject.

These results were partially generalized to higher dimensions in [V1].

The moduli space of stable bundles on a higher-dimensional hyperkähler manifold is no longer automatically smooth. However, it is a singular hyperkähler variety in the sense of [V2]. This implies, in particular, that it is hyperkähler near every smooth point. Moreover, a singular hyperkähler veriety can be canonically desingularized to a smooth hyperkähler manifold.

An outstanding problem in the theory of compact hyperkähler manifolds is to find an example of such a manifold which would be simply connected and not equivalent deformationally to a product of the ones constructed by Beauville. Several important results ([OGr]) on this subject have appeared recently, but the problem is still not completely closed. The first chapter of the present book describes a different approach to this subject. The results of Mukai and Verbitsky are extended and strengthened in a way that conjecturally leads to the hoped-for examples of compact hyperkähler manifolds belonging to a new deformation class.

Both the Calabi-Yau Theorem and the Bochner Theorem are results of a global nature and cannot be used to construct *non-compact* hyperkähler manifolds. Two general methods are known which can be used for this purpose. The first one uses the link between the theory of hyperkähler manifolds and the holomorphic geometry provided by the notion of the twistor space. The twistor space construction, introduced by R. Penrose, has a long and glorious history. The reader is referred to [HKLR] for a detailed exposition of this subject. Here we only mention that the twistor space X for a $4n$-dimensional hyperkähler manifold M is a holomorphic manifold of complex dimension $2n+2$ canonically associated to M, and that the holomorphic structure on the twistor space X embodies most of the differential-geometric properties of the hyperkähler manifold M.

There exists a theorem which allows one to reconstruct a hyperkähler manifold M from its twistor space X equipped with some additional structures. Therefore, if it is impossible to construct M explicitly, one can construct X instead. In this way one can, for example, construct an infinite-dimensional family of hyperkähler metrics on the vector space $\mathbb{C}^{2n}$ (see [HKLR]).

Another general method of constructing non-compact hyperkähler manifolds is the famous hyperkähler reduction technique introduced by Hitchin [Hi1], Hitchin et al. [HKLR]. It is this technique that led to recent discoveries of hyperkähler structures on many interesting manifolds. Fortunately, the hyperkähler reduction is well-covered in the literature (see, for example, [Hi3]). Therefore we will only list the most important examples of hyperkähler manifolds obtained by this method.

- One of the original examples of hyperkähler manifolds given by Calabi was the total space of the cotangent bundle to a complex projective space. This manifold admits an elementary construction via the hyperkähler reduction.

- Let $\Gamma \in SL(2, \mathbb{C})$ be a finite subgroup. The quotient $\mathbb{C}^2/\Gamma$ has an isolated singularity at 0 which can be blown up to a non-singular complex manifold. This manifold can be equipped with a hyperkähler metric by means of the hyperkähler reduction. The resulting metric is asymptotically locally Euclidean (ALE) at infinity. It was discovered by P.B. Kronheimer [Kr1] and generalized by Kronheimer and H. Nakajima [KN]. Nakajima [N] has recently generalized this example even further and obtained a whole family of hyperkähler manifolds called the quiver varieties.

- Let G be a compact Lie group, and let LG be the (infinite-dimensional) Lie group of maps from the unit cricle S^1 to G. A hyperkähler metric on the quotient LG/G has been constructed by S. Donaldson [D].

- Let S be a compact complex curve, and let $\mathcal{M}$ be the moduli space of bundles on S equipped with a flat connection. Hitchin [Hi2] has constructed a hyperkähler structure on the space $\mathcal{M}$. This construction has been recently generalized by C. Simpson [Sim] to the case when S is an arbitrary projective complex manifold.

A related group of examples is obtained by considering the solutions to a system of ordinary differential equations called the Nahm equations. These equations first appeared in the work of Schmid [Sch] on the variations of Hodge structures. They have been used by Kronheimer [Kr2],[Kr3] to obtain a hyperkähler metric on an orbit in the coadjoint representstion of an arbitrary semisimple complex ie group G. More recently Kronheimer's method was used in papers [BG],[DS] to obtain new examples of hyperkähler manifolds. This method is not directly related to the hyperkähler reduction but shares many features with it. Some of the metrics obtained by reduction can be also constructed via the Nahm equations, and vice versa.

Unfortunately, while hyperkähler reduction is a generous source of new hyperkähler metrics, it can only be pushed so far. One of the problems which seems to lie outside of the scope of this approach is that of constructing a hyperkähler metric on the total space of the cotangent bundle to a non-homogeneous Kähler manifold. The second chapter of the present book describes such a construction. This construction is local and works for an arbitrary Kähler manifold. The methods used are necessarily different from the ones already exploited in the literature and consist of explicit but cumbersome application of the deformation theory in the spirit of Kodaira and Spencer [Kod].

Bibliography

[Beau] A. Beauville, *Variétés Kähleriennes dont la première classe de Chern est nulle*, J. Diff. Geom. **19** (1983), 755–782.

[Ber] M. Berger, *Sur les groupes d'holonomie des variétés à connexion affine et des variétés riemanniennes*, Bull. Soc. Math. France, **83** (1955), 279–330.

[BG] O. Biquard and P. Gauduchon, *Hyperkähler metrics on cotangent bundles of Hermitian symmetric spaces*, in "Proc. of the Special Session an Geometry and Physics held at Aarhus University and of the Summer School at Odense University, Denmark, 1995", Lecture Notes in Pure and Applied Mathematics, volume 184, Marcel Dekker Inc. 1996, 768 pages.

[Boch] S. Bochner, *Curvature in Hermitian metric*, Bull. Amer. Math. Soc. **53** (1947), 179–195.

[C] E. Calabi, *Métriques kählериennes et fibrés holomorphes*, Ann. Ecol. Norm. Sup. **12** (1979), 269–294.

[DS] A. Dancer and A. Swann, *The structure of quaternionic kähler quotients*, in "Proc. of the Special Session an Geometry and Physics held at Aarhus University and of the Summer School at Odense University, Denmark, 1995", Lecture Notes in Pure and Applied Mathematics, volume 184, Marcel Dekker Inc. 1996, 768 pages.

[D] S.K. Donaldson, *Boundary value problems for Yang-Mills fields*, J. Geom. Phys. **8** (1992), 89–122.

[F] A. Fujiki, *On primitively symplectic compact Kähler V-manifolds of dimension four*, in Classification of algebraic and analytic manifolds, Progr. Math. **39** (1983), 71–250.

[Hi1] N.J. Hitchin, *Metrics on moduli spaces*, Proc. Lefschetz Centennial Conference (Mexico City 1984), Contemp. Math. **58**, Part I, AMS, Providence, RI, 1986.

[Hi2] N.J. Hitchin, *The self-duality equations on a Riemann surface*, Proc. London Math. Soc. **55** (1987), 59–126.

[Hi3] N.J. Hitchin, *Hyperkähler manifolds*, Séminaire Bourbaki, Exposé 748, Astérisque, **206** (1992), 137–166.

[HKLR] N.J. Hitchin, A. Karlhede, U. Lindström, M. Roček, *Hyperkähler metrics and supersymmetry*, Comm. Math. Phys (1987).

[Huy] D. Huybrechts, *Birational symplectic manifolds and their deformations*, alg-geom/9601015.

[Kod] K. Kodaira and D. Spencer, *On deformations of complex analytic structures I, II*, Ann. Math. **67** (1958), 328–466.

[Kob] S. Kobayashi, *Differential geometry of complex vector bundles*, Princeton University Press, 1987.

[Kr1] P.B. Kronheimer, *The construction of ALE spaces as hyper-Kaehler quotients*, J. Diff. Geom. **29** (1989), 665–683.

[Kr2] P.B. Kronheimer, *A hyper-Kählerian structure on coadjoint orbits of a semisimple complex group*, J. of LMS, **42** (1990), 193–208.

[Kr3] P.B. Kronheimer, *Instantons and the nilpotent variety*, J. Diff. Geom. **32** (1990), 473–490.

[KN] P.B. Kronheimer and H. Nakajima, *Yang-Mills instantons on ALE gravitational instantons*, Math. Ann. **288** (1990), 263–307.

[M1] S. Mukai, *Symplectic structure of the moduli space of sheaves on abelian or K3 surfaces*, Invent. Math. **77** (1984), 101–116.

[M2] S. Mukai, *Moduli of vector bundles on K3 surfaces, and symplectic manifolds*, Sugaku Expositions **1** (1988), 138–174.

[N] H. Nakajima, *Instantons on ALE spaces, quiver varieties, and Kac-Moody algebras*, Duke Math. J. **76** (1994), 365–416.

[OGr] Kieran G. O'Grady, *Desingularized moduli spaces of sheaves on a K3*, alg-geom/9708009.

[Sim] C. Simpson, *Moduli of representations of the fundamental group of a smooth projective variety*, I: Publ. Math. I.H.E.S. **79** (1994), 47–129; II: Publ. Math. I.H.E.S. **80** (1994), 5–79.

[Sch] W. Schmid, *Variations of Hodge structure: the singularities of the period mapping*, Invent. Math. **22** (1973), 211–319.

[V1] M. Verbitsky, *Hyperholomorphic bundles over a hyperkähler manifold*, alg-geom/9307008 (1993), also published in: Jour. of Alg. Geom. **5** (1996), 633–669.

[V2] M. Verbitsky, *Hypercomplex Varieties*, alg-geom/9703016 (1997) (to appear in Comm. in Anal. and Geom.)

Contents

Part I. Hyperholomorphic sheaves and new examples of hyperkähler manifolds

Misha Verbitsky

Given a compact hyperkähler manifold M and a holomorphic bundle B over M, we consider a Hermitian connection ∇ on B which is compatible with all complex structures on M induced by the hyperkähler structure. Such a connection is unique, because it is Yang-Mills. We call the bundles admitting such connections **hyperholomorphic bundles**. A bundle is hyperholomorphic if and only if its Chern classes c_1, c_2 are $SU(2)$-invariant, with respect to the natural $SU(2)$-action on the cohomology. For several years, it was known that the moduli space of stable hyperholomorphic bundles is singular hyperkähler. More recently, it was proven that singular hyperkähler varieties admit a canonical hyperkähler desingularization. In the present paper, we show that a moduli space of stable hyperholomorphic bundles is compact, given some assumptions on Chern classes of B and hyperkähler geometry of M (we also require $\dim_{\mathbb{C}} M > 2$). Conjecturally, this leads to new examples of hyperkähler manifolds. We develop the theory of hyperholomorphic sheaves, which are (intuitively speaking) coherent sheaves compatible with hyperkähler structure. We show that hyperholomorphic sheaves with isolated singularities can be canonically desingularized by a blow-up. This theory is used to study degenerations of hyperholomorphic bundles.

1 Introduction

For an introduction to basic results and the history of hyperkähler geometry, see [Bes].

This Introduction is independent from the rest of this paper.

1.1 An overview

Examples of hyperkähler manifolds

A Riemannian manifold M is called **hyperkähler** if the tangent bundle of M is equipped with an action of quaternian algebra, and its metric is Kähler with respect to the complex structures I_ι, for all embeddings $\mathbb{C} \stackrel{\iota}{\hookrightarrow} \mathbb{H}$. The complex structures I_ι are called **induced complex structures**; the corresponding Kähler manifold is denoted by (M, I_ι).

For a more formal definition of a hyperkähler manifold, see Definition 2.1. The notion of a hyperkähler manifold was introduced by E. Calabi ([C]).

Clearly, the real dimension of M is divisible by 4. For $\dim_{\mathbb{R}} M = 4$, there are only two classes of compact hyperkähler manifolds: compact tori and K3 surfaces.

Let M be a complex surface and $M^{(n)}$ be its n-th symmetric power, $M^{(n)} = M^n/S_n$. The variety $M^{(n)}$ admits a natural desingularization $M^{[n]}$, called **the Hilbert scheme of points.**

The manifold $M^{[n]}$ admits a hyperkähler metrics whenever the surface M is compact and hyperkähler ([Bea]). This way, Beauville constructed two series of examples of hyperkähler manifolds, associated with a torus (so-called "higher Kummer variety") and a K3 surface. It was conjectured that all compact hyperkähler manifolds M with $H^1(M) = 0$, $H^{2,0}(M) = \mathbb{C}$ are deformationally equivalent to one of these examples. In this paper, we study the deformations of coherent sheaves over higher-dimensional hyperkähler manifolds in order to construct counterexamples to this conjecture. A different approach to the construction of new examples of hyperkähler manifolds is found in the recent paper of K. O'Grady, who studies the moduli of semistable bundles over a K3 surface and resolves the singularities using methods of symplectic geometry ([O'G]).

Hyperholomorphic bundles

Let M be a compact hyperkähler manifold, and I an induced complex structure. It is well known that the differential forms and cohomology of M are equipped with a natural $SU(2)$-action (Lemma 2.5). In [V1], we studied the holomorphic vector bundles F on (M, I) which are compatible with a hyperkähler structure, in the sense that any of the following conditions hold:

$$\begin{array}{ll} \text{(i)} & \text{The bundle } F \text{ admits a Hermitian connection } \nabla \text{ with a} \\ & \text{curvature } \Theta \in \Lambda^2(M, \operatorname{End}(F)) \text{ which is of Hodge type} \\ & (1,1) \text{ with respect to any of induced complex structures.} \\ \text{(ii)} & \text{The bundle } F \text{ is a direct sum of stable bundles, and its} \\ & \text{Chern classes } c_1(F), c_2(F) \text{ are } SU(2)\text{-invariant.} \end{array} \quad (1.1)$$

These conditions are equivalent (Theorem 2.27). Moreover, the connection ∇ of (1.1) (i) is Yang-Mills (Proposition 2.25), and by Uhlenbeck–Yau theorem (Theorem 2.24), it is unique.

A holomorphic vector bundle satisfying any of the conditions of (1.1) is called **hyperholomorphic** ([V1]).

Clearly, a stable deformation of a hyperholomorphic bundle is again a hyperholomorphic bundle. In [V1], we proved that a deformation space of hyperholomorphic bundles is a singular hyperkähler variety. A recent development in the theory of singular hyperkähler varieties ([V-d], [V-d2], [V-d3]) gave a way to desingularize singular hyperkähler manifolds, in a canonical way. It was proven (Theorem 2.16) that a normalization of a singular hyperkähler variety (taken with respect to any induced complex structure I) is a smooth hyperkähler manifold.

This suggested a possibility of constructing new examples of compact hyperkähler manifolds, obtained as deformations of hyperholomorphic bundles. Two problems arise.

Problem 1. The deformation space of hyperholomorphic bundles is *a priori* non-compact and must be compactified.

Problem 2. The geometry of deformation spaces is notoriously hard to study. Even the dimension of a deformation space is difficult to compute, in simplest examples. How to find, for example, the dimension of the deformation space of a tangent bundle, on a Hilbert scheme of points on a K3 surface? The Betti numbers are even more difficult to compute. Therefore, there is no easy way to distinguish a deformation space of hyperholomorphic bundles from already known examples of hyperkähler manifolds.

In this paper, we address Problem 1. Problem 2 can be solved by studying the algebraic geometry of moduli spaces. It turns out that, for a generic deformation of a complex structure, the Hilbert scheme of points on a K3 surface has no closed complex subvarieties ([V5]; see also Theorem 2.17). It is possible to find a 21-dimensional family of deformations of the moduli space Def(B) of hyperholomorphic bundles, with all fibers having complex subvarieties (Lemma 10.28). Using this observation, it is possible to show that Def(B) is a new example of a hyperkähler manifold. Details of this approach are given in Subsection 10.3, and the complete proofs will be given in a forthcoming paper.

It was proven that a Hilbert scheme of a generic K3 surface has no trianalytic subvarieties.[1] Given a hyperkähler manifold M and an appropriate hyperholomorphic bundle B, denote the deformation space of hyperholomorphic connections on B by Def(B). Then the moduli of complex structures on M are locally embedded to a moduli of complex structures on Def(B) (Claim 10.26). Since the dimension of the moduli of complex structures on Def(B) is equal to its second Betti number minus 2 (Theorem 5.9), the second Betti number of Def(B) is no less than the second Betti number of M. The Betti numbers of Beauville's examples of simple hyperkähler manifolds are 23 (Hilbert scheme of points on a K3 surface) and 7 (generalized Kummer variety). Therefore, for M a generic deformation of a Hilbert scheme of points on K3, Def(B) is either a new manifold or a generic deformation of a Hilbert scheme of points on K3. It is easy to construct trianalytic subvarieties of the varieties Def(B), for hyperholomorphic B (see [V2], Appendix for details). This was the motivation of our work on trianalytic subvarieties of the Hilbert scheme of points on a K3 surface ([V5]).

For a generic complex structure on a hyperkähler manifold, all stable bundles are hyperholomorphic ([V2]). Nevertheless, hyperholomorphic bundles over higher-dimensional hyperkähler manifolds are in short supply. In fact, the only example to work with is the tangent bundle and its tensor powers, and their Chern classes are not prime. Therefore, there is no way to insure that a deformation of a stable bundle will remain stable (like it happens, for instance, in the case of deformations of stable bundles of rank 2 with odd first Chern class over a K3 surface). Even worse, a new kind of singularities may appear which never appears for 2-dimensional base manifolds: a deformation of a stable bundle can have a singular reflexization.

[1] Trianalytic subvariety (Definition 2.9) is a closed subset which is complex analytic with respect to any of induced complex structures.

We study the singularities of stable coherent sheaves over hyperkähler manifolds, using Yang-Mills theory for reflexive sheaves developed by S. Bando and Y.-T. Siu ([BS]).

Hyperholomorphic sheaves

A compactification of the moduli of hyperholomorphic bundles is the main purpose of this paper. We require the compactification to be singular hyperkähler. A natural approach to this problem requires one to study the coherent sheaves which are compatible with a hyperkähler structure, in the same sense as hyperholomorphic bundles are holomorphic bundles compatible with a hyperkähler structure. Such sheaves are called **hyperholomorphic sheaves** (Definition 3.11). Our approach to the theory of hyperholomorphic sheaves uses the notion of admissible Yang-Mills connection on a coherent sheaf ([BS]).

The equivalence of conditions (1.1) (i) and (1.1) (ii) is based on Uhlenbeck–Yau theorem (Theorem 2.24), which states that every stable bundle F with $\deg c_1(F) = 0$ admits a unique Yang–Mills connection, that is, a connection ∇ satisfying $\Lambda\nabla^2 = 0$ (see Subsection 2.5 for details). S. Bando and Y.-T. Siu developed a similar approach to the Yang–Mills theory on (possibly singular) coherent sheaves. Consider a coherent sheaf F and a Hermitian metric h on a locally trivial part of $F\Big|_U$. Then h is called admissible (Definition 3.5) if the curvature ∇^2 of the Hermitian connection on $F\Big|_U$ is square-integrable, and the section $\Lambda\nabla^2 \in \operatorname{End}(F\Big|_U)$ is uniformly bounded. The admissible metric is called **Yang-Mills** if $\Lambda\nabla^2 = 0$ (see Definition 3.6 for details). There exists an analogue of Uhlenbeck–Yau theorem for coherent sheaves (Theorem 3.8): a stable sheaf admits a unique admissible Yang–Mills metric, and conversely, a sheaf admitting a Yang–Mills metric is a direct sum of stable sheaves with the first Chern class of zero degree.

A coherent sheaf F is called **reflexive** if it is isomorphic to its second dual sheaf F^{**}. The sheaf F^{**} is always reflexive, and it is called **a reflexization** of F (Definition 3.1).

Applying the arguments of Bando and Siu to a reflexive coherent sheaf F over a hyperkähler manifold (M, I), we show that the following conditions are equivalent (Theorem 3.19).

(i) The sheaf F is stable and its Chern classes $c_1(F)$, $c_2(F)$ are $SU(2)$-invariant

(ii) F admits an admissible Yang–Mills connection, and its curvature is of type (1,1) with respect to all induced complex structures.

A reflexive sheaf satisfying any of the these conditions is called **reflexive stable hyperholomorphic**. An arbitrary torsion-free coherent sheaf is called **stable hyperholomorphic** if its reflexization is hyperholomorphic, and its second Chern class is $SU(2)$-invariant, and **semistable hyperholomorphic** if it is a successive extension of stable hyperholomorphic sheaves (see Definition 3.11 for details).

This paper is dedicated to the study of hyperholomorphic sheaves.

Deformations of hyperholomorphic sheaves

By Proposition 2.14, for an induced complex structure I of general type, **all** coherent sheaves are hyperholomorphic. However, the complex structures of general type are never algebraic, and in complex analytic situation, the moduli of coherent sheaves are, generaly speaking, non-compact. We study the flat deformations of hyperholomorphic sheaves over (M, I), where I is an algebraic complex structure.

A priori, a flat deformation of a hyperholomorphic sheaf will be no longer hyperholomorphic. We show that for some algebraic complex structures, called **C-restricted complex structures**, a flat deformation of a hyperholomorphic sheaf remains hyperholomorphic (Theorem 5.14). This argument is quite convoluted, and takes two sections (Sections 4 and 5).

Further on, we study the local structure of stable reflexive hyperholomorphic sheaves with isolated singularities. We prove the Desingularization Theorem for such hyperholomorphic sheaves (Theorem 6.1). It turns out that such a sheaf can be desingularized by a single blow-up. The proof of this result is parallel to the proof of Desingularization Theorem for singular hyperkähler varieties (Theorem 2.16).

The main idea of the desingularization of singular hyperkähler varieties ([V-d2]) is the following. Given a point x on a singular hyperkähler variety M and an induced complex structure I, the complex variety (M, I) admits a local $\mathbb{C}^*$-action which preserves x and acts as a dilatation on the Zariski tangent space of x. Here we show that any stable hyperholomorphic sheaf F is equivariant with respect to this $\mathbb{C}^*$-action (Theorem 6.6, Definition 6.11). Then an elementary algebro-geometric argument (Proposition 6.12) implies that F is desingularized by a blow-up.

Using the desingularization of hyperholomorphic sheaves, we prove that a hyperholomorphic deformation of a hyperholomorphic bundle is again a bundle (Theorem 9.3), assuming that it has isolated singularities. The proof of this result is conceptual but quite difficult, it takes 3 sections (Sections 7–9), and uses arguments of quaternionic-Kähler geometry ([Sw], [N2]) and twistor transform ([KV]).

In our study of deformations of hyperholomorphic sheaves, we usually assume that a deformation of a hyperholomorphic sheaf over (M, I) is again hyperholomorphic, i. e. that an induced complex structure I is C-restricted, for C sufficiently big (Definition 5.1). Since C-restrictness is a tricky condition, it is preferable to get rid of it. For this purpose, we use the theory of twistor paths, developed in [V3-bis], to show that the moduli spaces of hyperholomorphic sheaves are real analytic equivalent for different complex structures I on M (Theorem 10.14). This is done as follows.

A hyperkähler structure on M admits a 2-dimensional sphere of induced complex structures. This gives a rational curve in the moduli space *Comp* of complex structures on M, so-called **twistor curve**. A sequence of such rational curves connect any two points of *Comp* (Theorem 10.4). A sequence of connected twistor curves is called **a twistor path**. If the intersection points of these curves are generic, the twistor path is called **admissible** (Definition 10.6). It is known (Theorem 10.8) that an admissible twistor path induces a real analytic isomorphism of the moduli spaces of hyperholomorphic bundles. There exist admissible twistor paths connecting any two complex structures (Claim 10.13). Thus, if we prove that the moduli of deformations of hyperholomorphic bundles are compact for one generic hyperkähler structure, we prove a similar result for all generic hyperkähler structures (Theorem 10.14). Applying this argument to the moduli of deformations of a tangent bundle, we obtain the following theorem.

Theorem 1.1: Let M be a Hilbert scheme of points on a K3 surface, $\dim_{\mathbb{H}}(M) > 1$ and $\mathcal{H}$ a generic hyperkähler structure on M. Assume that for all induced complex structures I, except at most a finite many of, all semistable bundle deformations of the tangent bundle $T(M, I)$ are stable. Then, for all complex structures J on M and all generic polarizations ω on (M, J), the deformation space $\mathcal{M}_{J,\omega}(T(M, J))$ is singular hyperkähler and compact, and admits a smooth compact hyperkähler desingularization.

Proof: This is Theorem 10.20. ∎

In the course of this paper, we develop the theory of C-restricted complex structures (Sections 4 and 5) and another theory, which we called **the Swann's formalism for vector bundles** (Sections 7 and 8). These themes are of independent interest. We give a separate introduction to C-

restricted complex structures (Subsection 1.2) and Swann's formalism (Subsection 1.3).

1.2 C-restricted complex structures: an introduction

This part of the Introduction is highly non-precise. Our purpose is to clarify the intuitive meaning of C-restricted complex structure.

Consider a compact hyperkähler manifold M, which is **simple** (Definition 2.7), that is, satisfies $H^1(M) = 0$, $H^{2,0}(M) = \mathbb{C}$.

A **reflexive hyperholomorphic sheaf** is by definition a semistable sheaf which has a filtration of stable sheaves with $SU(2)$-invariant c_1 and c_2. A **hyperholomorphic sheaf** is a torsion-free sheaf which has hyperholomorphic reflexization and has $SU(2)$-invariant c_2 (Definition 3.11). If the complex structure I is of general type, all coherent sheaves are hyperholomorphic (Definition 2.13, Proposition 2.14), because all integer (p,p)-classes are $SU(2)$-invariant. However, for generic complex structures I, the corresponding complex manifold (M, I) is never algebraic. If we wish to compactify the moduli of holomorphic bundles, we need to consider algebraic complex structures, and if we want to stay in hyperholomorphic category, the complex structures must be generic. This paradox is reconciled by considering the C-restricted complex structures (Definition 5.1).

Given a generic hyperkähler structure $\mathcal{H}$, consider an algebraic complex structure I with $Pic(M, I) = \mathbb{Z}$. The group of rational (p,p)-cycles has form

$$\begin{aligned} H_I^{p,p}(M,\mathbb{Q}) =& H^{2p}(M,\mathbb{Q})^{SU(2)} \oplus a \cdot H^{2p}(M,\mathbb{Q})^{SU(2)} \\ & a^2 \cdot \oplus H^{2p}(M,\mathbb{Q})^{SU(2)} \oplus ... \end{aligned} \tag{1.2}$$

where a is a generator of $Pic(M, I) \subset H_I^{p,p}(M,\mathbb{Z})$ and $H^{2p}(M,\mathbb{Q})^{SU(2)}$ is the group of rational $SU(2)$-invariant cycles. This decomposition follows from an explicit description of the algebra of cohomology given by Theorem 4.6. Let

$$\Pi : H_I^{p,p}(M,\mathbb{Q}) \longrightarrow a \cdot H^{2p}(M,\mathbb{Q})^{SU(2)} \oplus a^2 \cdot H^{2p}(M,\mathbb{Q})^{SU(2)} \oplus ...$$

be the projection onto non-$SU(2)$-invariant part. Using Wirtinger's equality, we prove that a fundamental class $[X]$ of a complex subvariety $X \subset (M, I)$ is $SU(2)$-invariant unless $\deg \Pi([X]) \neq 0$ (Proposition 2.11). A similar result holds for the second Chern class of a stable bundle (Corollary 3.24,).

A C-restricted complex structure is, heuristically, a structure for which the decomposition (1.2) folds, and $\deg a > C$. For a C-restricted complex structure I, and a fundamental class $[X]$ of a complex subvariety $X \subset (M, I)$

of complex codimension 2, we have $\deg[X] > C$ or X is trianalytic. A version of Wirtinger's inequality for vector bundles (Corollary 3.24) implies that a stable vector bundle B over (M, I) is hyperholomorphic, unless $|\deg c_2(B)| > C$. Therefore, over a C-restricted (M, I), all torsion-free semistable coherent sheaves with bounded degree of the second Chern class are hyperholomorphic (Theorem 5.14).

The utility of C-restricted induced complex structures is that they are algebraic, but behave like generic induced complex structures with respect to the sheaves F with low $|\deg c_2(F)|$ and $|\deg c_1(F)|$.

We prove that a generic hyperkähler structure admits C-restricted induced complex structures for all C, and the set of C-restricted induced complex structures is dense in the set of all induced complex structures (Theorem 5.13). We prove this by studying the algebro-geometric properties of the moduli of hyperkähler structures on a given hyperkähler manifold (Subsection 5.2).

1.3 Quaternionic-Kähler manifolds and Swann's formalism

Quaternionic-Kähler manifolds (Subsection 7.3) are a beautiful subject of Riemannian geometry. We are interested in these manifolds because they are intimately connected with singularities of hyperholomorphic sheaves. A stable hyperholomorphic sheaf is equipped with a natural connection, which is called **hyperholomorphic connection**. By definition, a hyperholomorphic connection on a torsion-free coherent sheaf is a connection ∇ defined outside of singularities of F, with square-integrable curvature ∇^2 which is an $SU(2)$- invariant 2-form (Definition 3.15). We have shown that a stable hyperholomorphic sheaf admits a hyperholomorphic connection, and conversely, a reflexive sheaf admitting a hyperholomorphic connection is a direct sum of stable hyperholomorphic sheaves (Theorem 3.19).

Consider a reflexive sheaf F over (M, I) with an isolated singularity in $x \in M$. Let ∇ be a hyperholomorphic connection on F. We prove that F can be desingularized by a blow-up of its singular set. In other words, for $\pi : \widetilde{M} \longrightarrow (M, I)$ a blow-up of $x \in M$, the pull-back $\pi^* F$ is a bundle over $\widetilde{M}$.

Consider the restriction $\pi^* F\big|_C$ of $\pi^* F$ to the blow-up divisor

$$C = \mathbb{P}T_x M \cong \mathbb{C}P^{2n-1}.$$

To be able to deal with the singularities of F effectively, we need to prove that the bundle $\pi^* F\big|_C$ is semistable and satisfies $c_1\left(\pi^* F\big|_C\right) = 0$.

The following intuitive picture motivated our work with bundles over quaternionic-Kähler manifolds. The manifold $C = \mathbb{P}T_xM$ is has a quaternionic structure, which comes from the $SU(2)$-action on T_xM. We know that bundles which are compatible with a hyperkähler structure (hyperholomorphic bundles) are (semi-)stable. If we were able to prove that the bundle $\pi^*F\Big|_C$ is in some way compatible with quaternionic structure on C, we could hope to prove that it is (semi-)stable.

To give a precise formulation of these heuristic arguments, we need to work with the theory of quaternionic-Kähler manifolds, developed by Berard Bergery and Salamon ([Sal]). A quaternionic-Kähler manifold (Definition 7.8) is a Riemannian manifold Q equipped with a bundle W of algebras acting on its tangent bundle, and satisfying the following conditions. The fibers of W are (non-canonically) isomorphic to the quaternion algebra, the map $W \hookrightarrow \operatorname{End}(TQ)$ is compatible with the Levi-Civita connection, and the unit quaternions $h \in W$ act as orthogonal automorphisms on TQ. For each quaternionic-Kähler manifold Q, one has a twistor space $\operatorname{Tw}(Q)$ (Definition 7.10), which is a total space of a spherical fibration consisting of all $h \in W$ satisfying $h^2 = -1$. The twistor space is a complex manifold ([Sal]), and it is Kähler unless W is flat, in which case Q is hyperkähler. Further on, we shall use the term "quaternionic-Kähler" for manifolds with non-trivial W.

Consider the twistor space $\operatorname{Tw}(M)$ of a hyperkähler manifold M, equipped with a natural map

$$\sigma : \operatorname{Tw}(M) \longrightarrow M.$$

Let (B, ∇) be a bundle over M equipped with a hyperholomorphic connection. A pullback $(\sigma^*B, \sigma^*\nabla)$ is a holomorphic bundle on $\operatorname{Tw}(M)$ (Lemma 7.2), that is, the operator $\sigma^*\nabla^{0,1}$ is a holomorphic structure operator on σ^*B. This correspondence is called **the direct twistor transform**. It is invertible: from a holomorphic bundle $(\sigma^*B, \sigma^*\nabla^{0,1})$ on $\operatorname{Tw}(M)$ it is possible to reconstruct (B, ∇), which is unique ([KV]; see also Theorem 7.3).

A similar construction exists on quaternionic-Kähler manifolds, due to T. Nitta ([N1], [N2]). A bundle (B, ∇) on a quaternionic-Kähler manifold Q is called **a B_2-bundle** if its curvature ∇^2 is invariant with respect to the adjoint action of $\mathbb{H}^*$ on $\Lambda^2(M, \operatorname{End}(B))$ (Definition 7.12). An analogue of direct and inverse transform exists for B_2-bundles (Theorem 7.14). Most importantly, T. Nitta proved that on a quaternionic-Kähler manifold of positive scalar curvature a twistor transform of a B_2-bundle is a Yang-Mills bundle on $\operatorname{Tw}(Q)$ (Theorem 7.17). This implies that a twistor transform of a Hermitian B_2-bundle is a direct sum of stable bundles with $\deg c_1 = 0$.

In the situation described in the beginning of this Subsection, we have a manifold $C = \mathbb{P}T_xM \cong \mathbb{C}P^{2n-1}$ which is a twistor space of a quaternionic projective space

$$\mathbb{P}_{\mathbb{H}}T_xM = \left(T_xM\backslash 0\right)/\mathbb{H}^* \cong \mathbb{H}P^n.$$

To prove that $\pi^*F\Big|_C$ is stable, we need to show that $\pi^*F\Big|_C$ is obtained as twistor transform of some Hermitian B_2-bundle on $\mathbb{P}_{\mathbb{H}}T_xM$.

This is done using an equivalence between the category of quaternionic-Kähler manifolds of positive scalar curvature and the category of hyperkähler manifolds equipped with a special type of $\mathbb{H}^*$-action, constructed by A. Swann ([Sw]). Given a quaternionic-Kähler manifold Q, we consider a principal bundle $\mathcal{U}(Q)$ consisting of all quaternion frames on Q (7.4). Then $\mathcal{U}(Q)$ is fibered over Q with a fiber $\mathbb{H}/\{\pm 1\}$. It is easy to show that $\mathcal{U}(Q)$ is equipped with an action of quaternion algebra in its tangent bundle. A. Swann proved that if Q has with positive scalar curvature, then this action of quaternion algebra comes from a hyperkähler structure on $\mathcal{U}(M)$ (Theorem 7.24).

The correspondence $Q \longrightarrow \mathcal{U}(Q)$ induces an equivalence of appropriately defined categories (Theorem 7.25). We call this construction **Swann's formalism.**

The twistor space $\operatorname{Tw}(\mathcal{U}(Q))$ of the hyperkähler manifold $\mathcal{U}(Q)$ is equipped with a holomorphic action of $\mathbb{C}^*$. Every B_2-bundle corresponds to a $\mathbb{C}^*$-invariant holomorphic bundle on $\operatorname{Tw}(\mathcal{U}(Q))$ and this correspondence induces an equivalence of appropriately defined categories, called **Swann's formalism for budnles** (Theorem 8.5). Applying this equivalence to the $\mathbb{C}^*$-equivariant sheaf obtained as an associate graded sheaf of a hyperholomorphic sheaf, we obtain a B_2 bundle on $\mathbb{P}_{\mathbb{H}}T_xM$, and $\pi^*F\Big|_C$ is obtained from this B_2-bundle by a twistor transform.

The correspondence between B_2-bundles on Q and $\mathbb{C}^*$-invariant holomorphic bundles on $\operatorname{Tw}(\mathcal{U}(Q))$ is an interesting geometric phenomenon which is of independent interest. We construct it by reduction to $\dim Q = 0$, where it follows from an explicit calculation involving 2-forms over a flat manifold of real dimension 4.

1.4 Contents

The paper is organized as follows.

- Section 1 is an introduction. It is independent from the rest of this apper.

- Section 2 is an introduction to the theory of hyperkähler manifolds. We give a compenduum of results from hyperkähler geometry which are due to F. Bogomolov ([Bo]) and A. Beauville ([Bea]), and give an introduction to the results of [V1], [V-d3], [V2(II)].

- Section 3 contains a definition and basic properties of hyperholomorphic sheaves. We prove that a stable hyperholomorphic sheaf admits a hyperholomorphic connection, and conversely, a reflexive sheaf admitting a hyperholomorphic connection is stable hyperholomorphic (Theorem 3.19). This equivalence is constructed using Bando-Siu theory of Yang–Mills connections on coherent sheaves.

  We prove an analogue of Wirtinger's inequality for stable sheaves (Corollary 3.24), which states that for any induced complex structure $J \neq \pm I$, and any stable reflexive sheaf F on (M, I), we have

$$\deg_I\left(2c_2(F) - \frac{r-1}{r}c_1(F)^2\right) \geqslant \left|\deg_J\left(2c_2(F) - \frac{r-1}{r}c_1(F)^2\right)\right|,$$

  and the equality holds if and only if F is hyperholomorphic.

- Section 4 contains the preliminary material used for the study of C-restricted complex structures in Section 5. We give an exposition of various algebraic structures on the cohomology of a hyperkähler manifold, which were discovered in [V0] and [V3]. In the last Subsection, we apply the Wirtinger's inequality to prove that the fundamental classes of complex subvarieties and c_2 of stable reflexive sheaves satisfy a certain set of axioms. Cohomology classes satisfying these axioms are called classes of CA-type. This definition simplifies the work on C-restricted complex structures in Section 5.

- In Section 5 we define C-restricted complex structures and prove the following. Consider a compact hyperkähler manifold and an $SU(2)$-invariant class $a \in H^4(M)$. Then for all C-restricted complex structures I, with $C > \deg a$, and all semistable sheaves I on (M, I) with $c_2(F) = a$, the sheaf F is hyperholomorphic (Theorem 5.14). This is used to show that a deformation of a hyperholomorphic sheaf is again hyperholomorphic, over (M, I) with I a C-restricted complex structure, $c > \deg c_2(F)$.

  We define the moduli space of hyperkähler structures, and show that for a dense set $\mathcal{C}$ of hyperkähler structures, all $\mathcal{H} \in \mathcal{C}$ admit a dense set of C-induced complex structures, for all $C \in \mathbb{R}$ (Theorem 5.13).

- In Section 6 we give a proof of Desingularization Theorem for stable reflexive hyperholomorphic sheaves with isolated singularities (Theorem 6.1). We study the natural $\mathbb{C}^*$-action on a local ring of a hyperkähler manifold defined in [V-d2]. We show that a sheaf F admitting a hyperholomorphic connection is equivariant with respect to this $\mathbb{C}^*$-action. Then F can be desingularized by a blow-up, because any $\mathbb{C}^*$-equivariant sheaf with an isolated singularity can be desingularized by a blow-up (Proposition 6.12).

- Section 7 is a primer on twistor transform and quaternionic-Kähler geometry. We give an exposition of the works of A. Swann ([Sw]), T. Nitta ([N1], [N2]) on quaternionic-Kähler manifolds and explain the direct and inverse twistor transform over hyperkähler and quaternionic--Kähler manifolds.

- Section 8 gives a correspondence between B_2-bundles on a quaternionic--Kähler manifold, and $\mathbb{C}^*$-equivariant holomorphic bundles on the twistor space of the corresponding hyperkähler manifold constructed by A. Swann. This is called "Swann's formalism for vector bundles". We use this correspondence to prove that an associate graded sheaf of a hyperholomorphic sheaf is equipped with a natural connection which is compatible with quaternions. This implies polystability of the bundle $\pi^* F\Big|_C$ (see Subsection 1.3).

- In Section 9, we use the polystability of the bundle $\pi^* F\Big|_C$ to show that a hyperholomorphic deformation of a hyperholomorphic bundle is again a bundle. Together with results on C-restricted complex structures and Maruyama's compactification ([Ma2]), this implies that the moduli of semistable bundles are compact, under conditions of C-restrictness and non-existence of trianalytic subvarieties (Theorem 9.11).

- In Section 10, we apply these results to the hyperkähler geometry. Using the desingularization theorem for singular hyperkähler manifolds (Theorem 2.16), we prove that the moduli of stable deformations of a hyperholomorphic bundle has a compact hyperkähler desingularization (Theorem 10.17). We give an exposition of the theory of twistor paths, which allows one to identify the categories of stable bundles for different Kähler structures on the same hyperkähler manifold (Theorem 10.8). These results allow one to weaken the conditions necessary for compactness of the moduli spaces of vector bundles. Finally, we give a conjectural exposition of how these results can be used to obtain new examples of compact hyperkähler manifolds.

2 Hyperkähler manifolds

2.1 Hyperkähler manifolds

This subsection contains a compression of the basic and best known results and definitions from hyperkähler geometry, found, for instance, in [Bes] or in [Bea].

Definition 2.1: ([Bes]) A **hyperkähler manifold** is a Riemannian manifold M endowed with three complex structures I, J and K, such that the following holds.

(i) the metric on M is Kähler with respect to these complex structures and

(ii) I, J and K, considered as endomorphisms of a real tangent bundle, satisfy the relation $I \circ J = -J \circ I = K$.

The notion of a hyperkähler manifold was introduced by E. Calabi ([C]).

Clearly, a hyperkähler manifold has a natural action of the quaternion algebra $\mathbb{H}$ in its real tangent bundle TM. Therefore its complex dimension is even. For each quaternion $L \in \mathbb{H}$, $L^2 = -1$, the corresponding automorphism of TM is an almost complex structure. It is easy to check that this almost complex structure is integrable ([Bes]).

Definition 2.2: Let M be a hyperkähler manifold, and L a quaternion satisfying $L^2 = -1$. The corresponding complex structure on M is called **an induced complex structure.** The M, considered as a Kähler manifold, is denoted by (M, L). In this case, the hyperkähler structure is called **compatible with the complex structure** L.

Let M be a compact complex manifold. We say that M is **of hyperkähler type** if M admits a hyperkähler structure compatible with the complex structure.

Definition 2.3: Let M be a complex manifold and Θ a closed holomorphic 2-form over M such that $\Theta^n = \Theta \wedge \Theta \wedge ...$, is a nowhere degenerate section of a canonical class of M ($2n = dim_{\mathbb{C}}(M)$). Then M is called **holomorphically symplectic.**

Let M be a hyperkähler manifold; denote the Riemannian form on M by $< \cdot, \cdot >$. Let the form $\omega_I :=< I(\cdot), \cdot >$ be the usual Kähler form which is closed and parallel (with respect to the Levi-Civita connection). Analogously defined forms ω_J and ω_K are also closed and parallel.

A simple linear algebraic consideration ([Bes]) shows that the form $\Theta := \omega_J + \sqrt{-1}\omega_K$ is of type $(2,0)$ and, being closed, this form is also holomorphic. Also, the form Θ is nowhere degenerate, as another linear algebraic argument shows. It is called **the canonical holomorphic symplectic form of a manifold M.** Thus, for each hyperkähler manifold M, and an induced complex structure L, the underlying complex manifold (M, L) is holomorphically symplectic. The converse assertion is also true:

Theorem 2.4: ([Bea], [Bes]) Let M be a compact holomorphically symplectic Kähler manifold with the holomorphic symplectic form Θ, a Kähler class $[\omega] \in H^{1,1}(M)$ and a complex structure I. Let $n = \dim_{\mathbb{C}} M$. Assume that $\int_M \omega^n = \int_M (Re\Theta)^n$. Then there is a unique hyperkähler structure $(I, J, K, (\cdot, \cdot))$ over M such that the cohomology class of the symplectic form $\omega_I = (\cdot, I\cdot)$ is equal to $[\omega]$ and the canonical symplectic form $\omega_J + \sqrt{-1}\,\omega_K$ is equal to Θ.

Theorem 2.4 follows from the conjecture of Calabi, proven by Yau ([Y]). ■

Let M be a hyperkähler manifold. We identify the group $SU(2)$ with the group of unitary quaternions. This gives a canonical action of $SU(2)$ on the tangent bundle, and all its tensor powers. In particular, we obtain a natural action of $SU(2)$ on the bundle of differential forms.

Lemma 2.5: The action of $SU(2)$ on differential forms commutes with the Laplacian.

Proof: This is Proposition 1.1 of [V2(II)]. ■

Thus, for compact M, we may speak of the natural action of $SU(2)$ in cohomology.

Further in this article, we use the following statement.

Lemma 2.6: Let ω be a differential form over a hyperkähler manifold M. The form ω is $SU(2)$-invariant if and only if it is of Hodge type (p, p) with respect to all induced complex structures on M.

Proof: This is [V1], Proposition 1.2. ∎

2.2 Simple hyperkähler manifolds

Definition 2.7: ([Bea]) A connected simply connected compact hyperkähler manifold M is called **simple** if M cannot be represented as a product of two hyperkähler manifolds:

$$M \neq M_1 \times M_2, \text{ where } dim\, M_1 > 0 \text{ and } dim\, M_2 > 0$$

Bogomolov proved that every compact hyperkähler manifold has a finite covering which is a product of a compact torus and several simple hyperkähler manifolds. Bogomolov's theorem implies the following result ([Bea]):

Theorem 2.8: Let M be a compact hyperkähler manifold. Then the following conditions are equivalent.

(i) M is simple

(ii) M satisfies $H^1(M, \mathbb{R}) = 0$, $H^{2,0}(M) = \mathbb{C}$, where $H^{2,0}(M)$ is the space of $(2, 0)$-classes taken with respect to any of induced complex structures.

∎

2.3 Trianalytic subvarieties in hyperkähler manifolds.

In this subsection, we give a definition and basic properties of trianalytic subvarieties of hyperkähler manifolds. We follow [V2(II)], [V-d2].

Let M be a compact hyperkähler manifold, $\dim_{\mathbb{R}} M = 2m$.

Definition 2.9: Let $N \subset M$ be a closed subset of M. Then N is called **trianalytic** if N is a complex analytic subset of (M, L) for any induced complex structure L.

Let I be an induced complex structure on M, and $N \subset (M, I)$ be a closed analytic subvariety of (M, I), $dim_{\mathbb{C}} N = n$. Consider the homology class represented by N. Let $[N] \in H^{2m-2n}(M)$ denote the Poincare dual cohomology class, so called **fundamental class** of N. Recall that the hyperkähler structure induces the action of the group $SU(2)$ on the space $H^{2m-2n}(M)$.

Theorem 2.10: Assume that $[N] \in H^{2m-2n}(M)$ is invariant with respect to the action of $SU(2)$ on $H^{2m-2n}(M)$. Then N is trianalytic.

Proof: This is Theorem 4.1 of [V2(II)]. ■

The following assertion is the key to the proof of Theorem 2.10 (see [V2(II)] for details).

Proposition 2.11: (Wirtinger's inequality) Let M be a compact hyperkähler manifold, I an induced complex structure and $X \subset (M, I)$ a closed complex subvariety for complex dimension k. Let J be an induced complex structure, $J \neq \pm I$, and ω_I, ω_J the associated Kähler forms. Consider the numbers

$$\deg_I X := \int_X \omega_I^k, \quad \deg_J X := \int_X \omega_J^k$$

(these numbers are called **degree of a subvariety** X **with respect to** I, J. Then $\deg_I X \geqslant |\deg_J X|$, and the inequality is strict unless X is trianalytic.

■

Remark 2.12: Trianalytic subvarieties have an action of quaternion algebra in the tangent bundle. In particular, the real dimension of such subvarieties is divisible by 4.

Definition 2.13: Let M be a complex manifold admitting a hyperkähler structure $\mathcal{H}$. We say that M is **of general type** or **generic** with respect to $\mathcal{H}$ if all elements of the group

$$\bigoplus_p H^{p,p}(M) \cap H^{2p}(M, \mathbb{Z}) \subset H^*(M)$$

are $SU(2)$-invariant. We say that M is **Mumford–Tate generic** if for all $n \in \mathbb{Z}^{>0}$, all the cohomology classes

$$\alpha \in \bigoplus_p H^{p,p}(M^n) \cap H^{2p}(M^n, \mathbb{Z}) \subset H^*(M^n)$$

are $SU(2)$-invariant. In other words, M is Mumford–Tate generic if for all $n \in \mathbb{Z}^{>0}$, the n-th power M^n is generic. Clearly, Mumford–Tate generic implies generic.

Proposition 2.14: Let M be a compact manifold, $\mathcal{H}$ a hyperkähler structure on M and S be the set of induced complex structures over M. Denote by $S_0 \subset S$ the set of $L \in S$ such that (M, L) is Mumford-Tate generic with respect to $\mathcal{H}$. Then S_0 is dense in S. Moreover, the complement $S \backslash S_0$ is countable.

Proof: This is Proposition 2.2 from [V2(II)] ∎

Theorem 2.10 has the following immediate corollary:

Corollary 2.15: Let M be a compact holomorphically symplectic manifold. Assume that M is of general type with respect to a hyperkähler structure $\mathcal{H}$. Let $S \subset M$ be closed complex analytic subvariety. Then S is trianalytic with respect to $\mathcal{H}$.

∎

In [V-d3], [V-d], [V-d2], we gave a number of equivalent definitions of a singular hyperkähler and hypercomplex variety. We refer the reader to [V-d2] for the precise definition; for our present purposes it suffices to say that all trianalytic subvarieties are hyperkähler varieties. The following Desingularization Theorem is very useful in the study of trianalytic subvarieties.

Theorem 2.16: ([V-d2]) Let M be a hyperkähler variety, and I an induced complex structure. Consider the normalization

$$\widetilde{(M, I)} \xrightarrow{n} (M, I)$$

of (M, I). Then $\widetilde{(M, I)}$ is smooth and has a natural hyperkähler structure $\mathcal{H}$, such that the associated map $n: \widetilde{(M, I)} \longrightarrow (M, I)$ agrees with $\mathcal{H}$. Moreover, the hyperkähler manifold $\widetilde{M} := \widetilde{(M, I)}$ is independent from the choice of induced complex structure I.

∎

Let M be a K3 surface, and $M^{[n]}$ be a Hilbert scheme of points on M. Then $M^{[n]}$ admits a hyperkähler structure ([Bea]). In [V5], we proved the following theorem.

Theorem 2.17: Let M be a complex K3 surface without automorphisms. Assume that M is Mumford-Tate generic with respect to some hyperkähler structure. Consider the Hilbert scheme $M^{[n]}$ of points on M. Pick a hyperkähler structure on $M^{[n]}$ which is compatible with the complex structure. Then $M^{[n]}$ has no proper trianalytic subvarieties.

■

2.4 Hyperholomorphic bundles

This subsection contains several versions of a definition of hyperholomorphic connection in a complex vector bundle over a hyperkähler manifold. We follow [V1].

Let B be a holomorphic vector bundle over a complex manifold M, ∇ a connection in B and $\Theta \in \Lambda^2 \otimes End(B)$ be its curvature. This connection is called **compatible with a holomorphic structure** if $\nabla_X(\zeta) = 0$ for any holomorphic section ζ and any antiholomorphic tangent vector field $X \in T^{0,1}(M)$. If there exists a holomorphic structure compatible with the given Hermitian connection then this connection is called **integrable.**

One can define a **Hodge decomposition** in the space of differential forms with coefficients in any complex bundle, in particular, $End(B)$.

Theorem 2.18: Let ∇ be a Hermitian connection in a complex vector bundle B over a complex manifold. Then ∇ is integrable if and only if $\Theta \in \Lambda^{1,1}(M, \operatorname{End}(B))$, where $\Lambda^{1,1}(M, \operatorname{End}(B))$ denotes the forms of Hodge type (1,1). Also, the holomorphic structure compatible with ∇ is unique.

Proof: This is Proposition 4.17 of [Ko], Chapter I. ■

This result has the following more general version:

Proposition 2.19: Let ∇ be an arbitrary (not necessarily Hermitian) connection in a complex vector bundle B. Then ∇ is integrable if and only its $(0,1)$-part has square zero.

■

This proposition is a version of Newlander-Nirenberg theorem. For vector bundles, it was proven by Atiyah and Bott.

Definition 2.20: Let B be a Hermitian vector bundle with a connection ∇ over a hyperkähler manifold M. Then ∇ is called **hyperholomorphic** if ∇ is integrable with respect to each of the complex structures induced by the hyperkähler structure.

As follows from Theorem 2.18, ∇ is hyperholomorphic if and only if its curvature Θ is of Hodge type (1,1) with respect to any of complex structures induced by a hyperkähler structure.

As follows from Lemma 2.6, ∇ is hyperholomorphic if and only if Θ is a $SU(2)$-invariant differential form.

Example 2.21: (Examples of hyperholomorphic bundles)

(i) Let M be a hyperkähler manifold, and TM be its tangent bundle equipped with the Levi–Civita connection ∇. Consider a complex structure on TM induced from the quaternion action. Then ∇ is a Hermitian connection which is integrable with respect to each induced complex structure, and hence, is hyperholomorphic.

(ii) For B a hyperholomorphic bundle, all its tensor powers are also hyperholomorphic.

(iii) Thus, the bundles of differential forms on a hyperkähler manifold are also hyperholomorphic.

2.5 Stable bundles and Yang–Mills connections.

This subsection is a compendium of the most basic results and definitions from the Yang–Mills theory over Kähler manifolds, concluding in the fundamental theorem of Uhlenbeck–Yau [UY].

Definition 2.22: Let F be a coherent sheaf over an n-dimensional compact Kähler manifold M. We define **the degree** $\deg(F)$ (sometimes the degree is also denoted by $\deg c_1(F)$) as

$$\deg(F) = \int_M \frac{c_1(F) \wedge \omega^{n-1}}{vol(M)}$$

and slope(F) as

$$\operatorname{slope}(F) = \frac{1}{\operatorname{rank}(F)} \cdot \deg(F).$$

The number slope(F) depends only on a cohomology class of $c_1(F)$.

Let F be a coherent sheaf on M and $F' \subset F$ its proper subsheaf. Then F' is called **destabilizing subsheaf** if $\operatorname{slope}(F') \geqslant \operatorname{slope}(F)$

A coherent sheaf F is called **stable** [1] if it has no destabilizing subsheaves. A coherent sheaf F is called **semistable** if for all destabilizing subsheaves $F' \subset F$, we have $\operatorname{slope}(F') = \operatorname{slope}(F)$.

Later on, we usually consider the bundles B with $deg(B) = 0$.

Let M be a Kähler manifold with a Kähler form ω. For differential forms with coefficients in any vector bundle there is a Hodge operator $L : \eta \longrightarrow \omega \wedge \eta$. There is also a fiberwise-adjoint Hodge operator Λ (see [GH]).

Definition 2.23: Let B be a holomorphic bundle over a Kähler manifold M with a holomorphic Hermitian connection ∇ and a curvature $\Theta \in \Lambda^{1,1} \otimes End(B)$. The Hermitian metric on B and the connection ∇ defined by this metric are called **Yang-Mills** if

$$\Lambda(\Theta) = constant \cdot \operatorname{Id}\Big|_B,$$

where Λ is a Hodge operator and $\operatorname{Id}\Big|_B$ is the identity endomorphism which is a section of $End(B)$.

Further on, we consider only these Yang–Mills connections for which this constant is zero.

A holomorphic bundle is called **indecomposable** if it cannot be decomposed onto a direct sum of two or more holomorphic bundles.

The following fundamental theorem provides examples of Yang–Mills bundles.

Theorem 2.24: (Uhlenbeck-Yau) Let B be an indecomposable holomorphic bundle over a compact Kähler manifold. Then B admits a Hermitian Yang-Mills connection if and only if it is stable, and this connection is unique.

Proof: [UY]. ■

Proposition 2.25: Let M be a hyperkähler manifold, L an induced complex structure and B be a complex vector bundle over (M, L). Then every

[1] In the sense of Mumford-Takemoto

hyperholomorphic connection ∇ in B is Yang-Mills and satisfies $\Lambda(\Theta) = 0$ where Θ is a curvature of ∇.

Proof: We use the definition of a hyperholomorphic connection as one with $SU(2)$-invariant curvature. Then Proposition 2.25 follows from the

Lemma 2.26: Let $\Theta \in \Lambda^2(M)$ be a $SU(2)$-invariant differential 2-form on M. Then $\Lambda_L(\Theta) = 0$ for each induced complex structure L.[2]

Proof: This is Lemma 2.1 of [V1]. ∎

Let M be a compact hyperkähler manifold, I an induced complex structure. For any stable holomorphic bundle on (M, I) there exists a unique Hermitian Yang-Mills connection which, for some bundles, turns out to be hyperholomorphic. It is possible to tell when this happens.

Theorem 2.27: Let B be a stable holomorphic bundle over (M, I), where M is a hyperkähler manifold and I is an induced complex structure over M. Then B admits a compatible hyperholomorphic connection if and only if the first two Chern classes $c_1(B)$ and $c_2(B)$ are $SU(2)$-invariant.[3]

Proof: This is Theorem 2.5 of [V1]. ∎

2.6 Twistor spaces

Let M be a hyperkähler manifold. Consider the product manifold $X = M \times S^2$. Embed the sphere $S^2 \subset \mathbb{H}$ into the quaternion algebra $\mathbb{H}$ as the subset of all quaternions J with $J^2 = -1$. For every point $x = m \times J \in X = M \times S^2$ the tangent space $T_x X$ is canonically decomposed $T_x X = T_m M \oplus T_J S^2$. Identify $S^2 = \mathbb{C}P^1$ and let $I_J : T_J S^2 \to T_J S^2$ be the complex structure operator. Let $I_m : T_m M \to T_m M$ be the complex structure on M induced by $J \in S^2 \subset \mathbb{H}$.

The operator $I_x = I_m \oplus I_J : T_x X \to T_x X$ satisfies $I_x \circ I_x = -1$. It depends smoothly on the point x, hence defines an almost complex structure on X. This almost complex structure is known to be integrable (see [Sal]).

Definition 2.28: The complex manifold $\langle X, I_x \rangle$ is called *the twistor space* for the hyperkähler manifold M, denoted by $\operatorname{Tw}(M)$. This manifold

[2] By Λ_L we understand the Hodge operator Λ associated with the Kähler complex structure L.

[3] We use Lemma 2.5 to speak of action of $SU(2)$ in cohomology of M.

is equipped with a real analytic projection $\sigma : \operatorname{Tw}(M) \longrightarrow M$ and a complex analytic projection $\pi : \operatorname{Tw}(M) \longrightarrow \mathbb{C}P^1$.

The twistor space $\operatorname{Tw}(M)$ is not, generally speaking, a Kähler manifold. For M compact, it is easy to show that $\operatorname{Tw}(M)$ does not admit a Kähler metric. We consider $\operatorname{Tw}(M)$ as a Hermitian manifold with the product metric.

Definition 2.29: Let X be an n-dimensional Hermitian manifold and let $\sqrt{-1}\omega$ be the imaginary part of the metric on X. Thus ω is a real $(1,1)$-form. Assume that the form ω satisfies the following condition of Li and Yau ([LY]).

$$\omega^{n-2} \wedge d\omega = 0. \tag{2.1}$$

Such Hermitian metrics are called **metrics satisfying the condition of Li–Yau.**

For a closed real 2-form η let

$$\deg \eta = \int_X \omega^{n-1} \wedge \eta.$$

The condition (2.1) ensures that $\deg \eta$ depends only on the cohomology class of η. Thus it defines a degree functional $\deg : H^2(X, \mathbb{R}) \to \mathbb{R}$. This functional allows one to repeat verbatim the Mumford-Takemoto definitions of stable and semistable bundles in this more general situation. Moreover, the Hermitian Yang-Mills equations also carry over word-by-word. Li and Yau proved a version of Uhlenbeck–Yau theorem in this situation ([LY]; see also Theorem 3.8).

Proposition 2.30: Let M be a hyperkähler manifold and $\operatorname{Tw}(M)$ its twistor space, considered as a Hermitian manifold. Then $\operatorname{Tw}(M)$ satisfies the conditions of Li–Yau.

Proof: [KV], Proposition 4.5. ∎

3 Hyperholomorphic sheaves

3.1 Stable sheaves and Yang-Mills connections

In [BS], S. Bando and Y.-T. Siu developed the machinery allowing one to apply the methods of Yang-Mills theory to torsion-free coherent sheaves. In the course of this paper, we apply their work to generalise the results of [V1]. In this Subsection, we give a short exposition of their results.

Definition 3.1: Let X be a complex manifold, and F a coherent sheaf on X. Consider the sheaf $F^* := \mathcal{H}om_{\mathcal{O}_X}(F, \mathcal{O}_X)$. There is a natural functorial map $\rho_F : F \longrightarrow F^{**}$. The sheaf F^{**} is called **a reflexive hull**, or **reflexization** of F. The sheaf F is called **reflexive** if the map $\rho_F : F \longrightarrow F^{**}$ is an isomorphism.

Remark 3.2: For all coherent sheaves F, the map $\rho_{F^*} : F^* \longrightarrow F^{***}$ is an isomorphism ([OSS], Ch. II, the proof of Lemma 1.1.12). Therefore, a reflexive hull of a sheaf is always reflexive.

Claim 3.3: Let X be a Kähler manifold, and F a torsion-free coherent sheaf over X. Then F (semi)stable if and only if F^{**} is (semi)stable.

Proof: This is [OSS], Ch. II, Lemma 1.2.4. ∎

Definition 3.4: Let X be a Kähler manifold, and F a coherent sheaf over X. The sheaf F is called **polystable** if F is a direct sum of stable sheaves.

The admissible Hermitian metrics, introduced by Bando and Siu in [BS], play the role of the ordinary Hermitian metrics for vector bundles.

Let X be a Kähler manifold. In Hodge theory, one considers the operator $\Lambda : \Lambda^{p,q}(X) \longrightarrow \Lambda^{p-1,q-1}(X)$ acting on differential forms on X, which is adjoint to the multiplication by the Kähler form. This operator is defined on differential forms with coefficient in every bundle. Considering a curvature Θ of a bundle B as a 2-form with coefficients in $\operatorname{End}(B)$, we define the expression $\Lambda\Theta$ which is a section of $\operatorname{End}(B)$.

Definition 3.5: Let X be a Kähler manifold, and F a reflexive coherent sheaf over X. Let $U \subset X$ be the set of all points at which F is locally trivial. By definition, the restriction $F\big|_U$ of F to U is a bundle. An **admissible metric** on F is a Hermitian metric h on the bundle $F\big|_U$ which satisfies the following assumptions

(i) the curvature Θ of (F, h) is square integrable, and

(ii) the corresponding section $\Lambda\Theta \in \operatorname{End}(F\big|_U)$ is uniformly bounded.

Definition 3.6: Let X be a Kähler manifold, F a reflexive coherent sheaf over X, and h an admissible metric on F. Consider the corresponding Hermitian connection ∇ on $F\Big|_U$. The metric h and the connection ∇ are called **Yang-Mills** if its curvature satisfies

$$\Lambda\Theta \in \operatorname{End}(F\Big|_U) = c \cdot \mathsf{id}$$

where c is a constant and id the unit section $\mathsf{id} \in \operatorname{End}(F\Big|_U)$.

Further in this paper, we shall only consider Yang-Mills connections with $\Lambda\Theta = 0$.

Remark 3.7: By Gauss-Bonnet formule, the constant c is equal to $\deg(F)$, where $\deg(F)$ is the degree of F (Definition 2.22).

One of the main results of [BS] is the following analogue of Uhlenbeck–Yau theorem (Theorem 2.24).

Theorem 3.8: Let M be a compact Kähler manifold, or a compact Hermitian manifold satisfying conditions of Li-Yau (Definition 2.29), and F a coherent sheaf without torsion. Then F admits an admissible Yang–Mills metric is and only if F is polystable. Moreover, if F is stable, then this metric is unique, up to a constant multiplier.

Proof: In [BS], Theorem 3.8 is proved for Kähler M ([BS], Theorem 3). It is easy to adapt this proof for Hermitian manifolds satisfying conditions of Li–Yau. ∎

Remark 3.9: Clearly, the connection, corresponding to a metric on F, does not change when the metric is multiplied by a scalar. The Yang–Mills metric on a polystable sheaf is unique up to a componentwise multiplication by scalar multipliers. Thus, the Yang–Mills connection of Theorem 3.8 is unique.

Another important theorem of [BS] is the following.

Theorem 3.10: Let (F, h) be a holomorphic vector bundle with a Hermitian metric h defined on a Kähler manifold X (not necessary compact nor complete) outside a closed subset S with locally finite Hausdorff measure of real co-dimension 4. Assume that the curvature tensor of F is locally square integrable on X. Then F extends to the whole space X as a reflexive sheaf $\mathcal{F}$. Moreover, if the metric h is Yang-Mills, then h can be smoothly extended as a Yang-Mills metric over the place where $\mathcal{F}$ is locally free.

Proof: This is [BS], Theorem 2. ∎

3.2 Stable and semistable sheaves over hyperkähler manifolds

Let M be a compact hyperkähler manifold, I an induced complex structure, F a torsion-free coherent sheaf over (M, I) and F^{**} its reflexization. Recall that the cohomology of M are equipped with a natural $SU(2)$-action (Lemma 2.5). The motivation for the following definition is Theorem 2.27 and Theorem 3.8.

Definition 3.11: Assume that the first two Chern classes of the sheaves F, F^{**} are $SU(2)$-invariant. Then F is called **stable hyperholomorphic** if F is stable, and **semistable hyperholomorphic** if F can be obtained as a successive extension of stable hyperholomorphic sheaves.

Remark 3.12: The slope of a hyperholomorphic sheaf is zero, because a degree of an $SU(2)$-invariant 2-form is zero (Lemma 2.26).

Claim 3.13: Let F be a semistable coherent sheaf over (M, I). Then the following conditions are equivalent.

(i) F is stable hyperholomorphic

(ii) Consider the support S of the sheaf F^{**}/F as a complex subvariety of (M, I). Let $X_1, \ldots, X_n$ be the set of irreducible components of S of codimension 2. Then X_i is trianalytic for all i, and the sheaf F^{**} is stable hyperholomorphic.

Proof: Consider an exact sequence

$$0 \longrightarrow F \longrightarrow F^{**} \longrightarrow F^{**}/F \longrightarrow 0.$$

Let $[F/F^{**}] \in H^4(M)$ be the fundamental class of the union of all codimension-2 components of support of the sheaf F/F^{**}, taken with appropriate multiplicities. Then, $c_2(F^{**}/F) = -[F/F^{**}]$. From the product formula for Chern classes, it follows that

$$c_2(F) = c_2(F_i^{**}) + c_2(F^{**}/F) = c_2(F_i^{**}) - [F/F^{**}]. \tag{3.1}$$

Clearly, if all X_i are trianalytic then the class $[F/F^{**}]$ is $SU(2)$-invariant. Thus, if the sheaf F^{**} is hyperholomorphic and all X_i are trianalytic, then the second Chern class of F is $SU(2)$-invariant, and F is hyperholomorphic. Conversely, assume that F is hyperholomorphic. We need to show that all X_i are trianalytic. By definition,

$$[F/F^{**}] = \sum_i \lambda_i [X_i]$$

where $[X_i]$ denotes the fundamental class of X_i, and λ_i is the multiplicity of F/F^{**} at X_i. By (3.1), (F hyperholomorphic) implies that the class $[F/F^{**}]$ is $SU(2)$-invariant. Since $[F/F^{**}]$ is $SU(2)$-invariant, we have

$$\sum_i \lambda_i \deg_J(X_i) = \sum_i \lambda_i \deg_I(X_i).$$

By Wirtinger's inequality (Proposition 2.11),

$$\deg_J(X_i) \leqslant \deg_I(X_i),$$

and the equality is reached only if X_i is trianalytic. By definition, all the numbers λ_i are positive. Therefore,

$$\sum_i \lambda_i \deg_J(X_i) \leqslant \sum_i \lambda_i \deg_I(X_i).$$

and the equality is reached only if all the subvarieties X_i are trianalytic. This finishes the proof of Claim 3.13. ■

Claim 3.14: Let M be a compact hyperkähler manifold, and I an induced complex structure of general type. Then all torsion-free coherent sheaves over (M, I) are semistable hyperholomorphic.

Proof: Let F be a torsion-free coherent sheaf over (M, I). Clearly from the definition of induced complex structure of general type, the sheaves F and F^{**} have $SU(2)$-invariant Chern classes. Now, all $SU(2)$-invariant 2-forms have degree zero (Lemma 2.26), and thus, F is semistable. ■

3.3 Hyperholomorphic connection in torsion-free sheaves

Let M be a hyperkähler manifold, I an induced complex structure, and F a torsion-free sheaf over (M, I). Consider the natural $SU(2)$-action in the bundle $\Lambda^i(M, B)$ of the differential i-forms with coefficients in a vector bundle B. Let $\Lambda^i_{inv}(M, B) \subset \Lambda^i(M, B)$ be the bundle of $SU(2)$-invariant i-forms.

Definition 3.15: Let $X \subset (M, I)$ be a complex subvariety of codimension at least 2, such that $F\big|_{M\backslash X}$ is a bundle, h be an admissible metric on $F\big|_{M\backslash X}$ and ∇ the associated connection. Then ∇ is called **hyperholomorphic** if its curvature

$$\Theta_\nabla = \nabla^2 \in \Lambda^2\left(M, \operatorname{End}\left(F\big|_{M\backslash X}\right)\right)$$

is $SU(2)$-invariant, i. e. belongs to $\Lambda^2_{inv}\left(M, \operatorname{End}\left(F\big|_{M\backslash X}\right)\right)$.

Claim 3.16: The singularities of a hyperholomorphic connection form a trianalytic subvariety in M.

Proof: Let J be an induced complex structure on M, and U the set of all points of (M, I) where F is non-singular. Clearly, (F, ∇) is a bundle with admissible connection on (U, J). Therefore, the holomorphic structure on $F\big|_{(U,J)}$ can be extended to (M, J). Thus, the singular set of F is holomorphic with respect to J. This proves Claim 3.16. ∎

Proposition 3.17: Let M be a compact hyperkähler manifold, I an induced complex structure and F a reflexive sheaf admitting a hyperholomorphic connection. Then F is a polystable hyperholomorphic sheaf.

Proof: By Remark 3.20 and Theorem 3.8, F is polystable. We need only to show that the Chern classes $c_1(F)$ and $c_2(F)$ are $SU(2)$-invariant. Let $U \subset M$ be the maximal open subset of M such that $F\big|_U$ is locally trivial. By Theorem 3.10, the metric h and the connection ∇ can be extended to U. Let $\operatorname{Tw} U \subset \operatorname{Tw} M$ be the corresponding twistor space, and σ : $\operatorname{Tw} U \longrightarrow U$ the standard map. Consider the bundle $\sigma^* F\big|_U$, equipped with a connection $\sigma^*\nabla$. It is well known [1] that $\sigma^* F\big|_U$ is a bundle with an admissible Yang-Mills metric (we use Yang-Mills in the sense of Li-Yau, see Definition 2.29). By Theorem 3.10, $\sigma^* F\big|_U$ can be extended to a reflexive sheaf $\mathcal{F}$ on $\operatorname{Tw} M$. Clearly, this extension coincides with the push-forward of $\sigma^* F\big|_U$. The singular set $\widetilde{S}$ of $\mathcal{F}$ is a pull-back of the singular set S of

[1] See for instance the section "Direct and inverse twistor transform" in [KV].

F. Thus, S is trianalytic. By desingularization theorem (Theorem 2.16), S can be desingularized to a hyperkähler manifold in such a way that its twistors form a desingularization of $\mathcal{S}$. From the exact description of the singularities of $\mathcal{S}$, provided by the desingularization theorem, we obtain that the standard projection $\pi : \mathcal{S} \longrightarrow \mathbb{C}P^1$ is flat. By the following lemma, the restriction of $\mathcal{F}$ to the fiber $(M, I) = \pi^{-1}(\{I\})$ of π coincides with F.

Lemma 3.18: Let $\pi : X \longrightarrow Y$ be a map of complex varieties, and $S \hookrightarrow X$ a subvariety of X of codimension at least 2, which is flat over Y. Denote by $U \stackrel{j}{\hookrightarrow} X$ the complement $U = (X \backslash S)$. Let F be a vector bundle over U, and $j_* F$ its push-forward. Then the restriction of $j_* F$ to the fibers of π is reflexive.

Proof: Let $Z = \pi^{-1}(\{y\})$ be a fiber of π. Since S is flat over Y and of codimension at least 2, we have $j_*(\mathcal{O}_{Z \cap U}) = \mathcal{O}_Z$. Clearly, for an open embedding $\gamma : T_1 \longrightarrow T_2$ and coherent sheaves A, B on T_1, we have $\gamma_*(A \otimes B) = \gamma_* A \otimes \gamma_* B$. Thus, for all coherent sheaves A on U, we have

$$j_* A \otimes \mathcal{O}_Z = j_*(A \otimes \mathcal{O}_{Z \cap U}). \tag{3.2}$$

This implies that $j_*(F\big|_Z) = j_* F\big|_Z$. It is well known ([OSS], Ch. II, 1.1.12; see also Lemma 9.2) that a push-forward of a reflexive sheaf under an open embedding γ is reflexive, provided that the complement of the image of γ has codimension at least 2. Therefore, $j_* F\big|_Z$ is a reflexive sheaf over Z. This proves Lemma 3.18. ∎

Return to the proof of Proposition 3.17. Consider the sheaf $\mathcal{F}$ on the twistor space constructed above. Since $\mathcal{F}$ is reflexive, its singularities have codimension at least 3 ([OSS], Ch. II, 1.1.10). Therefore, $\mathcal{F}$ is flat in codimension 2, and the first two Chern classes of $F = \mathcal{F}\big|_{\pi^{-1}(I)}$ can be obtained by restricting the first two Chern classes of $\mathcal{F}$ to the subvariety $(M, I) = \pi^{-1}(I) \subset \operatorname{Tw}(M)$. It remains to show that such restriction is $SU(2)$-invariant. Clearly, $H^2((M, I)) = H^2((M, I) \backslash S)$, and $H^4((M, I)) = H^4((M, I) \backslash S)$. Therefore,

$$c_1\left(\mathcal{F}\Big|_{(M,I)}\right) = c_1\left(\mathcal{F}\Big|_{(M,I)\backslash S}\right)$$

and

$$c_2\left(\mathcal{F}\Big|_{(M,I)}\right) = c_2\left(\mathcal{F}\Big|_{(M,I)\backslash S}\right).$$

On the other hand, the restriction $\mathcal{F}\Big|_{\operatorname{Tw}(M)\backslash S}$ is a bundle. Therefore, the classes

$$c_1\left(\mathcal{F}\Big|_{(M,I)\backslash S}\right),\quad c_2\left(\mathcal{F}\Big|_{(M,I)\backslash S}\right)$$

are independent from $I \in \mathbb{C}P^1$. On the other hand, these classes are of type (p,p) with respect to all induced complex structures $I \in \mathbb{C}P^1$. By Lemma 2.6, this implies that the classes $c_1(\mathcal{F}\Big|_{(M,I)})$, $c_1(\mathcal{F}\Big|_{(M,I)})$ are $SU(2)$-invariant. As we have shown above, these two classes are equal to the first Chern classes of F. Proposition 3.17 is proven. ∎

3.4 Existence of hyperholomorphic connections

The following theorem is the main result of this section.

Theorem 3.19: Let M be a compact hyperkähler manifold, I an induced complex structure and F a reflexive sheaf on (M,I). Then F admits a hyperholomorphic connection if and only if F is polystable hyperholomorphic in the sense of Definition 3.11.

Remark 3.20: From Lemma 2.26, it is clear that a hyperholomorphic connection is always Yang-Mills. Therefore, such a connection is unique (Theorem 3.8).

The "only if" part of Theorem 3.19 is Proposition 3.17. The proof of "if" part of Theorem 3.19 takes the rest of this subsection.

Let I be an induced complex structure. We denote the corresponding Hodge decomposition on differential forms by $\Lambda^*(M) = \oplus\Lambda_I^{p,q}(M)$, and the standard Hodge operator by $\Lambda_I : \Lambda_I^{p,q}(M) \longrightarrow \Lambda_I^{p-1,q-1}(M)$. All these structures are defined on the differential forms with coefficients in a bundle. Let $\deg_I \eta := \int_M Tr(\Lambda_I)^k(\eta)$, for $\eta \in \Lambda^k(M, \operatorname{End} B)$. The following claim is implied by an elementary linear-algebraic computation.

Claim 3.21: Let M be a hyperkähler manifold, B a Hermitian vector bundle over M, and Θ a 2-form on M with coefficients in $\mathfrak{su}(B)$. Assume that

$$\Lambda_I\Theta = 0,\quad \Theta \in \Lambda_I^{1,1}(M, \operatorname{End} B)$$

for some induced complex structure I. Assume, moreover, that Θ is square-integrable. Let J be another induced complex structure, $J \neq \pm I$. Then

$$\deg_I \Theta^2 \geqslant |\deg_J \Theta^2|,$$

and the equality is reached only if Θ is $SU(2)$-invariant.

Proof: The following general argument is used.

Sublemma 3.22: Let M be a Kähler manifold, B a Hermitian vector bundle over M, and Ξ a square-integrable 2-form on M with coefficients in $\mathfrak{su}(B)$. Then:

(i) For

$$\Lambda_I \Xi = 0, \quad \Xi \in \Lambda_I^{1,1}(M, \operatorname{End} B)$$

we have

$$\deg_I \Xi^2 = C \int_M |\Xi|^2 \operatorname{Vol} M,$$

where $C = (4\pi^2 n(n-1))^{-1}\ M$.

(ii) For

$$\Xi \in \Lambda_I^{2,0}(M, \operatorname{End} B) \oplus \Lambda_I^{0,2}(M, \operatorname{End} B)$$

we have

$$\deg_I \Xi^2 = -C \int_M |\Xi|^2 \operatorname{Vol} M,$$

where C is the same constant as appeared in (i).

Proof: The proof is based on a linear-algebraic computation (so-called Lübcke-type argument). The same computation is used to prove Hodge-Riemann bilinear relations. ∎

Return to the proof of Claim 3.21. Let $\Theta = \Theta_J^{1,1} + \Theta_J^{2,0} + \Theta_J^{0,2}$ be the Hodge decomposition associated with J. The following Claim shows that $\Theta_J^{1,1}$ satisfies conditions of Sublemma 3.22 (i).

Claim 3.23: Let M be a hyperkähler manifold, I, L induced complex structures and Θ a 2-form on M satisfying

$$\Lambda_I \Theta = 0, \quad \Theta \in \Lambda_I^{1,1}(M).$$

Let $\Theta_L^{1,1}$ be the $(1,1)$-component of Θ taken with respect to L. Then $\Lambda_L \Theta_L^{1,1} = 0$.

Proof: Clearly, $\Lambda_L\Theta_L^{1,1} = \Lambda_L\Theta$. Consider the natural Hermitian structure on the space of 2-forms. Since Θ is of type $(1,1)$ with respect to I, Θ is fiberwise orthogonal to the holomorphic symplectic form $\Omega = \omega_J + \sqrt{-1}\Omega_K \in \Lambda_I^{2,0}(M)$. By the same reason, Θ is orthogonal to $\overline{\Omega}$. Therefore, Θ is orthogonal to ω_J and ω_K. Since $\Lambda_I\Theta = 0$, Θ is also orthogonal to ω_I. The map Λ_L is a projection to the form ω_L which is a linear combination of ω_I, ω_J and ω_K. Since Θ is fiberwise orthogonal to ω_L, we have $\Lambda_L\Theta = 0$. ∎

By Sublemma 3.22, we have

$$\deg_J \left(\Theta_J^{1,1}\right)^2 = C\int_M |\Theta_J^{1,1}|^2$$

and

$$\deg_J \left(\Theta_J^{2,0} + \Theta_J^{0,2}\right)^2 = -C\int_M |\Theta_J^{2,0} + \Theta_J^{0,2}|^2.$$

Thus,

$$\deg_J \Theta^2 = C\int_M \left|\Theta_J^{1,1}\right| 2 - C\int_M \left|\Theta_J^{2,0} + \Theta_J^{0,2}\right| 2.$$

On the other hand,

$$\deg_I \Theta^2 = C\int_M |\Theta|\ 2 = C\int_M \left|\Theta_J^{1,1}\right| 2 + C\int_M \left|\Theta_J^{2,0} + \Theta_J^{0,2}\right| 2.$$

Thus, $\deg_I \Theta^2 > |\deg_J \Theta^2|$ unless $\Theta_J^{2,0} + \Theta_J^{0,2} = 0$. On the other hand, $\Theta_J^{2,0} + \Theta_J^{0,2} = 0$ means that Θ is of type $(1,1)$ with respect to J. Consider the standard $U(1)$-action on differential forms associated with the complex structures I and J. These two $U(1)$-actions generate the whole Lie group $SU(2)$ acting on $\Lambda^2(M)$ (here we use that $I \neq \pm J$). Since Θ is of type $(1,1)$ with respect to I and J, this form is $SU(2)$-invariant. This proves Claim 3.21.∎

Return to the proof of Theorem 3.19. Let ∇ be the admissible Yang-Mills connection in F, and Θ its curvature. Recall that the form $Tr\Theta^2$ represents the cohomology class $2c_2(F) - \frac{r-1}{r}c_1(F)^2$, where c_i are Chern classes of F. Since the form $Tr\Theta^2$ is square-integrable, the integral

$$\deg_J \Theta^2 = \int_M Tr\Theta^2\omega_J^{n-2}$$

makes sense. In [BS], it was shown how to approximate the connection ∇ by smooth connections, via the heat equation. This argument, in particular, was used to show that the value of integrals like $\int_M Tr\Theta^2\omega_J^{n-2}$ can be computed through cohomology classes and the Gauss–Bonnet formula

$$Tr\Theta^2 = 2c_2(F) - \frac{r-1}{r}c_1(F)^2.$$

Since the classes $c_2(F)$, $c_1(F)$ are $SU(2)$-invariant, we have

$$\deg_I \Theta^2 = \deg_J \Theta^2$$

for all induced complex structures I, J. By Claim 3.21, this implies that Θ is $SU(2)$-invariant. Theorem 3.19 is proven. ∎

The same argument implies the following corollary.

Corollary 3.24: Let M be a compact hyperkähler manifold, I an induced complex structure, F a stable reflexive sheaf on (M, I), and J be an induced complex structure, $J \neq \pm I$. Then

$$\deg_I\left(2c_2(F) - \frac{r-1}{r}c_1(F)^2\right) \geqslant \left|\deg_J\left(2c_2(F) - \frac{r-1}{r}c_1(F)^2\right)\right|,$$

and the equality holds if and only if F is hyperholomorphic.

∎

3.5 Tensor category of hyperholomorphic sheaves

This subsection is extraneous. Further on, we do not use the tensor structure on the category of hyperholomorphic sheaves. However, we need the canonical identification of the categories of hyperholomorphic sheaves associated with different induced complex structures.

From Bando-Siu (Theorem 3.8) it follows that on a compact Kähler manifold a tensor product of stable reflexive sheaves is polystable. Similarly, Theorem 3.19 implies that a tensor product of polystable hyperholomorphic sheaves is polystable hyperholomorphic. We define the following category.

Definition 3.25: Let M be a compace hyperkähler manifold and I an induced complex structure. Let $\mathcal{F}_{st}(M, I)$ be a category with objects reflexive polystable hyperholomorphic sheaves and morphisms as in category of coherent sheaves. This category is obviously additive. The tensor product on $\mathcal{F}_{st}(M, I)$ is induced from the tensor product of coherent sheaves.

Claim 3.26: The category $\mathcal{F}_{st}(M, I)$ is abelian. Moreover, it is a Tannakian tensor category.

Proof: Let $\varphi : F_1 \longrightarrow F_2$ be a morphism of hyperholomorphi sheaves. In Definition 2.22 , we introduced **a slope** of a coherent sheaf. Clearly, $sl(F_1) \leqslant sl(\operatorname{im}\varphi) \leqslant sl(F_2)$. All hyperholomorphic sheaves have slope 0 by Remark 3.12. Thus, $sl(\operatorname{im}\varphi) = 0$ and the subsheaf $\operatorname{im}\varphi \subset F_2$ is destabilizing. Since F_2 is polystable, this sheaf is decomposed:

$$F_2 = \operatorname{im}\varphi \oplus \operatorname{coker}\varphi.$$

A similar argument proves that $F_1 = \ker\varphi \oplus \operatorname{coim}\varphi$, with all summands hyperholomorphic. This proves that $\mathcal{F}_{st}(M, I)$ is abelian. The Tannakian properties are clear. ∎

The category $\mathcal{F}_{st}(M, I)$ does not depend from the choice of induced complex structure I:

Theorem 3.27: Let M be a compact hyperkähler manifold, I_1, I_2 induced complex structures and $\mathcal{F}_{st}(M, I_1)$, $\mathcal{F}_{st}(M, I_2)$ the associated categories of polystable reflexive hyperholomorphic sheaves. Then, there exists a natural equivalence of tensor categories

$$\Phi_{I_1, I_2} : \mathcal{F}_{st}(M, I_1) \longrightarrow \mathcal{F}_{st}(M, I_2).$$

Proof: Let $F \in \mathcal{F}_{st}(M, I_1)$ be a reflexive polystable hyperholomorphic sheaf and ∇ the canonical admissible Yang-Mills connection. Consider the sheaf $\mathcal{F}$ on the twistor space $\operatorname{Tw}(M)$ constructed as in the proof of Proposition 3.17. Restricting $\mathcal{F}$ to $\pi^{-1}(I_2) \subset \operatorname{Tw}(M)$, we obtain a coherent sheaf F' on (M, I_2). As we have shown in the proof of Proposition 3.17, the sheaf $(F')^{**}$ is polystable hyperholomorphic. Let $\Phi_{I_1, I_2}(F) := (F')^{**}$. It is easy to check that thus constructed map of objects gives a functor

$$\Phi_{I_1, I_2} : \mathcal{F}_{st}(M, I_1) \longrightarrow \mathcal{F}_{st}(M, I_2),$$

and moreover, $\Phi_{I_1, I_2} \circ \Phi_{I_2, I_1} = Id$. This shows that Φ_{I_1, I_2} is an equivalence. Theorem 3.27 is proven. ∎

Definition 3.28: By Theorem 3.27, the category $\mathcal{F}_{st}(M, I_1)$ is independent from the choice of induced complex structure. We call this category **the category of polystable hyperholomorphic reflexive sheaves on**

M and denote it by $\mathcal{F}(M)$. The objects of $\mathcal{F}(M)$ are called **hyperholomorphic sheaves on** M. For a hyperholomorphic sheaf on M, we denote by F_I the corresponding sheaf from $\mathcal{F}_{st}(M, I_1)$.

Remark 3.29: Using the same argument as proves Theorem 10.8 (ii), it is easy to check that the category $\mathcal{F}(M)$ is a deformational invariant of M. That is, for two hyperkähler manifolds M_1, M_2 which are deformationally equivalent, the categories $\mathcal{F}(M_i)$ are also equivalent, assuming that $Pic(M_1) = Pic(M_2) = 0$. The proof of this result is essentially contained in [V3-bis].

Remark 3.30: As Deligne proved ([D]), for a each Tannakian category $\mathcal{C}$ equipped with a fiber functor, there exists a natural pro-algebraic group G such that $\mathcal{C}$ is a group of representations of G. For $\mathcal{F}(M)$, there are several natural fiber functors. The simplest one is defined for each induced complex structure I such that (M, I) is algebraic (such complex structures always exist, as proven in [F]; see also [V-a] and Subsection 4.1). Let $\mathcal{K}(M, I)$ is the space of rational functions on (M, I). For $F \in \mathcal{F}_{st}(M, I)$, consider the functor $F \longrightarrow \eta_I(F)$, where $\eta_I(F)$ is the space of global sections of $F \otimes \mathcal{K}(M, I)$. This is clearly a fiber functor, which associates to $\mathcal{F}(M)$ the group G_I. The corresponding pro-algebraic group G_I is a deformational, that is, topological, invariant of the hyperkähler manifold.

4 Cohomology of hyperkähler manifolds

This section contains a serie of preliminary results which are used further on to define and study the C-restricted complex structures.

4.1 Algebraic induced complex structures

This subsection contains a recapitulation of results of [V-a].

A more general version of the following theorem was proven by A. Fujiki ([F], Theorem 4.8 (2)).

Theorem 4.1: Let M be a compact simple hyperkähler manifold and $\mathcal{R}$ be the set of induced complex structures $\mathcal{R} \cong \mathbb{C}P^1$. Let $\mathcal{R}_{alg} \subset \mathcal{R}$ be the set of all algebraic induced complex structures. Then $\mathcal{R}_{alg}$ is countable and dense in $\mathcal{R}$.

Proof: This is [V-a], Theorem 2.2. ∎

In the proof of Theorem 4.1, the following important lemma was used.

Lemma 4.2:

(i) Let $\mathcal{O} \subset H^2(M, \mathbb{R})$ be the set of all cohomology classes which are Kähler with respect to some induced comples structure. Then $\mathcal{O}$ is open in $H^2(M, \mathbb{R})$. Moreover, for all $\omega \in \mathcal{O}$, the class ω is *not* $SU(2)$-invariant.

(ii) Let $\eta \in H^2(M, \mathbb{R})$ be a cohomology class which is not $SU(2)$-invariant. Then there exists a unique up to a sign induced complex structure $I \in \mathcal{R}/\{\pm 1\}$ such that η belongs to $H^{1,1}_I(M)$.

Proof: This statement is a form of [V-a], Lemma 2.3. ■

4.2 The action of $\mathfrak{so}(5)$ on the cohomology of a hyperkähler manifold

This subsection is a recollection of data from [V0] and [V2(II)].

Let M be a hyperkähler manifold. For an induced complex structure R over M, consider the Kähler form $\omega_R = (\cdot, R\cdot)$, where $(\cdot, \cdot)$ is the Riemannian form. As usually, L_R denotes the operator of exterior multiplication by ω_R, which is acting on the differential forms $A^*(M, \mathbb{C})$ over M. Consider the adjoint operator to L_R, denoted by Λ_R.

One may ask oneself, what algebra is generated by L_R and Λ_R for all induced complex structures R? The answer was given in [V0], where the following theorem was proven.

Theorem 4.3: ([V0]) Let $M, \mathcal{H}$ be a hyperkähler manifold, and $\mathfrak{a}_{\mathcal{H}}$ be a Lie algebra generated by L_R and Λ_R for all induced complex structures R over M. Then the Lie algebra $\mathfrak{a}_{\mathcal{H}}$ is isomorphic to $\mathfrak{so}(4, 1)$. ■

The following facts about a structure of $\mathfrak{a}_{\mathcal{H}}$ were also proven in [V0].

Let I, J and K be three induced complex structures on M, such that $I \circ J = -J \circ I = K$. For an induced complex structure R, consider an operator adR on cohomology, acting on (p, q)-forms as a multiplication by $(p - q)\sqrt{-1}$. The operators adR generate a 3-dimensional Lie algebra $\mathfrak{g}_{\mathcal{H}}$, which is isomorphic to $\mathfrak{su}(2)$. This algebra coincides with the Lie algebra associated to the standard $SU(2)$-action on $H^*(M)$. The algebra $\mathfrak{a}_{\mathcal{H}}$ contains

$\mathfrak{g}_{\mathcal{H}}$ as a subalgebra, as follows:

$$[\Lambda_J, L_K] = [L_J, \Lambda_K] = ad\, I \text{ (etc)}. \tag{4.1}$$

The algebra $\mathfrak{a}_{\mathcal{H}}$ is 10-dimensional. It has the following basis: $L_R, \Lambda_R, ad\, R$ $(R = I, J, K)$ and the element $H = [L_R, \Lambda_R]$. The operator H is a standard Hodge operator; it acts on r-forms over M as multiplication by a scalar $n - r$, where $n = dim_{\mathbb{C}} M$.

Definition 4.4: Let $\mathfrak{g}$ be a semisimple Lie algebra, V its representation and $V = \oplus V_\alpha$ a $\mathfrak{g}$-invariant decompostion of V, such that for all α, V_α is a direct sum of isomorphic finite-dimensional representations W_α of V, and all W_α are distinct. Then the decomposition $V = \oplus V_\alpha$ is called **the isotypic decomposition of** V.

It is clear that for all finite-dimensional representations, isotypic decomposition always exists and is unique.

Let M be a compact hyperkähler manifold. Consider the cohomology space $H^*(M)$ equipped with the natural action of $\mathfrak{a}_{\mathcal{H}} = \mathfrak{so}(5)$. Let $H^*_o \subset H^*(M)$ be the isotypic component containing $H^0(M) \subset H^*(M)$. Using the root system written explicitly for $\mathfrak{a}_{\mathcal{H}}$ in [V0], [V3], it is easy to check that $H^*_o(M)$ is an irreducible representation of $\mathfrak{so}(5)$. Let $p: H^*(M) \longrightarrow H^*_o(M)$ be the unique $\mathfrak{so}(5)$-invariant projection, and $i: H^*_o(M) \hookrightarrow H^*(M)$ the natural embedding.

Let M be a compact hyperkähler manifold, I an induced complex structure, and ω_I the corresponding Kähler form. Consider the **degree map** $\deg_I: H^{2p}(M) \longrightarrow \mathbb{C}$, $\eta \longrightarrow \int_M \eta \wedge \omega_I^{n-p}$, where $n = \dim_{\mathbb{C}} M$.

Proposition 4.5: The space

$$H^*_o(M) \subset H^*(M)$$

is a subalgebra of $H^*(M)$, which is invariant under the $SU(2)$-action. Moreover, for all induced complex structures I, the degree map

$$\deg_I: H^*(M) \longrightarrow \mathbb{C}$$

satisfies

$$\deg_I(\eta) = \deg_I(i(p(\eta))),$$

where $i: H^*_o(M) \hookrightarrow H^*(M)$, $p: H^*(M) \longrightarrow H^*_o(M)$ are the $\mathfrak{so}(5)$-invariant maps defined above. And finally, the projection $p: H^*(M) \longrightarrow H^*_o(M)$ is $SU(2)$-invariant.

Proof: The space $H^*_o(M)$ is generated from $\mathbf{1} \in H^0(M)$ by operators L_R, Λ_R. To prove that $H^*_o(M)$ is closed under multiplication, we have to show that $H^*_o(M)$ is generated (as a linear space) by expressions of type $L_{r_1} \circ L_{R_2} \circ ... \circ \mathbf{1}$. By (4.1), the commutators of L_R, Λ_R map such expressions to linear combinations of such expressions. On the other hand, the operators Λ_R map $\mathbf{1}$ to zero. Thus, the operators Λ_R map expressions of type $L_{r_1} \circ L_{R_2} \circ ... \circ \mathbf{1}$ to linear combinations of such expressions. This proves that $H^*_o(M)$ is closed under multiplication. The second statement of Proposition 4.5 is clear (see, e. g. [V2(II)], proof of Proposition 4.5). It remains to show that $H^*_o(M) \subset H^*(M)$ is an $SU(2)$-invariant subspace and that $p: H^*(M) \longrightarrow H^*_o(M)$ is compatible with the $SU(2)$-action. From (4.1), we obtain that the Lie group G_A associated with $\mathfrak{a}_{\mathcal{H}} \cong \mathfrak{so}(1,4)$ contains $SU(2)$ acting in a standard way on $H^*(M)$. Since the map $p: H^*(M) \longrightarrow H^*_o(M)$ commutes with G_A-action, p also commutes with $SU(2)$-action. We proved Proposition 4.5. ∎

4.3 Structure of the cohomology ring

In [V3] (see also [V3-bis]), we have computed explicitly the subalgebra of cohomology of M generated by $H^2(M)$. This computation can be summed up as follows.

Theorem 4.6: ([V3], Theorem 15.2) Let M be a compact hyperkähler manifold, $H^1(M) = 0$, $\dim_{\mathbb{C}} M = 2n$, and $H^*_r(M)$ the subalgebra of cohomology of M generated by $H^2(M)$. Then

$$\begin{cases} H^{2i}_r(M) \cong S^i H^2(M) & \text{for } i \leqslant n, \text{ and} \\ H^{2i}_r(M) \cong S^{2n-i} H^2(M) & \text{for } i \geqslant n \end{cases}$$

∎

Theorem 4.7: Let M be a simple hyperkähler manifold. Consider the group G generated by a union of all $SU(2)$ for all hyperkähler structures on M. Then the Lie algebra of G is isomorphic to $\mathfrak{so}(H^2(M))$, for a certain natural integer bilinear symmetric form on $H^2(M)$, called Bogomolov-Beauville form.

Proof: [V3] (see also [V3-bis]). ∎

The key element in the proof of Theorem 4.6 and Theorem 4.7 is the following algebraic computation.

Theorem 4.8: Let M be a simple hyperkähler manifold, and $\mathcal{H}$ a hyperkähler structure on M. Consider the Lie subalgebra

$$\mathfrak{a}_{\mathcal{H}} \subset \operatorname{End}(H^*(M)), \quad \mathfrak{a}_{\mathcal{H}} \cong \mathfrak{so}(1,4),$$

associated with the hyperkähler structure (Subsection 4.2). Let

$$\mathfrak{g} \subset \operatorname{End}(H^*(M))$$

be the Lie algebra generated by subalgebras $\mathfrak{a}_{\mathcal{H}} \subset \operatorname{End}(H^*(M))$, for all hyperkäher structures $\mathcal{H}$ on M. Then

(i) The algebra $\mathfrak{g}$ is naturally isomorphic to the Lie algebra $\mathfrak{so}(V \oplus \mathfrak{H})$, where V is the linear space $H^2(M,\mathbb{R})$ equipped with the Bogomolov–Beauville pairing, and $\mathfrak{H}$ is a 2-dimensional vector space with a quadratic form of signature $(1,-1)$.

(ii) The space $H^*_r(M)$ is invariant under the action of $\mathfrak{g}$. Moreover,

$$H^*_r(M) \subset H^*(M)$$

is an isotypic [1] component of the space $H^*(M)$ considered as a representation of $\mathfrak{g}$.

Proof: [V3] (see also [V3-bis]). ■

As one of the consequences of Theorem 4.6, we obtain the following lemma, which will be used further on in this paper.

Lemma 4.9: Let M be a simple hyperkähler manifold, $\dim_{\mathbb{H}} M = n$, and $p : H^*(M) \longrightarrow H^*_o(M)$ the map defined in Subsection 4.2. Then, for all $x, y \in H^*_r(M)$, we have

$$p(x)p(y) = p(xy), \quad \text{whenever} \quad xy \in \bigoplus_{i \leqslant 2n} H^i(M).$$

Proof: Let $\omega_I, \omega_J, \omega_K, x_1, ..., x_n$ be an orthonormal basis in $H^2(M)$. Clearly, the vectors $x_1, ..., x_k$ are $SU(2)$-invariant. Therefore, these vectors

[1] See Definition 4.4 for the definition of isotypic decomposition.

are highest vectors of the corresponding $\mathfrak{a}_{\mathcal{H}}$-representations, with respect to the root system and Cartan subalgebra for $\mathfrak{a}_{\mathcal{H}}$ which is written in [V0] or [V3]. We obtain that the monomials

$$P_{k_1,k_2,k_3,\{n_i\}} = \omega_I^{k_1}\omega_J^{k_2}\omega_K^{k_3}\prod x_i^{n_i}, \sum n_i = N, \quad P_{k_1,k_2,k_3,\{n_i\}} \in \bigoplus_{i\leqslant 2n} H^i(M)$$

belong to the different isotypical components for different N's. By Theorem 4.6, a product of two such monomials $P_{k_1,k_2,k_3,\{n_i\}}$ and $P_{k_1',k_2',k_3',\{n_i'\}}$ is equal to $P_{k_1+k_1',k_2+k_2',k_3+k_3',\{n_i+n_i'\}}$, assuming that

$$P_{k_1,k_2,k_3,\{n_i\}} P_{k_1',k_2',k_3',\{n_i'\}} \in \bigoplus_{i\leqslant 2n} H^i(M).$$

Thus, the isotypical decomposition associated with the $\mathfrak{a}_{\mathcal{H}}$-action is compatible with multiplicative structure on $H^*(M)$, for low-dimensional cycles. This implies Lemma 4.9. ∎

We shall use the following corollary of Lemma 4.9.

Corollary 4.10: Let M be a simple hyperkähler manifold, $\dim_{\mathbb{H}} M > 1$, and $\omega_1, \omega_2 \in H^2(M)$ cohomology classes which are $SU(2)$-invariant. Then, for all induced complex structures I, we have $\deg_I(\omega_1\omega_2) = 0$.

Proof: By definition, the classes ω_1, ω_2 satisfy $\omega_i \in \ker p$. By Lemma 4.9, we have $\omega_1\omega_2 \in \ker p$. By Proposition 4.5, $\deg_I \omega_1\omega_2 = 0$. ∎

Let ω be a rational Kähler form. The corresponding $\mathfrak{sl}(2)$-action on $H^*(M)$ is clearly compatible with the rational structure on $H^*(M)$. It is easy to see (using, for instance, Lemma 4.2) that $\mathfrak{g}$ is generated by $\mathfrak{sl}(2)$-triples associated with rational Kähler forms ω. Therefore, the action of $\mathfrak{g}$ on $H^*(M)$ is compatible with the rational structure on $H^*(M)$. Using the isotypic decomposition, we define a natural $\mathfrak{g}$-invariant map $r: H^*(M) \longrightarrow H^*_r(M)$. Further on, we shall use the following properties of this map.

Claim 4.11:

(i) The map $r: H^*(M) \longrightarrow H^*_r(M)$ is compatible with the rational structure on $H^*(M)$.

(ii) For every $x \in \ker r$, and every hyperkähler structure $\mathcal{H}$, the corresponding map $p: H^*(M) \longrightarrow H^*_o(M)$ satisfies $p(x) = 0$.

(iii) For every $x \in \ker r$, every hyperkähler structure $\mathcal{H}$, and every induced complex structure I on M, we have $\deg_I x = 0$.

Proof: Claim 4.11 (i) is clear, because the action of $\mathfrak{g}$ on $H^*(M)$ is compatible with the rational structure on $H^*(M)$. To prove Claim 4.11 (ii), we notice that the space $H_r^*(M)$ is generated from $H^0(M)$ by the action of $\mathfrak{g}$, and $H_o^*(M)$ is generated from $H^0(M)$ by the action of $\mathfrak{a}_{\mathcal{H}}$. Since $\mathfrak{a}_{\mathcal{H}}$ is by definition a subalgebra in $\mathfrak{g}$, we have $H_o^*(M) \subset H_r^*(M)$. The isotypic projection $r : H^*(M) \longrightarrow H_r^*(M)$ is by definition compatible with the $\mathfrak{g}$-action. Since $\mathfrak{a}_{\mathcal{H}} \subset \mathfrak{g}$, the map r is also compatible with the $\mathfrak{a}_{\mathcal{H}}$-action. Therefore, $\ker r \subset \ker p$. Claim 4.11 (iii) is implied by Claim 4.11 (ii) and Proposition 4.5. ∎

Let x_i be an basis in $H^2(M, \mathbb{Q})$ which is rational and orthonormal with respect to Bogomolov-Beauville pairing, $(x_i, x_i)_{\mathcal{B}} = \varepsilon_i = \pm 1$. Consider the cohomology class $\theta' := \varepsilon_i x_i^2 \in H^4(M, \mathbb{Q})$. Let $\theta \in H^4(M, \mathbb{Z})$ be a non-zero integer cohomology class which is proportional to θ'. From results of [V3] (see also [V3-bis]), the following proposition can be easily deduced.

Proposition 4.12: The cohomology class $\theta \in H^4(M, \mathbb{Z})$ is $SU(2)$-invariant for all hyperkähler structures on M. Moreover, for a generic hyperkähler structure, the group of $SU(2)$-invariant integer classes $\alpha \in H_r^4(M)$ has rank one, where $H_r^*(M)$ is the subalgebra of cohomology generated by $H^2(M)$.

Proof: Clearly, if an integer class α is $SU(2)$-invariant for a generic hyperkähler structure, then α is G-invariant, where G is the group defined in Theorem 4.7. On the other hand, $H_r^4(M) \cong S^2(H^2(M))$, as follows from Theorem 4.6. Clearly, the vector $\theta \in H_r^4(M) \cong S^2(H^2(M))$ is $\mathfrak{so}(H^2(M))$-invariant. Moreover, the space of $\mathfrak{so}(H^2(M))$-invariant vectors in $S^2(H^2(M))$ is one-dimensional. Finally, from an explicit computation of G it follows that G acts on $H^4(M)$ as $SO(H^2(M))$, and thus, the Lie algebra invariants coincide with invariants of G. We found that the space of G-invariants in $H_r^4(M)$ is one-dimensional and generated by θ. This proves Proposition 4.12. ∎

Remark 4.13: It is clear how to generalize Proposition 4.12 from dimension 4 to all dimensions. The space $H_r^{2d}(M)^G$ of G-invariants in $H_r^{2d}(M)$ is 1-dimensional for d even and zero-dimensional for d odd.

4.4 Cohomology classes of CA-type

Let M be a compact hyperkähler manifold, and I an induced complex structure. All cohomology classes which appear as fundamental classes of complex subvarieties of (M, I) satisfy certain properties. Classes satisfying these properties are called classes of CA-type, from Complex Analytic. Here is the definition of CA-type.

Definition 4.14: Let $\eta \in H_I^{2,2}(M) \cap H^4(M, \mathbb{Z})$ be an integer (2,2)-class. Assume that for all induced complex structures J, satisfying $I \circ J = -J \circ I$, we have $\deg_I(\eta) \geqslant \deg_J(\eta)$, and the equality is reached only if η is $SU(2)$-invariant. Assume, moreover, that $\deg_I(\eta) \geqslant |\deg_J(\eta)|$. Then η is called **a class of CA-type.**

Theorem 4.15: Let M be a simple hyperkähler manifold, of dimension $\dim_{\mathbb{H}} M > 1$, I an induced complex structure, and $\eta \in H_I^{2,2}(M) \cap H^4(M, \mathbb{Z})$ an integer (2,2)-class. Assume that one of the following conditions holds.

(i) There exists a complex subvariety $X \subset (M, I)$ such that η is the fundamental class of X

(ii) There exists a stable coherent torsion-free sheaf F over (M, I), such that the first Chern class of F is zero, and $\eta = c_2(F)$.

Then η is of CA-type.

Proof: Theorem 4.15 (i) is a direct consequence of Wirtinger's inequality (Proposition 2.11). It remains to prove Theorem 4.15 (ii).

We assume, temporarily, that F is reflexive. By Corollary 3.24, we have

$$\deg_I(2c_2(F) - \frac{r-1}{r} c_1(F)^2) \geqslant \left|\deg_J(2c_2(F) - \frac{r-1}{r} c_1(F)^2)\right|, \tag{4.2}$$

and the equality happens only if F is hyperholomorphic. Since $c_1(F)$ is $SU(2)$-invariant, we have $\deg_I(c_1(F)^2) = \deg_J(c_1(F)^2) = 0$ (Corollary 4.10). Thus, (4.2) implies that

$$\deg_I 2c_2(F) \geqslant |\deg_J 2c_2(F)|$$

and the inequality is strict unless F is hyperholomorphic, in which case, the class $c_2(F)$ is $SU(2)$-invariant by definition. We have proven Theorem 4.15 (ii) for the case of reflexive F.

For F not necessary reflexive sheaf, we have shown in the proof of Claim 3.13 that

$$c_2(F) = c_2(F^{**}) + \sum n_i[X_i],$$

where n_i are positive integers, and $[X_i]$ are the fundamental classes of irreducible components of support of the sheaf F^{**}/F. Therefore, the class $c_2(F)$ is a sum of classes of CA-type. Clearly, a sum of cohomology classes of CA-type is again a class of CA-type. This proves Theorem 4.15. ∎

5 C-restricted complex structures on hyperkähler manifolds

5.1 Existence of C-restricted complex structures

We assume from now till the end of this section that the hyperkähler manifold M is simple (Definition 2.7). This assumption can be avoided, but it simplifies notation.

We assume from now till the end of this section that the hyperkähler manifold M is compact of real dimension $\dim_{\mathbb{R}} M \geqslant 8$, i. e. $\dim_{\mathbb{H}} M \geqslant 2$. This assumption is absolutely necessary. The case of hyperkähler surfaces with $\dim_{\mathbb{H}} M = 1$ (torus and K3 surface) is trivial and for our purposes not interesting. It is not difficult to extend our definitions and results to the case of a compact hyperkähler manifold which is a product of simple hyperkähler manifolds with $\dim_{\mathbb{H}} M \geqslant 2$.

Definition 5.1: Let M be a compact hyperkähler manifold, and I an induced complex structure. As usually, we denote by $\deg_I : H^{2p}(M) \longrightarrow \mathbb{C}$ the associated degree map, and by $H^*(M) = \oplus H_I^{p,q}(M)$ the Hodge decomposition. Assume that I is algebraic. Let C be a positive real number. We say that the induced complex structure I is C-**restricted** if the following conditions hold.

(i) For all non-$SU(2)$-invariant cohomology classes classes $\eta \in H_I^{1,1}(M) \cap H^2(M,\mathbb{Z})$, we have $|\deg_I(\eta)| > C$.

(ii) Let $\eta \in H_I^{2,2}(M)$ be a cohomology class of CA-type which is not $SU(2)$-invariant. Then $|\deg_I(\eta)| > C$.

The heuristic (completely informal) meaning of this definition is the following. The degree map plays the role of the metric on the cohomology. Cohomology classes with small degrees are "small", the rest is "big". Under reasonably strong assumptions, there are only finitely many "small" integer classes, and the rest is "big". For each non-$SU(2)$-invariant cohomology class η there exists at most two induced complex structures for which η is of type (p,p). Thus, for most induced complex structures, all non-$SU(2)$-invariant integer (p,p) classes are "big". Intuitively, the C-restriction means that all non-$SU(2)$-invariant integer (1,1) and (2,2)-cohomology classes are "big". This definition is needed for the study of first and second Chern classes of sheaves. The following property of C-restricted complex structures is used (see Theorem 4.15): for every subvariety $X \subset (M,I)$ of complex codimension 2, either X is trianalytic or $\deg_I(X) > C$.

Definition 5.2: Let M be a compact manifold, and $\mathcal{H}$ a hyperkähler structure on M. We say that $\mathcal{H}$ **admits C-restricted complex structures** if for all $C > 0$, the set of all C-restricted algebraic complex structures is dense in the set $R_{\mathcal{H}} = \mathbb{C}P^1$ of all induced complex structures.

Proposition 5.3: Let M be a compact simple hyperkähler manifold, $\dim_{\mathbb{H}}(M) > 1$, and $r: H^4(M) \longrightarrow H^4_r(M)$ be the map defined in Claim 4.11. Assume that for all algebraic induced complex structures I, the group $H^{1,1}_I(M) \cap H^2(M,\mathbb{Z})$ has rank one, and the group

$$H^{2,2}_I(M) \cap H^4(M,\mathbb{Z})/(\ker r)$$

has rank 2. Then M admits C-restricted complex structures.

Proof: The proof of Proposition 5.3 takes the rest of this section.

Denote by $\mathcal{R}$ the set $\mathcal{R} \cong \mathbb{C}P^1$ of all induced complex structures on M. Consider the set $\mathcal{R}/\{\pm 1\}$ of induced complex structures up to a sign (Lemma 4.2). Let $\alpha \in H^2(M)$ be a cohomology class which is not $SU(2)$-invariant. According to Lemma 4.2, there exists a unique element $c(\alpha) \in \mathcal{R}/\{\pm 1\}$ such that $\alpha \in H^{1,1}_{c(\alpha)}(M)$. This defines a map

$$c: \ (H^2(M,\mathbb{R}) \backslash H^2_{inv}(M)) \longrightarrow \mathcal{R}/\{\pm 1\},$$

where $H^2_{inv}(M) \subset H^2(M)$ is the set of all $SU(2)$-invariant cohomology classes. For induced complex structures I and $-I$, and $\eta \in H^{2p}(M)$, the degree maps satisfy

$$\deg_I(\eta) = (-1)^p \deg_{-I}(\eta). \tag{5.1}$$

Thus, the number $|\deg_I(\eta)|$ is independent from the sign of I.

Let $\eta \in H^*(M, \mathbb{Z})$ be a cohomology class. The **largest divisor** of η is the biggest positive integer number k such that the cohomology class ηk is also integer.

Let $\alpha \in H^2(M, \mathbb{Z})$ be an integer cohomology class, which is not $SU(2)$-invariant, k its largest divisor and $\tilde{\alpha} := \alpha k$ the corresponding integer class. Denote by $\widetilde{\deg}(\alpha)$ the number

$$\widetilde{\deg}(\alpha) := \left|\deg_{c(\alpha)}(\tilde{\alpha})\right|.$$

The induced complex structure $c(\alpha)$ is defined up to a sign, but from (5.1) it is clear that $\widetilde{\deg}(\alpha)$ is independent from the choice of a sign.

Lemma 5.4: Let M be a compact hyperkähler manifold, and I be an algebraic induced complex structure, such that the group $H_I^{1,1}(M) \cap H^2(M, \mathbb{Z})$ has rank one, and the group $H_I^{2,2}(M) \cap H^4(M, \mathbb{Z})/(\ker r)$ has rank 2. Denote by α the generator of $H_I^{1,1}(M) \cap H^2(M, \mathbb{Z})$. Since the class α is proportional to a Kähler form, α is not $SU(2)$-invariant (Lemma 4.2, (i)). Let $d := \widetilde{\deg}\alpha$. Then, there exists a positive real constant A depending on volume of M, its topology and its dimension, such that I is $d{\cdot}A$-restricted.

Proof: This lemma is a trivial calculation based on results of [V3] (see also [V3-bis] and Subsection 4.3).

Since $H_I^{1,1}(M) \cap H^2(M, \mathbb{Z})$ has rank one, for all $\eta \in H_I^{1,1}(M) \cap H^2(M, \mathbb{Z})$, $\eta \neq 0$, we have $|\deg_I \eta| \geqslant d$. This proves the first condition of Definition 5.1.

Let θ be the $SU(2)$-invariant integer cycle $\theta \in H^4(M)$ defined in Proposition 4.12. By Lemma 2.6, $\theta \in H_I^{2,2}(M)$. Consider $\alpha^2 \in H_I^{2,2}(M)$, where α is the generator of $H_I^{1,1}(M) \cap H^2(M, \mathbb{Z})$.

Sublemma 5.5: Let J be an induced complex structure, $J \circ I = -J \circ I$, and $\deg_I$, $\deg_J$ the degree maps associated with I, J. Then

$$\deg_I \alpha^2 > 0, \deg_J \alpha^2 = 0, \deg_I \theta = \deg_J \theta > 0.$$

Proof: Since α is a Kähler class with respect to I, we have $\deg_I \alpha^2 > 0$. Since the cohomology class θ is $SU(2)$-invariant, and $SU(2)$ acts transitively on the set of induced complex structures, we have $\deg_I \theta = \deg_J \theta$. It remains to show that $\deg_J \alpha^2 = 0$ and $\deg_J \theta > 0$. The manifold M is by our assumptions simple; thus, $\dim H^{2,0}(M) = 1$ ([Bes]). Therefore, in the natural $SU(2)$-invariant decomposition

$$H^2(M) = H^2_{inv}(M) \oplus H^2_+(M), \tag{5.2}$$

we have $\dim H^2_+(M) = 3$. In particular, the intersection $H^2_+(M) \cap H^{1,1}_I(M)$ is 1-dimensional. Consider the decomposition of α, associated with (5.2): $\alpha = \alpha_+ + \alpha_{inv}$. Since α is of type $(1,1)$ with respect to I, the class α_+ is proportional to the Kähler class ω_I, with positive coefficient. A similar argument leads to the following decomposition for θ:

$$\theta = \omega_I^2 + \omega_J^2 + \omega_K^2 + \sum x_i^2,$$

where $K = I \circ J$ is an induced complex structure, and the classes x_i belong to $H^2_{inv}(M)$. From Corollary 4.10, we obtain that the classes x_i^2 satisfy $\deg_I(x_i^2) = 0$ (here we use $\dim_{\mathbb{H}}(M) > 1$). Thus,

$$\deg_I(\theta) = \deg_I(\omega_I^2 + \omega_J^2 + \omega_K^2) = \deg_I(\omega_I^2) > 0.$$

Similarly one checks that

$$\deg_J(\alpha^2) = \deg_J((\alpha_+ + \alpha_{inv})^2) = \deg_J(\alpha_+^2) = \deg_J(c^2\omega_I) = 0.$$

This proves Sublemma 5.5. ∎

Return to the proof of Lemma 5.4. Since $\deg_I \alpha^2 \neq \deg_J \alpha^2$, the class α^2 is *not* $SU(2)$-invariant. Since θ *is* $SU(2)$-invariant, θ is not collinear with α^2. The degrees $\deg_I$ of θ and α^2 are non-zero; we have $\deg_I(\theta) = \deg_J(\theta)$, $\deg_I(\alpha^2) \neq \deg_J(\alpha^2)$ for J an induced complex structure, $J \neq \pm I$. By Proposition 4.5, no non-trivial linear combination of θ, α^2 belongs to $\ker p$. By Claim 4.11 (ii), the classes θ, α^2 generate a 2-dimensional subspace in $H^4(M, \mathbb{Q}) / \ker r$.

By assumptions of Lemma 5.4, the group $H^{2,2}_I(M) \cap H^4(M, \mathbb{Z}) / (\ker r)$ has rank 2. Therefore ω and α^2 generate the space

$$H^{2,2}_I(M) \cap H^4(M, \mathbb{Q}) / (\ker r).$$

To prove Lemma 5.4 it suffices to show that for all integer classes

$$\beta = a\alpha^2 + b\theta, \quad a \in \mathbb{Q}\backslash 0, \quad \deg_I \beta \geqslant \deg_J \beta,$$

we have $|\deg_I \beta| > A \cdot d$, for a constant A depending only on volume, topology and dimension of M. Since $\deg_I \beta \geqslant |\deg_J \beta|$, and $\deg_J \alpha^2 = 0$ (Sublemma 5.5), we have

$$\deg_I(a\alpha^2 + b\theta) \geqslant |\deg_J b\theta|.$$

Therefore, either a and b have the same sign, or $\deg_I(a\alpha^2) > 2\deg_I(b\theta)$. In both cases,

$$|\deg_I \beta| \geqslant \frac{1}{2}\deg_I(a\alpha^2). \tag{5.3}$$

Let $x \in \mathbb{Q}^{>0}$ be the smallest positive rational value of a for which there exists an integer class $\beta = a\alpha^2 + b\theta$. We have an integer lattice L_1 in $H^4_r(M)$ provided by the products of integer classes; the integer lattice $L_2 \supset L_1$ provided by integer cycles might be different from that one. Clearly, x is greater than determinant $\det(L_1/L_2)$ of L_1 over L_2, and this determinant is determined by the topology of M.

Form the definition of x and (5.3), we have $|\deg_I \beta| > x^2 \deg_I(\alpha^2)$. On the other hand, $\deg_I(\alpha^2) > C\deg_I(\alpha)$, where C is a constant depending on volume and dimension of M. Setting $A := x^2 \cdot C$, we obtain $|\deg_I \beta| > x^2 \cdot C \cdot d$. This proves Lemma 5.4. ∎

Consider the maps

$$\widetilde{\deg}: \ H^2(M,\mathbb{Z})\backslash H^2_{inv}(M) \longrightarrow \mathbb{R},$$

$$c: \ H^2(M)\backslash H^2_{inv}(M) \longrightarrow \mathcal{R}/\{\pm 1\}$$

introduced in the beginning of the proof of Proposition 5.3.

Lemma 5.6: In assumptions of Proposition 5.3, let

$$\mathcal{O} \subset H^2(M,\mathbb{R})\backslash H^2_{inv}(M)$$

be an open subset of $H^2(M,\mathbb{R})$, such that for all $x \in \mathcal{O}$, $k \in \mathbb{R}^{>0}$, we have $k \cdot x \in \mathcal{O}$. Assume that $\mathcal{O}$ contains the Kähler class ω_I for all induced complex structures $I \in \mathcal{R}$. For a positive number $C \in \mathbb{R}^{>0}$, consider the set $X_C \subset \mathcal{O}$

$$X_C := \left\{\alpha \in \mathcal{O} \cap H^2(M,\mathbb{Z}) \ | \ \widetilde{\deg}(\alpha) \geqslant C\right\}.$$

Then $c(X_C)$ is dense in $\mathcal{R}/\{\pm 1\}$ for all $C \in \mathbb{R}^{>0}$.

Proof: The map $\widetilde{\deg}$ can be expressed in the following wey. We call an integer cohomology class $\alpha \in H^2(M,\mathbb{Z})$ **indivisible** if its largest divisor is 1, that is, there are no integer classes α', and numbers $k \in \mathbb{Z}$, $k > 1$, such that $\alpha = k\alpha'$.

Sublemma 5.7: Let $\alpha \in H^2(M)$ be an non-$SU(2)$-invariant cohomology class and $\alpha = \alpha_{inv} + \alpha_+$ be a decomposition associated with (5.2). Assume that α is indivisible. Then

$$\widetilde{\deg}(\alpha) = C\sqrt{((\alpha_+, \alpha_+)_{\mathcal{B}})}, \tag{5.4}$$

where $(\cdot,\cdot)_{\mathcal{B}}$ is the Bogomolov-Beauville pairing on $H^2(M)$ ([V3-bis]; see also Theorem 4.7), and C a constant depending on $\dim M$, $\operatorname{Vol} M$.

Proof: By Proposition 4.5,

$$\deg_I(\alpha) = \deg_I(\alpha_+)$$

(clearly, $p(\alpha) = \alpha_+$). By definition of $(\cdot,\cdot)_{\mathcal{B}}$, we have

$$\deg_I(\alpha_+) = (\alpha_+, \omega_{c(\alpha)})_{\mathcal{B}}$$

On the other hand, α_+ is collinear with $\omega_{c(\alpha)}$ by definition of the map c. Now (5.4) follows trivially from routine properties of bilinear forms. ∎

Let I be an induced complex structure such that the cohomology class ω_I is irrational: $\omega_I \notin H^2(M,\mathbb{Q})$.

To prove Lemma 5.6, we have to produce a sequence $x_i \in \mathcal{O} \cap H^2(M,\mathbb{Z})$ such that

(i) $c(x_i)$ converges to I,

(ii) and $\lim \widetilde{\deg}(x_i) = \infty$. (5.5)

We introduce a metric $(\cdot,\cdot)_{\mathcal{H}}$ on $H^2(M,\mathbb{R})$,

$$(\alpha,\beta)_{\mathcal{H}} := (\alpha_+,\beta_+)_{\mathcal{B}} - (\alpha_{inv},\beta_{inv})_{\mathcal{B}}.$$

It is easy to check that $(\cdot,\cdot)_{\mathcal{H}}$ is positive definite ([V3]). For every ε, there exists a rational class $\omega_\varepsilon \in H^2(M,\mathbb{Q})$ which approximates ω_I with precision

$$(\omega_\varepsilon - \omega_I, \omega_\varepsilon - \omega_I)_{\mathcal{H}} < \varepsilon.$$

Since $\mathcal{O}$ is open and contains ω_I, we may assume that ω_ε belongs to $\mathcal{O}$. Take a sequence ε_i converging to 0, and let $\widetilde{x}_i := \omega_{\varepsilon_i}$ be the corresponding sequence of rational cohomology cycles. Let $x_i := \lambda_i\widetilde{x}_i$ be the minimal positive integer such that $x_i \in H^2(M,\mathbb{Z})$. We are going to show that the sequence x_i satisfies the conditions of (5.5). First of all, $\widetilde{x}_i$ converges to ω_I, and the map

$$c: H^2(M)\backslash H^2_{inv}(M) \longrightarrow \mathcal{R}/\{\pm 1\}$$

is continuous. Therefore, $\lim c(\widetilde{x}_i) = c(\omega_I) = I$. By construction of c, c satisfies $c(x) = c(\lambda x)$, and thus, $c(x_i) = c(\widetilde{x}_i)$. This proves the condition (i) of (5.5). On the other hand, since ω_I is irrational, the sequence λ_i goes to infinity. Therefore,

$$\lim(x_i, x_i)_{\mathcal{H}} = \infty.$$

It remains to compare $(x_i, x_i)_{\mathcal{H}}$ with $\widetilde{\deg} x_i$. By (5.4),

$$\widetilde{\deg} x_i = \sqrt{((x_i)_+, (x_i)_+)_{\mathcal{B}}}.$$

On the other hand, since $(x_i)_+ \in H^2_+(M)$, we have

$$((x_i)_+, (x_i)_+)_{\mathcal{B}} = ((x_i)_+, (x_i)_+)_{\mathcal{H}}.$$

To prove (5.5) (ii), it remains to show that

$$\lim((x_i)_+, (x_i)_+)_{\mathcal{H}} = \lim(x_i, x_i)_{\mathcal{H}}.$$

Since the cohomology class $\widetilde{x}_i \in H^2(M, \mathbb{Q})$ ε-approximates ω_I, and ω_I belongs to $H^2_+(M)$, we have

$$(\widetilde{x}_i - (\widetilde{x}_i)_+, \widetilde{x}_i - (\widetilde{x}_i)_+)_{\mathcal{H}} < \varepsilon_i.$$

Therefore,

$$(x_i - (x_i)_+, x_i - (x_i)_+)_{\mathcal{H}} < \lambda_i \varepsilon_i. \tag{5.6}$$

On the other hand, for i sufficiently big, the cohomology class $\widetilde{x}_i$ approaches ω_I, and

$$(x_i, x_i)_{\mathcal{H}} > \frac{1}{2}\lambda_i(\omega_I, \omega_I)_{\mathcal{H}} \tag{5.7}$$

Comparing (5.6) and (5.7) and using the distance property for the distance given by $\sqrt{(\cdot, \cdot)_{\mathcal{H}}}$, we find that

$$\sqrt{(x_i)_+, (x_i)_+} > \sqrt{\frac{1}{2}\lambda_i(\omega_I, \omega_I)_{\mathcal{H}}} - \sqrt{\lambda_i \varepsilon_i} = \sqrt{\lambda_i} \cdot \left(\sqrt{\frac{1}{2}(\omega_I, \omega_I)_{\mathcal{H}}} - \sqrt{\varepsilon_i}\right). \tag{5.8}$$

Since ε_i converges to 0 and λ_i converges to infinity, the right hand side of (5.8) converges to infinity. On the other hand, by (5.4) the left hand side of (5.8) is equal constant times $\widetilde{\deg} x_i$, so $\lim \widetilde{\deg} x_i = \infty$. This proves the second condition of (5.5). Lemma 5.6 is proven. ■

We use Lemma 5.4 and Lemma 5.6 in order to finish the proof of Proposition 5.3.

Let M be a compact hyperkähler manifold, and $\mathcal{O} \subset H^2(M,\mathbb{R})$ be the set of all Kähler classes for the Kähler metrics compatible with one of induced complex structures. By Lemma 4.2, $\mathcal{O}$ is open in $H^2(M,\mathbb{R})$. Applying Lemma 5.6 to $\mathcal{O}$, we obtain the following. In assumptions of Proposition 5.3, let $Y_C \subset \mathcal{R}$ be the set of all algebraic induced complex structures I with $\deg\alpha > C$, where α is a rational Kähler class, $\alpha \in H^{1,1}(M) \cap H^2(M,\mathbb{Z})$. Then Y_C is dense in $\mathcal{R}$. Now, Lemma 5.4, implies that for all $I \in Y_C$, the induced complex structure I is $A \cdot C$-restricted, where A is the universal constant of Lemma 5.4. Thus, for all C the set of C-restricted induced complex structures is dense in $\mathcal{R}$. This proves that M admits C-restricted complex structures. We finished the proof of Proposition 5.3. ∎

5.2 Hyperkähler structures admitting C-restricted complex structures

Let M be a compact complex manifold admitting a hyperkähler structure $\mathcal{H}$. Assume that $(M,\mathcal{H})$ is a simple hyperkähler manifold of dimension $\dim_{\mathbb{H}} M > 1$. The following definition of (coarse, marked) moduli space for complex and hyperkähler structures on M is standard.

Definition 5.8: Let M_{C^∞} be the M considered as a differential manifold, $\widetilde{Comp}$ be the set of all integrable complex structures, and $\widetilde{Hyp}$ be the set of all hyperkähler structures on M_{C^∞}. The set $\widetilde{Hyp}$ is equipped with a natural topology. Let $\widetilde{Hyp}^0$ be a connected component of $\widetilde{Hyp}$ containing $\mathcal{H}$ and $\widetilde{Comp}^0$ be a set of all complex structures $I \in \widetilde{Comp}$ which are compatible with some hyperkähler structure $\mathcal{H}_1 \in \widetilde{Hyp}^0$. Let *Diff* be the group of diffeomorphisms of M which act trivially on the cohomology. The coarse, marked moduli *Hyp* of hyperkähler structures on M is the quotient $Hyp := \widetilde{Hyp}^0/Diff$ equipped with a natural topology. The coarse, marked moduli *Comp* of complex structures on M is defined as $Comp := \widetilde{Comp}^0/Diff$. For a detailed discussion of various aspects of this definition, see [V3].

Consider the variety

$$X \subset \mathbb{P}H^2(M,\mathbb{C}),$$

consisting of all lines $l \in \mathbb{P}H^2(M,\mathbb{C})$ which are isotropic with respect to the

Bogomolov-Beauville's pairing:

$$X := \{l \in H^2(M, \mathbb{C}) \mid (l, l)_{\mathcal{B}} = 0\}.$$

Since M is simple, $\dim H^{2,0}(M, I) = 1$ for all induced complex structures. Let $P_c : Comp \longrightarrow \mathbb{P}H^2(M, \mathbb{C})$ map I to the line $H_I^{2,0}(M) \subset H^2(M, \mathbb{C})$. The map P_c is called **the period map**. It is well known that $Comp$ is equipped with a natural complex structure. From general properties of the period map it follows that P_c is compatible with this complex structure. Clearly from the definition of Bogomolov-Beauville's form, $P_c(I) \in X$ for all induced complex structures $I \in Comp$ (see [Bea] for details).

Theorem 5.9: [Bes] (Bogomolov) The complex analytic map

$$P_c : Comp \longrightarrow X$$

is locally an etale covering. [1]

■

It is possible to formulate a similar statement about hyperkähler structures. For a hyperkähler structure $\mathcal{H}$, consider the set $\mathcal{R}_{\mathcal{H}} \subset Comp$ of all induced complex structures associated with this hyperkähler structure. The subset $\mathcal{R}_{\mathcal{H}} \subset Comp$ is a complex analytic subvariety, which is isomorphic to $\mathbb{C}P^1$. Let $S := P_c(\mathcal{R}_{\mathcal{H}})$ be the corresponding projective line in X, and $\overline{L}(X)$ be the space of smooth deformations of S in X. The points of $L(X)$ correspond to smooth rational curves of degree 2 in $\mathbb{P}H^2(M, \mathbb{C})$. For every such curve s, there exists a unique 3-dimensional plane $L(s) \subset H^2(M, \mathbb{C})$, such that s is contained in $\mathbb{P}L$. Let Gr be the Grassmanian manifold of all 3-dimensional planes in $H^2(M, \mathbb{C})$ and $Gr_0 \subset Gr$ the set of all planes $L \in Gr$ such that the restriction of the Bogomolov-Beauville form to L is non-degenerate. Let $L(X) \subset \overline{L}(X)$ be the space of all rational curves $s \in \overline{L}(X)$ such that the restriction of the Bogomolov-Beauville form to $L(s)$ is non-degenerate: $L(s) \in Gr_0$. The correspondence $s \longrightarrow L(s)$ gives a map $\kappa : L(X) \longrightarrow Gr_0$.

Lemma 5.10: The map $\kappa : L(X) \longrightarrow Gr_0$ is an isomorphism of complex varieties.

---

[1] The space $Comp$ is smooth, as follows from Theorem 5.9. This space is, however, in most cases not separable ([H]). The space Hyp has no natural complex structures, and can be odd-dimensional.

Proof: For every plane $L \in Gr_0$, consider the set $s(L)$ of all isotropic lines $l \in L$, that is, lines satisfying $(l,l)_B = 0$. Since $(\cdot,\cdot)_B\Big|_L$ is non-degenerate, the set $s(L)$ is a rational curve in $\mathbb{P}L$. Clearly, this curve has degree 2. Therefore, $s(L)$ belongs to $X(L)$. The map $L \longrightarrow s(L)$ is inverse to κ. ■

Consider the standard anticomplex involution

$$\iota: H^2(M,\mathbb{C}) \longrightarrow H^2(M,\mathbb{C}), \quad \eta \longrightarrow \overline{\eta}.$$

Clearly, ι is compatible with the Bogomolov-Beauville form. Therefore, ι acts on $L(X)$ as an anticomplex involution. Let $L(X)_\iota \subset L(X)$ be the set of all $S \in L(X)$ fixed by ι.

Every hyperkähler structure

$$\mathcal{H} \in Hyp$$

gives a rational curve $\mathcal{R}_{\mathcal{H}} \subset Comp$ with points corresponding all induced complex structures. Let $P_h(\mathcal{H}) \subset X$ be the line $P_c(\mathcal{R}_{\mathcal{H}})$. Clearly from the definition, $P_h(\mathcal{H})$ belongs to $L(X)_\iota$. We have constructed a map $P_h : Hyp \longrightarrow L(X)_\iota$. Let $L(Comp)$ be the space of deformations if $\mathcal{R}_{\mathcal{H}}$ in $Comp$. Denote by

$$\gamma: Hyp \longrightarrow L(Comp)$$

the map $\mathcal{H} \longrightarrow \mathcal{R}_{\mathcal{H}}$. The following result gives a hyperkähler analogue of Bogomolov's theorem (Theorem 5.9).

Theorem 5.11: The map $\gamma: Hyp \longrightarrow L(Comp)$ is an embedding. The map $P_h : Hyp \longrightarrow L(X)_\iota$ is locally a covering.

Proof: The first claim is an immediate consequence of Calabi-Yau Theorem (Theorem 2.4). Now, Theorem 5.11 follows from the Bogomolov's theorem (Theorem 5.9) and dimension count. ■

Let $I \in Comp$ be a complex structure on M. Consider the groups

$$H_h^2(M,I) := H^{1,1}(M,I) \cap H^2(M,\mathbb{Z})$$

and

$$H_h^2(M,I) := H_r^{2,2}(M,I) \cap H^4(M,\mathbb{Z}).$$

For a general I, $H_h^2(M,I) = 0$ and $H_h^4(M,I) = \mathbb{Z}$ as follows from Proposition 4.12. Therefore, the set of all I with $\operatorname{rk} H_h^2(M,I) = 1$, $\operatorname{rk} H_h^4(M,I) = 2$ is a union of countably many subvarieties of codimension 1 in *Comp*. Similarly, the set $V \subset Comp$ of all I with $\operatorname{rk} H^2_h(M, I) > 1$, $\operatorname{rk} H^4_h(M, I) > 2$ is a union of countably many subvarieties of codimension more than 1. Together with Theorem 5.11, this implies the following.

Claim 5.12: Let $U \subset Hyp$ be the set of all $\mathcal{H} \in Hyp$ such that $\mathcal{R}_{\mathcal{H}}$ does not intersect V. Then U is dense in *Hyp*.

Proof: Consider a natural involution i of *Comp* which is compatible with the involution $\iota : X \longrightarrow X$ inder the period map $P_c : Comp \longrightarrow X$. This involution maps the complex structure I to $-I$.

$$\text{By Theorem 5.11, } Hyp \text{ is identified with an open subset in the set } L(X)_\iota \text{ of real points of } L(Comp). \tag{5.9}$$

Let $L_U \subset L(Comp)$ be the set of all lines which do not intersect V. Since V is a union of subvarieties of codimension at least 2, a general rational line $l \in L(Comp)$ does not intersect V. Therefore, L_U is dense in $L(Comp)$. Thus, the set of real points of L_U is dense $L(X)_\iota$. Using the identification (5.9), we obtain the statement of Claim 5.12. ∎

Claim 5.12 together with Proposition 5.3 imply the following theorem.

Theorem 5.13: Let M be a compact simple hyperkähler manifold, $\dim_{\mathbb{H}} M > 1$, and *Hyp* its coarse marked moduli of hyperkähler structures. Let $U \subset Hyp$ be the set of all hyperkähler structures which admit C-restricted complex structures (Definition 5.2). Then U is dense in *Hyp*.
∎

5.3 Deformations of coherent sheaves over manifolds with C-restricted complex structures

The following theorem shows that a semistable deformation of a hyperholomorphic sheaf on (M, I) is again hyperholomorphic, provided that I is a C-restricted complex structure and C is sufficiently big.

Theorem 5.14: Let M be a compact hyperkähler manifold, and $\mathcal{F} \in \mathcal{F}(M)$ a polystable hyperholomorphic sheaf on M (Definition 3.28). Let I

be a C-restricted induced complex structure, for $C = \deg_I c_2(\mathcal{F})$,[2] and F' be a semistable torsion-free coherent sheaf on (M, I) with the same rank and Chern classes as $\mathcal{F}$. Then the sheaf F' is hyperholomorphic.

Proof: Let $F_1, ..., F_n$ be the Jordan-Hölder series for the sheaf F'. Since $\mathcal{F}$ is hyperholomorphic, we have $\operatorname{slope}(\mathcal{F}) = 0$ (Remark 3.12). Therefore, $\operatorname{slope}(F_i) = 0$, and $\deg_I(c_1(F_i)) = 0$. By Definition 5.1 (i), then, the class $c_1(F_i)$ is $SU(2)$ invariant for all i. To prove that F' is hyperholomorphic it remains to show that the classes $c_2(F_i)$, $c_2(F_i^{**})$ are $SU(2)$-invariant for all i.

Consider an exact sequence

$$0 \longrightarrow F_i \longrightarrow F_i^{**} \longrightarrow F_i/F_i^{**} \longrightarrow 0.$$

Let $[F_i/F_i^{**}] \in H^4(M)$ be the fundamental class of the union of all components of $Sup(F_i/F_i^{**})$ of complex codimension 2, taken with appropriate multiplicities. Clearly, $c_2(F_i) = c_2(F_i^{**}) + [F_i/F_i^{**}]$. Since $[F_i/F_i^{**}]$ is an effective cycle, $\deg_I([F_i/F_i^{**}]) \geqslant 0$. By the Bogomolov-Miyaoka-Yau inequality (see Corollary 3.24), we have $\deg_I(c_2(F_i^{**}) \geqslant 0$. Therefore,

$$\deg_I c_2(F_i) \geqslant \deg_I c_2(F_i^{**}) \geqslant 0. \tag{5.10}$$

Using the product formula for Chern classes, we obtain

$$c_2(F) = \sum_i c_2(F_i) + \sum_{i,j} c_2(F_i) \wedge c_2(F_j). \tag{5.11}$$

By Corollary 4.10, $\deg_I(\sum_{i,j} c_2(F_i) \wedge c_2(F_j)) = 0$. Since the numbers $\deg_I c_2(F_i)$ are non-negative, we have $\deg_I c_2(F_i) \leqslant \deg_I c_2(F) = C$. By Theorem 4.15, the classes $c_2(F_i)$, $c_2(F_i^{**})$ are of CA-type. By Definition 5.1 (ii), then, the inequality $\deg_I c_2(F_i) \leqslant C$ implies that the class $c_2(F_i)$ is $SU(2)$-invariant. By (5.10), $\deg_I c_2(F_i^{**}) \leqslant \deg_I c_2(F_i)$, so the class $c_2(F_i^{**})$ is also $SU(2)$-invariant. Theorem 5.14 is proven. ∎

6 Desingularization of hyperholomorphic sheaves

The aim of this section is the following theorem.

[2] Clearly, since $\mathcal{F}$ is hyperholomorphic, the class $c_2(\mathcal{F})$ is $SU(2)$-invariant, and the number $\deg_I c_2(\mathcal{F})$ independent from I.

Theorem 6.1: Let M be a hyperkähler manifold, not necessarily compact, I an induced complex structure, and F a reflexive coherent sheaf over (M, I) equipped with a hyperholomorphic connection (Definition 3.15). Assume that F has isolated singularities. Let $\widetilde{M} \xrightarrow{\sigma} M$ be a blow-up of (M, I) in the singular set of F, and $\sigma^* F$ the pullback of F. Then $\sigma^* F$ is a locally trivial sheaf, that is, a holomorphic vector bundle.

We prove Theorem 6.1 in Subsection 6.4.

The idea of the proof is the following. We apply to F the methods used in the proof of Desingularization Theorem (Theorem 2.16). The main ingredient in the proof of Desingularization Theorem is the existence of a natural $\mathbb{C}^*$-action on the completion $\hat{\mathcal{O}}_x(M, I)$ of the local ring $\mathcal{O}_x(M, I)$, for all $x \in M$. This $\mathbb{C}^*$-action identifies $\hat{\mathcal{O}}_x(M, I)$ with a completion of a graded ring. Here we show that a sheaf F is $\mathbb{C}^*$-equivariant. Therefore, a germ of F at x has a grading, which is compatible with the natural $\mathbb{C}^*$-action on $\hat{\mathcal{O}}_x(M, I)$. Singularities of such reflexive sheaves can be resolved by a single blow-up.

6.1 Twistor lines and complexification

Further on, we need the following definition.

Definition 6.2: Let X be a real analytic variety, which is embedded to a complex variety $X_{\mathbb{C}}$. Assume that the sheaf of complex-valued real analytic functions on X coincides with the restriction of $\mathcal{O}_{X_{\mathbb{C}}}$ to $X \subset X_{\mathbb{C}}$. Then $X_{\mathbb{C}}$ is called **a complexification of** X.

For more details on complexification, the reader is referred to [GMT]. There are the most important properties.

Claim 6.3: In a neighbourhood of X, the manifold $X_{\mathbb{C}}$ has an anticomplex involution. The variety X is identified with the set of fixed points of this involution, considered as a real analytic variety.

Let Y be a complex variety, and X the underlying real analytic variety. Then the product of Y and its complex conjugate is a complexification of X, with embedding $X \hookrightarrow Y \times \overline{Y}$ given by the diagonal.

The complexification is unique in the following weak sense. For $X_{\mathbb{C}}$, $X'_{\mathbb{C}}$ complexifications of C, the complex manifolds $X_{\mathbb{C}}$, $X'_{\mathbb{C}}$ are naturally identified in a neighbourhood of X.

■

Let M be a hyperkähler manifold, $\operatorname{Tw}(M)$ its twistor space, and $\pi : \operatorname{Tw}(M) \longrightarrow \mathbb{C}P^1$ the twistor projection. Let $l \subset \operatorname{Tw}(M)$ be a rational curve, such that the restriction of π to l is an identity. Such a curve gives a section of π, and vice versa, every section of π corresponds to such a curve. The set of sections of the projection π is called **the space of twistor lines**, denoted by Lin, or $\operatorname{Lin}(M)$. This space is equipped with complex structure, by Douady ([Do]).

Let $m \in M$ be a point. Consider a twistor line $s_m : I \longrightarrow (I \times m) \in \mathbb{C}P^1 \times M = \operatorname{Tw}$. Then s_m is called **a horizontal twistor line**. The space of horizontal twistor lines is a real analytic subvariety in Lin, denoted by Hor, or $\operatorname{Hor}(M)$. Clearly, the set Hor is naturally identified with M.

Proposition 6.4: (Hitchin, Karlhede, Lindström, Roček) Let M be a hyperkähler manifold, $\operatorname{Tw}(M)$ its twistor space, I, $J \in \mathbb{C}P^1$ induced complex structures, and Lin the space of twistor lines. The complex manifolds (M, I) and (M, J) are naturally embedded to $\operatorname{Tw}(M)$:

$$(M, I) = \pi^{-1}(I), \quad (M, J) = \pi^{-1}(J).$$

Consider a point $s \in \operatorname{Lin}$, $s : \mathbb{C}P^1 \longrightarrow \operatorname{Tw}(M)$. Let

$$ev_{I,J} : \operatorname{Lin}(M) \longrightarrow (M, I) \times (M, J)$$

be the map defined by $ev_{I,J}(s) = (s(I), s(J))$. Assume that $I \neq J$. Then there exists a neighbourhood U of $\operatorname{Hor} \subset \operatorname{Lin}$, such that the restriction of $ev_{I,J}$ to U is an open embedding.

Proof: [HKLR], [V-d3]. ■

Consider the anticomplex involution i of $\mathbb{C}P^1 \cong S^2$ which corresponds to the central symmetry of S^2. Let $\iota : \operatorname{Tw} \longrightarrow \operatorname{Tw}$ be the corresponding involution of the twistor space $\operatorname{Tw}(M) = \mathbb{C}P^1 \times M$, $(x, m) \longrightarrow (i(x), m)$. It is clear that ι maps holomorphic subvarieties of $\operatorname{Tw}(M)$ to holomorphic subvarieties. Therefore, ι acts on Lin as an anticomplex involution. For $J = -I$, we obtain a local identification of Lin in a neighbourhood of Hor with $(M, I) \times (M, -I)$, that is, with (M, I) times its complex conjugate. Therefore, the space of twistor lines is a complexification of (M, I). The natural anticomplex involution of Claim 6.3 coincides with ι. This gives an identification of Hor and the real analytic manifold underlying (M, I).

We shall explain how to construct the natural $\mathbb{C}^*$-action on a local ring of a hyperkähler manifold, using the machinery of twistor lines.

Fix a point $x_0 \in M$ and induced complex structures I, J, such that $I \neq \pm J$. Let V_1, V_2 be neighbourhoods of $s_{x_0} \in \operatorname{Lin}$, and U_1, U_2 be neighbourhoods of (x_0, x_0) in $(M, I) \times (M, -I)$, $(M, J) \times (M, -J)$, such that the evaluation maps $\epsilon v_{I,-I}$, $\epsilon v_{J,-J}$ induce isomorphisms

$$\epsilon v_{I,-I} : V_1 \xrightarrow{\sim} U_1, \quad \epsilon v_{J,-J} : V_2 \xrightarrow{\sim} U_2.$$

Let B be an open neighbourhood of $x_0 \in M$, such that $(B, I) \times (B, -I) \subset U_1$ and $(B, I) \times (B, -I) \subset U_2$. Denote by $V_I \subset V_1$ be the preimage of $(B, I) \times (B, -I)$ under $\epsilon v_{I,-I}$, and by $V_J \subset V_2$ be the preimage of $(B, J) \times (B, -J)$ under $\epsilon v_{J,-J}$. Let $p_I : V_I \longrightarrow (B, I)$ be the evaluation, $s \longrightarrow s(I)$, and $e_I : (B, I) \longrightarrow V_1$ the map associating to $x \in B$ the unique twistor line passing through $(x, x_0) \subset (B, I) \times (B, -I)$. In the same fashion, we define e_J and p_J. We are interested in the composition

$$\Psi_{I,J} := e_I \circ p_J \circ e_J \circ p_I : (B_0, I) \longrightarrow (B, I)$$

which is defined in a smaller neighbourhood $B_0 \subset B$ of $x_0 \in M$.

The following proposition is the focal point of this Subsection: we explain the map $\Psi_{I,J}$ of [V-d2], [V-d3] is geometric terms (in [V-d2], [V-d3] this map was defined algebraically).

Proposition 6.5: Consider the map $\Psi_{I,J} : (B_0, I) \longrightarrow (B, I)$ defined above. By definition, $\Psi_{I,J}$ preserves the point $x_0 \in B_0 \subset B$. Let $d\Psi_{I,J}$ be the differential of $\Psi_{I,J}$ acting on the tangent space $T_{x_0}B_0$. Assume that $I \neq \pm J$. Then $d\Psi_{I,J}$ is a multiplication by a scalar $\lambda \in \mathbb{C}$, $0 < |\lambda| < 1$.

Proof: The map $\Psi_{I,J}$ was defined in [V-d2], [V-d3] using the identifications between the real analytic varieties underlying (M, I) and (M, J). We proved that $\Psi_{I,J}$ defined this way acts on $T_{x_0}B_0$ as a multiplication by the scalar $\lambda \in \mathbb{C}$, $0 < |\lambda| < 1$. It remains to show that the map $\Psi_{I,J}$ defined in [V-d2], [V-d3] coincides with $\Psi_{I,J}$ defined above.

Consider the natural identification

$$(B, I) \times (B, -I) \sim (B, J) \times (B, -J),$$

which is defined in a neighbourhood $B_{\mathbb{C}}$ of (x_0, x_0). There is a natural projection $a_I : B_{\mathbb{C}} \longrightarrow (M, I)$. Consider the embedding $b_I : (B, I) \longrightarrow B_{\mathbb{C}}$, $x \longrightarrow (x, x_0)$, defined in a neighbourhood of $x_0 \in (B, I)$. In a similar way we define a_J, b_J. In [V-d2], [V-d3] we defined $\Psi_{I,J}$ as a composition $b_I \circ a_J \circ b_J \circ a_I$. Earlier in this Subsection, we described a local identification of $(B, I) \times (B, -I)$ and $\operatorname{Lin}(B)$. Clearly, under this identification, the maps a_I, b_I correspond to p_I, ϵ_I. Therefore, the definition of $\Psi_{I,J}$ given in this paper is equivalent to the definition given in [V-d2], [V-d3]. ■

6.2 The automorphism $\Psi_{I,J}$ acting on hyperholomorphic sheaves

In this section, we prove that hyperholomorphic sheaves are equivariant with respect to the map $\Psi_{I,J}$, considered as an automorphism of the local ring $\mathcal{O}_{x_0}(M, I)$.

Theorem 6.6: Let M be a hyperkähler manifold, not necessarily compact, $x_0 \in M$ a point, I an induced complex structure and F a reflexive sheaf over (M, I) equipped with a hyperholomorphic connection. Let $J \neq \pm I$ be another induced complex structure, and B_0, B the neighbourhoods of $x_0 \in M$ for which the map $\Psi_{I,J} : B_0 \longrightarrow B$ was defined in Proposition 6.5. Assume that $\Psi_{I,J} : B_0 \longrightarrow B$ is an isomorphism. Then there exists a canonical functorial isomorphism of coherent sheaves

$$\Psi^F_{I,J} : F\Big|_{B_0} \longrightarrow \Psi^*_{I,J}(F\Big|_B).$$

Proof: Return to the notation introduced in Subsection 6.1. Let $W := V_I \cap V_J$. By definition of V_I, V_J, the evaluation maps produce open embeddings

$$\epsilon v_{I,-I} : \operatorname{Lin}(W) \hookrightarrow (W, I) \times (W, -I),$$

and

$$\epsilon v_{J,-J} : \operatorname{Lin}(W) \hookrightarrow (W, J) \times (W, -J),$$

Let $S \subset W$ be the singular set of $F\Big|_W$, $\operatorname{Tw}(S) \subset \operatorname{Tw}(W)$ the corresponding embedding, and $L_0 \subset \operatorname{Lin}(W)$ be the set of all lines $l \in \operatorname{Lin}(W)$ which do not intersect $\operatorname{Tw}(S)$. Consider the maps

$$p_I : L_0 \hookrightarrow (W, I) \backslash S$$

and

$$p_J : L_0 \hookrightarrow (W, J) \backslash S$$

obtained by restricting the evaluation map $p_I : \operatorname{Lin}(M) \longrightarrow (M, I)$ to $L_0 \subset \operatorname{Lin}(M)$. Since F is equipped with a hyperholomorphic connection, the vector bundle $F\Big|_{(M,J)\backslash S}$ has a natural holomorphic structure. Let $\underline{F}_1 := p_I^*\left(F\Big|_{(M,I)\backslash S}\right)$ and $\underline{F}_2 := p_J^*\left(F\Big|_{(M,J)\backslash S}\right)$ be the corresponding pullback sheaves over L_0, and F_1, F_2 the sheaves on $\operatorname{Lin}(W)$ obtained as direct images of $\underline{F}_1$, $\underline{F}_2$ under the open embedding $L_0 \hookrightarrow \operatorname{Lin}(W)$.

Lemma 6.7: Under these assumptions, the sheaves F_1, F_2 are coherent reflexive sheaves. Moreover, there exists a natural isomorphism of coherent sheaves $\Psi_{1,2} : F_1 \longrightarrow F_2$.

Proof: The complex codimension of the singular set S in (M, I) is at least 3, because F is reflexive ([OSS], Ch. II, 1.1.10). Since S is trianalytic (Claim 3.16), this codimension is even. Thus, $\operatorname{codim}_{\mathbb{C}}(S, (M, I)) \geqslant 4$. Therefore,

$$\operatorname{codim}_{\mathbb{C}}(\operatorname{Tw}(S), \operatorname{Tw}(M)) \geqslant 4.$$

Consider the set L_S of all twistor lines $l \in \operatorname{Lin}(W)$ passing through $\operatorname{Tw}(S)$. For generic points $x, y \in \operatorname{Tw}(W)$, there exists a unique line $l \in \operatorname{Lin}(W)$ passing through x, y. Therefore,

$$\operatorname{codim}_{\mathbb{C}}(L_S, \operatorname{Lin}(W)) = \operatorname{codim}_{\mathbb{C}}(\operatorname{Tw}(S), \operatorname{Tw}(M)) - 1 \geqslant 3.$$

By definition, $L_0 := \operatorname{Lin}(W)\backslash L_S$. Since F_1, F_2 are direct images of bundles $\underline{F}_1$, $\underline{F}_2$ over a subvariety L_S of codimension 3, these sheaves are coherent and reflexive ([OSS], Ch. II, 1.1.12; see also Lemma 9.2). To show that they are naturally isomorphic it remains to construct an isomorphism between $\underline{F}_1$ and $\underline{F}_2$.

Let $\mathcal{F}$ be a coherent sheaf on $\operatorname{Tw}(W)$ obtained from $F\Big|_W$ as in the proof of Proposition 3.17. The singular set of $\mathcal{F}$ is $\operatorname{Tw}(S) \subset \operatorname{Tw}(W)$. Therefore, the restriction $\mathcal{F}\Big|_{\operatorname{Tw}(W)\backslash\operatorname{Tw}(S)}$ is a holomorphic vector bundle. For all horizontal twistor lines $l_x \subset \operatorname{Tw}(W)\backslash\operatorname{Tw}(S)$, the restriction $\mathcal{F}\Big|_{l_x}$ is clearly a trivial vector bundle over $l_x \cong \mathbb{C}P^1$. A small deformation of a trivial vector bundle is again trivial. Shrinking W if necessary, we may assume that for all lines $l \in L_0$, the restriction of $\mathcal{F}$ to $l \cong \mathbb{C}P^1$ is a trivial vector bundle.

The isomorphism $\underline{\Psi}_{1,2} : \underline{F}_1 \longrightarrow \underline{F}_2$ is constructed as follows. Let $l \in L_0$ be a twistor line. The restriction $\mathcal{F}\Big|_l$ is trivial. Consider l as a map $l : \mathbb{C}P^1 \longrightarrow \operatorname{Tw}(M)$. We identify $\mathbb{C}P^1$ with the set of induced complex structures on M. By definition, the fiber of F_1 in l is naturally identified with the space $\mathcal{F}\Big|_{l(I)}$; and the fiber of F_2 in l is identifies with $\mathcal{F}\Big|_{l(J)}$. Since $\mathcal{F}\Big|_l$ is trivial, the fibers of the bundle $\mathcal{F}\Big|_l$ are naturally identified. This provides a vector bundle isomorphism $\underline{\Psi}_{1,2} : \underline{F}_1 \longrightarrow \underline{F}_2$ mapping $\underline{F}_1\Big|_l = \mathcal{F}\Big|_{l(I)}$ to $\underline{F}_2\Big|_l = \mathcal{F}\Big|_{l(J)}$. It remains to show that this isomorphism is compatible with the holomorphic structure.

Since the bundle $\mathcal{F}\Big|_l$ is trivial, we have an identification

$$\mathcal{F}\Big|_{l(I)} \cong \mathcal{F}\Big|_{l(J)} = \Gamma(\mathcal{F}\Big|_l),$$

where $\Gamma(\mathcal{F}\Big|_l)$ is the space of global sections of $\mathcal{F}\Big|_l$. Thus, $F_i\Big|_l = \Gamma(\mathcal{F}\Big|_l)$, and this identification is clearly holomorphic. This proves Lemma 6.7. ■

We return to the proof of Theorem 6.6. Denote by F_J the restriction of $\mathcal{F}$ to $(M, J) = \pi^{-1}(J) \subset \mathrm{Tw}(M)$. The map $\Psi_{I,J}$ was defined as a composition $\epsilon_I \circ p_J \circ \epsilon_J \circ p_I$. The sheaf $p_I^* F$ is by definition isomorphic to F_1, and $p_J^* F_J$ to F_2. On the other hand, clearly, $\epsilon_J^* F_2 = F_J$. Therefore, $(p_J \circ \epsilon_J)^* F_2 \cong F_2$. Using the isomorphism $F_1 \cong F_2$, we obtain $(p_J \circ \epsilon_J)^* F_1 \cong F_1$. To sum it up, we have the following isomorphisms:

$$\begin{aligned} p_I^* F &\cong F_1, \\ (p_J \circ \epsilon_J)^* F_1 &\cong F_1, \\ \epsilon_I^* F_1 &\cong F. \end{aligned}$$

A composition of these isomorphisms gives an isomorphism

$$\Psi_{I,J}^F : F\Big|_{B_0} \longrightarrow \Psi_{I,J}^*(F\Big|_B).$$

This proves Theorem 6.6. ■

6.3 A $\mathbb{C}^*$-action on a local ring of a hyperkähler manifold

Let M be a hyperkähler manifold, non necessarily compact, $x \in M$ a point and I, J induced complex structures, $I \neq J$. Consider the complete local ring $\mathcal{O}_{x,I} := \hat{\mathcal{O}}_x(M, I)$. Throughout this section we consider the map $\Psi_{I,J}$ (Proposition 6.5) as an automorphism of the ring $\mathcal{O}_{x,I}$. Let $\mathfrak{m}$ be the maximal ideal of $\mathcal{O}_{x,I}$, and $\mathfrak{m}/\mathfrak{m}^2$ the Zariski cotangent space of (M, I) in x.

By Proposition 6.5, $\Psi_{I,J}$ acts on $\mathfrak{m}/\mathfrak{m}^2$ as a multiplication by a number $\lambda \in \mathbb{C}$, $0 < |\lambda| < 1$. (6.1)

Let V_{λ^n} be the eigenspace corresponding to the eigenvalue λ^n,

$$V_{\lambda^n} := \{v \in \mathcal{O}_{x,I} \mid \Psi_{I,J}(v) = \lambda^n v\}.$$

Clearly, $\oplus V_{\lambda^i}$ is a graded subring in $\mathcal{O}_{x,I}$. In [V-d2], (see also [V-d3]) we proved that the ring $\oplus V_{\lambda^i}$ is dense in $\mathcal{O}_{x,I}$ with respect to the adic topology. Therefore, the ring $\mathcal{O}_{x,I}$ is identified with the adic completion of $\oplus V_{\lambda^i}$.

Consider at action of $\mathbb{C}^*$ on $\oplus V_{\lambda^i}$, with $z \in \mathbb{C}^*$ acting on V_{λ^i} as a multiplication by z^i. This $\mathbb{C}^*$-action is clearly continuous, with respect to the adic topology. Therefore, it can be extended to

$$\mathcal{O}_{x,I} = \widehat{\oplus V_{\lambda^i}}.$$

Definition 6.8: Let M, I, J, x, $\mathcal{O}_{x,I}$ be as in the beginning of this Subsection. Consider the $\mathbb{C}^*$-action

$$\Psi_{I,J}(z) : \mathcal{O}_{x,I} \longrightarrow \mathcal{O}_{x,I}$$

constructed as above. Then $\Psi_{I,J}(z)$ is called **the canonical $\mathbb{C}^*$-action associated with M, I, J, x.**

In the above notation, consider a reflexive sheaf F on (M, I) equipped with a hyperholomorphic connection. Denote the germ of F at x by F_x, $F_x := F \otimes_{\mathcal{O}_{(M,I)}} \mathcal{O}_{x,I}$. From Theorem 6.6, we obtain an isomorphism $F_x \cong \Psi_{I,J}^* F_x$. This isomorphism can be interpreted as an automorphism

$$\Psi_{I,J}^F : F_x \longrightarrow F_x$$

satisfying

$$\Psi_{I,J}^F(av) = \Psi_{I,J}(a)v, \tag{6.2}$$

for all $a \in \mathcal{O}_{x,I}$, $v \in F_x$.

By (6.2), the automorphism $\Psi_{I,J}^F$ respects the filtration

$$F_x \supset \mathfrak{m}F_x \supset \mathfrak{m}^2 F_x \supset \ldots$$

Thus, it makes sense to speak of $\Psi_{I,J}^F$-action on $\mathfrak{m}^i F_x / \mathfrak{m}^{i+1} F_x$.

Lemma 6.9: The automorphism $\Psi_{I,J}^F$ acts on $\mathfrak{m}^i F_x / \mathfrak{m}^{i+1} F_x$ as a multiplication by λ^i, where $\lambda \in \mathbb{C}$ is the number considered in (6.1).

Proof: By (6.2), it suffices to prove Lemma 6.9 for $i = 0$. In other words, we have to show that $\Psi^F_{I,J}$ acts as identity on $F_x/\mathfrak{m}F_x$. We reduced Lemma 6.9 to the following claim.

Claim 6.10: In the above assumptions, the automorphism $\Psi^F_{I,J}$ acts as identity on $F_x/\mathfrak{m}F_x$.

Proof: In the course of defining the map $\Psi^F_{I,J}$, we identified the space $\operatorname{Lin}(M)$ with a complexification of (M, I), and defined the maps

$$p_I : \operatorname{Lin}(M) \longrightarrow (M, I), \quad p_J : \operatorname{Lin}(M) \longrightarrow (M, I)$$

(these maps are smooth, in a neighbourhood of $\operatorname{Hor} \subset \operatorname{Lin}(M)$, by Proposition 6.4), and

$$\epsilon_I : (B, I) \longrightarrow \operatorname{Lin}(M), \quad \epsilon_J : (B, J) \longrightarrow \operatorname{Lin}(M)$$

(these maps are locally closed embeddings). Consider (M, J) as a subvariety of $\operatorname{Tw}(M)$, $(M, J) = \pi^{-1}(J)$. Let $\mathcal{F}$ be the lift of F to $\operatorname{Tw}(M)$ (see the proof of Proposition 3.17 for details). Denote the completion of $\mathcal{O}_x(M, J)$ by $\mathcal{O}_{x,J}$. Let F_J denote the $\mathcal{O}_{x,J}$-module $\left(\mathcal{F}\big|_{(M,J)}\right) \otimes_{\mathcal{O}_{(M,J)}} \hat{\mathcal{O}}_{x,J}$. Consider the horizontal twistor line $l_x \in \operatorname{Lin}(M)$. Let $\operatorname{Lin}_x(M)$ be the spectre of the completion $\mathcal{O}_{x,\mathrm{Lin}}$ of the local ring of holomorphic functions on $\operatorname{Lin}(M)$ in l_x. The maps p_I, p_J, ϵ_I, ϵ_J can be considered as maps of corresponding formal manifolds:

$$\begin{aligned} p_I &: \operatorname{Lin}_x(M) \longrightarrow \operatorname{Spec}(\mathcal{O}_{x,I}), \\ p_J &: \operatorname{Lin}_x(M) \longrightarrow \operatorname{Spec}(\mathcal{O}_{x,J}), \\ \epsilon_I &: \operatorname{Spec}(\mathcal{O}_{x,I}) \longrightarrow \operatorname{Lin}_x(M), \\ \epsilon_J &: \operatorname{Spec}(\mathcal{O}_{x,J}) \longrightarrow \operatorname{Lin}_x(M), \end{aligned}$$

As in Subsection 6.2, we consider the $\mathcal{O}_{x,\mathrm{Lin}}$-modules $F_1 := p_I^* F_x$ and $F_2 := p_J^* F_J$. By Lemma 6.7, there exists a natural isomorphism $\Psi_{1,2} : F_1 \longrightarrow F_2$.

Let $\mathfrak{m}_{l_x}$ be the maximal ideal of $\mathcal{O}_{x,\mathrm{Lin}}$. Since the morphism p_I is smooth, the space $F_1/\mathfrak{m}_{l_x}F_1$ is naturally isomorphic to $F_x/\mathfrak{m}F_x$. Similarly, the space $F_2/\mathfrak{m}_{l_x}F_2$ is isomorphic to $F_J/\mathfrak{m}_J F_J$, where $\mathfrak{m}_J$ is the maximal ideal of $\mathcal{O}_{x,J}$.

We have a chain of isomorphisms

$$\begin{aligned} F_x/\mathfrak{m}F_x &\xrightarrow{p_I^*} F_1/\mathfrak{m}_{l_x}F_1 \xrightarrow{\Psi_{1,2}} F_2/\mathfrak{m}_{l_x}F_2 \\ &\xrightarrow{\epsilon_J^*} F_J/\mathfrak{m}_J F_J \xrightarrow{p_J^*} F_2/\mathfrak{m}_{l_x}F_2 \\ &\xrightarrow{\Psi_{1,2}^{-1}} F_1/\mathfrak{m}_{l_x}F_1 \xrightarrow{\epsilon_I^*} F_x/\mathfrak{m}F_x. \end{aligned} \tag{6.3}$$

By definition, for any $f \in F_x/\mathfrak{m}F_x$, the value of $\Psi_{I,J}^F(f)$ is given by the composide map of (6.3) applied to f. The composition

$$F_2/\mathfrak{m}_{l_x}F_2 \xrightarrow{\epsilon_J^*} F_J/\mathfrak{m}_J F_J \xrightarrow{p_J^*} F_2/\mathfrak{m}_{l_x}F_2 \tag{6.4}$$

is identity, because the spaces $F_2/\mathfrak{m}_{l_x}F_2$ and $F_J/\mathfrak{m}_J F_J$ are canonically identified, and this identification can be performed via ϵ_J^* or p_J^*. Thus, the map (6.3) is a composition

$$F_x/\mathfrak{m}F_x \xrightarrow{p_I^*} F_1/\mathfrak{m}_{l_x}F_1 \xrightarrow{\Psi_{1,2}} F_2/\mathfrak{m}_{l_x}F_2 \xrightarrow{\Psi_{1,2}^{-1}} F_1/\mathfrak{m}_{l_x}F_1 \xrightarrow{\epsilon_I^*} F_x/\mathfrak{m}F_x.$$

This map is clearly equivalent to a composition

$$F_x/\mathfrak{m}F_x \xrightarrow{p_I^*} F_1/\mathfrak{m}_{l_x}F_1 \xrightarrow{\epsilon_I^*} F_x/\mathfrak{m}F_x,$$

which is identity according to the same reasoning which proved that (6.4) is identity. We proved Claim 6.10 and Lemma 6.9. ∎

Consider the λ^n-eigenspaces F_{λ^n} of F_x. Consider the $\oplus V_{\lambda^n}$-submodule $\oplus F_{\lambda^n} \subset F_x$, where $\oplus V_{\lambda^n} \subset \mathcal{O}_{x,I}$ is the ring defined in Subsection 6.3. From Claim 6.10 and (6.1) it follows that $\oplus F_{\lambda^n}$ is dense in F_x, with respect to the adic topology on F_x. For $z \in \mathbb{C}^*$, let $\Psi_{I,J}^F(z) : \oplus F_{\lambda^n} \longrightarrow \oplus F_{\lambda^n}$ act on F_{λ^n} as a multiplication by z^n. As in Definition 6.8, we extend $\Psi_{I,J}^F(z)$ to $F_x = \widehat{\oplus F_{\lambda^n}}$. This automorphism makes F_x into a $\mathbb{C}^*$-equivariant module over $\mathcal{O}_{x,I}$

Definition 6.11: The constructed above $\mathbb{C}^*$-equivariant structure on F_x is called **the canonical $\mathbb{C}^*$-equivariant structure on F_x associated with J.**

6.4 Desingularization of $\mathbb{C}^*$-equivariant sheaves

Let M be a hyperkähler manifold, I an induced complex structure and F a reflexive sheaf with isolated singularities over (M, I), equipped with a hyperholomorphic connection. We have shown that the sheaf F admits a $\mathbb{C}^*$-equivariant structure compatible with the canonical $\mathbb{C}^*$-action on the local ring of (M, I). Therefore, Theorem 6.1 is implied by the following proposition.

Proposition 6.12: Let B be a complex manifold, $x \in B$ a point. Assume that there is an action $\Psi(z)$ of $\mathbb{C}^*$ on B which fixes x and acts on T_xB be dilatations. Let F be a reflexive coherent sheaf on B, which is locally trivial outside of x. Assume that the germ F_x of F in x is equipped with a $\mathbb{C}^*$-equivariant structure, compatible with $\Psi(z)$. Let $\widetilde{B}$ be a blow-up of B in x, and $\pi : \widetilde{B} \longrightarrow B$ the standard projection. Then the pullback sheaf $\widetilde{F} := \pi^* F$ is locally trivial on $\widetilde{B}$.

Proof: Let $C := \pi^{-1}(x)$ be the singular locus of π. The sheaf F is locally trivial outside of x. Let d be the rank of $F\Big|_{B\backslash x}$. To prove that $\widetilde{F}$ is locally trivial, we need to show that for all points $y \in \widetilde{B}$, the fiber $\pi^* F\Big|_y$ is d-dimensional. Therefore, to prove Proposition 6.12 it suffices to show that $\pi^* F\Big|_C$ is a vector bundle of dimension d.

The variety C is naturally identified with the projectivization $\mathbb{P}T_xB$ of the tangent space T_xB. The total space of T_xB is equipped with a natural action of $\mathbb{C}^*$, acting by dilatations. Clearly, coherent sheaves on $\mathbb{P}T_xB$ are in one-to-one correspondence with $\mathbb{C}^*$-equivariant coherent sheaves on T_xB. Consider a local isomorphism $\varphi : T_xB \longrightarrow B$ which is compatible with $\mathbb{C}^*$-action, maps $0 \in T_xB$ to x and acts as identity on the tangent space $T_0(T_xB) = T_xB$. The sheaf $\varphi^* F$ is $\mathbb{C}^*$-equivariant. Clearly, the corresponding sheaf on $\mathbb{P}T_xB$ is canonically isomorphic with $\pi^* F\Big|_C$. Let $l \in T_xB$ be a line passing through 0, and $l\backslash 0$ its complement to 0. Denote the corresponding point of $\mathbb{P}T_xB$ by y. The restriction $\varphi^* F\Big|_{l\backslash 0}$ is a $\mathbb{C}^*$-equivariant vector bundle. The $\mathbb{C}^*$-equivariant structure identifies all the fibers of the bundle $\varphi^* F\Big|_{l\backslash 0}$. Let F_l be one of these fibers. Clearly, the fiber of $\pi^* F\Big|_C$ in y is canonically isomorphic to F_l. Therefore, the fiber of $\pi^* F\Big|_C$ in y is d-dimensional. We proved that $\pi^* F$ is a bundle. ∎

7 Twistor transform and quaternionic-Kähler geometry

This Section is a compilation of results known from the literature. Subsection 7.1 is based on [KV] and the results of Subsection 7.2 are implicit in [KV]. Subsection 7.3 is based on [Sal], [N1] and [N2], and Subsection 7.4 is a recapitulation of the results of A. Swann ([Sw]).

7.1 Direct and inverse twistor transform

In this Subsection, we recall the definition and the main properties of the direct and inverse twistor transform for bundles over hyperkähler manifolds ([KV]).

The following definition is a non-Hermitian analogue of the notion of a hyperholomorphic connection.

Definition 7.1: Let M be a hyperkähler manifold, not necessarily compact, and (B,∇) be a vector bundle with a connection over M, not necessarily Hermitian. Assume that the curvature of ∇ is contained in the space $\Lambda^2_{inv}(M,\operatorname{End}(B))$ of $SU(2)$-invariant 2-forms with coefficients in $\operatorname{End}(B)$. Then (B,∇) is called **an autodual bundle**, and ∇ **an autodual connection.**

Let $\operatorname{Tw}(M)$ be the twistor space of M, equipped with the standard maps $\pi : \operatorname{Tw}(M) \longrightarrow \mathbb{C}P^1$, $\sigma : \operatorname{Tw}(M) \longrightarrow M$.

We introduce the direct and inverse twistor transforms which relate autodual bundles on the hyperkähler manifold M and holomorphic bundles on its twistor space $\operatorname{Tw}(M)$.

Let B be a complex vector bundle on M equipped with a connection ∇. The pullback σ^*B of B to $\operatorname{Tw}(M)$ is equipped with a pullback connection $\sigma^*\nabla$.

Lemma 7.2: ([KV], Lemma 5.1) The connection ∇ is autodual if and only if the connection $\sigma^*\nabla$ has curvature of Hodge type $(1,1)$.

Proof: Follows from Lemma 2.6. ∎

In assumptions of Lemma 7.2, consider the $(0,1)$-part $(\sigma^*\nabla)^{0,1}$ of the connection $\sigma^*\nabla$. Since $\sigma^*\nabla$ has curvature of Hodge type $(1,1)$, we have

$$\left((\sigma^*\nabla)^{0,1}\right)^2 = 0,$$

and by Proposition 2.19, this connection is integrable. Consider $(\sigma^*\nabla)^{0,1}$ as a holomorphic structure operator on σ^*B.

Let $\mathcal{A}$ be the category of autodual bundles on M, and $\mathcal{C}$ the category of holomorphic vector bundles on $\operatorname{Tw}(M)$. We have constructed a functor

$$(\sigma^*\bullet)^{0,1} : \mathcal{A} \longrightarrow \mathcal{C},$$

$\nabla \longrightarrow (\sigma^*\nabla)^{0,1}$. Let $s \in \operatorname{Hor} \subset \operatorname{Tw}(M)$ be a horizontal twistor line (Subsection 6.1). For any $(B,\nabla) \in \mathcal{A}$, consider corresponding holomorphic vector bundle $(\sigma^*B, (\sigma^*\nabla)^{0,1})$. The restriction of $(\sigma^*B, (\sigma^*\nabla)^{0,1})$ to $s \cong \mathbb{C}P^1$ is a trivial vector bundle. A converse statement is also true. Denote by $\mathcal{C}_0$ the category of holomorphic vector bundles C on $\operatorname{Tw}(M)$, such that the restriction of C to any horizontal twistor line is trivial.

Theorem 7.3: Consider the functor

$$(\sigma^*\bullet)^{0,1} : \mathcal{A} \longrightarrow \mathcal{C}_0$$

constructed above. Then it is an equivalence of categories.

Proof: [KV], Theorem 5.12. ∎

Definition 7.4: Let M be a hyperkähler manifold, $\operatorname{Tw}(M)$ its twistor space and $\mathcal{F}$ a holomorphic vector bundle. We say that $\mathcal{F}$ is **compatible with twistor transform** if the restriction of C to any horizontal twistor line $s \in \operatorname{Tw}(M)$ is a trivial bundle on $s \cong \mathbb{C}P^1$.

Recall that a connection ∇ in a vector bundle over a complex manifold is called $(1,1)$-connection if its curvature is of Hodge type $(1,1)$.

Remark 7.5: Let $\mathcal{F}$ be a holomorphic bundle over $\operatorname{Tw}(M)$ which is compatible with twistor transform. Then $\mathcal{F}$ is equipped with a natural $(1,1)$-connection $\nabla_{\mathcal{F}} = \sigma^*\nabla$, where (B,∇) is the corresponding autodual bundle over M. The connection $\nabla_{\mathcal{F}}$ is not, generally speaking, Hermitian, or compatible with a Hermitian structure.

7.2 Twistor transform and Hermitian structures on vector bundles

Results of this Subsection were implicit in [KV], but in this presentation, they are new.

Let M be a hyperkähler manifold, not necessarily compact, and $\operatorname{Tw}(M)$ its twistor space. In Subsection 7.1, we have shown that certain holomorphic vector bundles over $\operatorname{Tw}(M)$ admit a canonical (1,1)-connection $\nabla_{\mathcal{F}}$ (Remark 7.5). This connection can be non-Hermitian. Here we study the Hermitian structures on $(\mathcal{F}, \nabla_{\mathcal{F}})$ in terms of holomorphic properties of $\mathcal{F}$.

Definition 7.6: Let F be a real analytic complex vector bundle over a real analytic manifold $X_{\mathbb{R}}$, and $h: F \times F \longrightarrow \mathbb{C}$ a $\mathcal{O}_{X_{\mathbb{R}}}$-linear pairing on F. Then h is called **semilinear** if for all $\alpha \in \mathcal{O}_{X_{\mathbb{R}}} \otimes_{\mathbb{R}} \mathbb{C}$, we have

$$h(\alpha x, y) = \alpha \cdot h(x, y), \quad \text{and} \quad h(x, \alpha y) = \overline{\alpha} \cdot h(x, y).$$

For X a complex manifold and F a holomorphic vector bundle, by a semilinear pairing on F we understand a semilinear pairing on the underlying real analytic bundle. Clearly, a real analytic Hermitian metric is always semilinear.

Let X be a complex manifold, $I: TX \longrightarrow TX$ the complex structure operator, and $i: X \longrightarrow X$ a real analytic map. We say that X is anticomplex if the induced morphism of tangent spaces satisfies $i \circ I = -I \circ I$. For a complex vector bundle F on X, consider the complex adjoint vector bundle $\overline{F}$, which coincides with F as a real vector bundle, with $\mathbb{C}$-action which is conjugate to that defined on F. Clearly, for every holomorphic vector bundle F, and any anticomplex map $i: X \longrightarrow X$, the bundle $i^*\overline{F}$ is equipped with a natural holomorphic structure.

Let M be a hyperkähler manifold, and $\operatorname{Tw}(M)$ its twistor space. Recall that $\operatorname{Tw}(M) = \mathbb{C}P^1 \times M$ is equipped with a canonical anticomplex involution ι, which acts as identity on M and as central symmetry $I \longrightarrow -I$ on $\mathbb{C}P^1 = S^2$. For any holomorphic bundle $\mathcal{F}$ on $\operatorname{Tw}(M)$, consider the corresponding holomorphic bundle $\iota^*\overline{\mathcal{F}}$.

Let M be a hyperkähler manifold, I an induced complex structure, F a vector bundle over M, equipped with an autodual connection ∇, and $\mathcal{F}$ the corresponding holomorphic vector bundle over $\operatorname{Tw}(M)$, equipped with a canonical connection $\nabla_{\mathcal{F}}$. As usually, we identify (M, I) and the fiber $\pi^{-1}(I)$

of the twistor projection $\pi : \operatorname{Tw}(M) \longrightarrow \mathbb{C}P^1$. Let $\nabla_{\mathcal{F}} = \nabla_I^{1,0} + \nabla_I^{0,1}$ be the Hodge decomposition of ∇ with respect to I.

> Clearly, the operator $\nabla_I^{1,0}$ can be considered as a holomorphic structure operator on F, considered as a complex vector bundle over $(M, -I)$. (7.1)

Then the holomorphic structure operator on $\mathcal{F}\Big|_{(M,I)}$ is equal to $\nabla_I^{0,1}$, and the holomorphic structure operator on $\mathcal{F}\Big|_{(M,-I)}$ is equal to $\nabla_I^{1,0}$.

Assume that the bundle $(\mathcal{F}, \nabla_{\mathcal{F}})$ is equipped with a non-degenerate semilinear pairing h which is compatible with the connection. Consider the natural connection $\nabla_{\mathcal{F}^*}$ on the dual bundle to $\mathcal{F}$, and its Hodge decomposition (with respect to I)

$$\nabla_{\mathcal{F}^*} = \nabla_{\mathcal{F}^*}^{1,0} + \nabla_{\mathcal{F}^*}^{0,1}.$$

Clearly, the pairing h gives a C^∞-isomorphism of $\mathcal{F}$ and the complex conjugate of its dual bundle, denoted as $\overline{\mathcal{F}}^*$. Since h is semilinear and compatible with the connection, it maps the holomorphic structure operator $\nabla_I^{0,1}$ to the complex conjugate of $\nabla_{\mathcal{F}^*}^{1,0}$. On the other hand, the operator $\nabla_{\mathcal{F}^*}^{1,0}$ is a holomorphic structure operator in $\mathcal{F}^*\Big|_{(M,-I)}$, as (7.1) claims. We obtain that the map h can be considered as an isomorphism of holomorphic vector bundles

$$h : \mathcal{F} \longrightarrow (\iota^* \overline{\mathcal{F}})^*.$$

This correspondence should be thought of as a (direct) twistor transform for bundles with a semilinear pairing.

Proposition 7.7: (direct and inverse twistor transform for bundles with semilinear pairing) Let M be a hyperkähler manifold, and $\mathcal{C}_{sl}$ the category of autodual bundles over M equipped with a non-degenerate semilinear pairing. Consider the category $\mathcal{C}_{hol,sl}$ of holomorphic vector bundles $\mathcal{F}$ on $\operatorname{Tw}(M)$, compatible with twistor transform and equipped with an isomorphism

$$\mathfrak{h} : \mathcal{F} \longrightarrow (\iota^* \overline{\mathcal{F}})^*.$$

Let $\mathcal{T} : \mathcal{C}_{sl} \longrightarrow \mathcal{C}_{hol,sl}$ be the functor constructed above. Then $\mathcal{T}$ is an isomorphism of categories.

Proof: Given a pair $\mathcal{F}, \mathfrak{h} : \mathcal{F} \longrightarrow (\iota^* \overline{\mathcal{F}})^*$, we need to construct a non-degenerate semilinear pairing h on $\mathcal{F}\Big|_{(M,I)}$, compatible with a connection.

Since $\mathcal{F}$ is compatible with twistor transform, it is a pullback of a bundle (F, ∇) on M. This identifies the real analytic bundles $\mathcal{F}\Big|_{(M,I')}$, for all induced complex structures I'. Taking $I' = \pm I$, we obtain an identification of the C^∞-bundles $\mathcal{F}\Big|_{(M,I)}$, $\mathcal{F}\Big|_{(M,-I)}$. Thus, $\mathfrak{h}$ can be considered as an isomorphism of $F = \mathcal{F}\Big|_{(M,I)}$ and $(\overline{\mathcal{F}})^*\Big|_{(M,I)}$. This allows one to consider $\mathfrak{h}$ as a semilinear form h on F. We need only to show that h is compatible with the connection ∇. Since $\nabla_{\mathcal{F}}$ is an invariant of holomorphic structure, the map $\mathfrak{h} : \mathcal{F} \longrightarrow (\iota^*\overline{\mathcal{F}})^*$ is compatible with the connection $\nabla_{\mathcal{F}}$. Thus, the obtained above form h is compatible with the connection $\nabla_{\mathcal{F}}\Big|_{(M,I)} = \nabla$. This proves Proposition 7.7. ∎

7.3 B_2-bundles on quaternionic-Kähler manifolds

Definition 7.8: ([Sal], [Bes]) Let M be a Riemannian manifold. Consider a bundle of algebras $\operatorname{End}(TM)$, where TM is the tangent bundle to M. Assume that $\operatorname{End}(TM)$ contains a 4-dimensional bundle of subalgebras $W \subset \operatorname{End}(TM)$, with fibers isomorphic to a quaternion algebra $\mathbb{H}$. Assume, moreover, that W is closed under the transposition map $\perp : \operatorname{End}(TM) \longrightarrow \operatorname{End}(TM)$ and is preserved by the Levi-Civita connection. Then M is called **quaternionic-Kähler.**

Example 7.9: Consider the quaternionic projective space

$$\mathbb{H}P^n = (\mathbb{H}^n \backslash 0)/\mathbb{H}^*.$$

It is easy to see that $\mathbb{H}P^n$ is a quaternionic-Kähler manifold. For more examples of quaternionic-Kähler manifolds, see [Bes].

A quaternionic-Kähler manifold is Einstein ([Bes]), i. e. its Ricci tensor is proportional to the metric: $Ric(M) = c \cdot g$, with $c \in \mathbb{R}$. When $c = 0$, the manifold M is hyperkähler, and its restricted holonomy group is $Sp(n)$; otherwise, the restricted holonomy is $Sp(n) \cdot Sp(1)$. The number c is called **the scalar curvature** of M. Further on, we shall use the term *quaternionic-Kähler manifold* for manifolds with non-zero scalar curvature.

The quaternionic projective space $\mathbb{H}P^n$ has positive scalar curvature.

The quaternionic projective space is the only example of quaternionic-Kähler manifold which we need, in the course of this paper. However, the formalism of quaternionic-Kähler manifolds is very beautiful and significantly simplifies the arguments, so we state the definitions and results for a general quaternionic-Kähler manifold whenever possible.

Let M be a quaternionic-Kähler manifold, and $W \subset \operatorname{End}(TM)$ the corresponding 4-dimensional bundle. For $x \in M$, consider the set $\mathcal{R}_x \subset W\Big|_x$, consisting of all $I \in W\Big|_x$ satisfying $I^2 = -1$. Consider $\mathcal{R}_x$ as a Riemannian submanifold of the total space of $W\Big|_x$. Clearly, $\mathcal{R}_x$ is isomorphic to a 2-dimensional sphere. Let $\mathcal{R} = \cup_x \mathcal{R}_x$ be the corresponding spherical fibration over M, and $\operatorname{Tw}(M)$ its total space. The manifold $\operatorname{Tw}(M)$ is equipped with an almost complex structure, which is defined in the same way as the almost complex structure for the twistor space of a hyperkähler manifold. This almost complex structure is known to be integrable (see [Sal]).

Definition 7.10: ([Sal], [Bes]) Let M be a quaternionic-Kähler manifold. Consider the complex manifold $\operatorname{Tw}(M)$ constructed above. Then $\operatorname{Tw}(M)$ is called **the twistor space of M.**

Note that (unlike in the hyperkähler case) the space $\operatorname{Tw}(M)$ is Kähler. For quaternionic-Kähler manifolds with positive scalar curvature, the anticanonical bundle of $\operatorname{Tw}(M)$ is ample, so $\operatorname{Tw}(M)$ is a Fano manifold.

Quaternionic-Kähler analogue of a twistor transform was studied by T. Nitta in a serie of papers ([N1], [N2] etc.) It turns out that the picture given in [KV] for Kähler manifolds is very similar to that observed by T. Nitta.

A role of $SU(2)$-invariant 2-forms is played by the so-called B_2-forms.

Definition 7.11: Let $SO(TM) \subset \operatorname{End}(TM)$ be a group bundle of all orthogonal automorphisms of TM, and $G_M := W \cap SO(TM)$. Clearly, the fibers of G_M are isomorphic to $SU(2)$. Consider the action of G_M on the bundle of 2-forms $\Lambda^2(M)$. Let $\Lambda^2_{inv}(M) \subset \Lambda^2(M)$ be the bundle of G_M-invariants. The bundle $\Lambda^2_{inv}(M)$ is called **the bundle of B_2-forms.** In a similar fashion we define B_2-forms with coefficients in a bundle.

Definition 7.12: In the above assumptions, let (B, ∇) be a bundle with connection over M. The bundle B is called **a B_2-bundle**, and ∇ is called **a B_2-connection**, if its curvature is a B_2-form.

Consider the natural projection $\sigma : \operatorname{Tw}(M) \longrightarrow M$. The proof of the following claim is completely analogous to the proof of Lemma 2.6 and Lemma 7.2.

Claim 7.13:

(i) Let ω be a 2-form on M. The pullback $\sigma^*\omega$ is of type $(1,1)$ on $\operatorname{Tw}(M)$ if and only if ω is a B_2-form on M.

(ii) Let B be a complex vector bundle on M equipped with a connection ∇, not necessarily Hermitian. The pullback σ^*B of B to $\operatorname{Tw}(M)$ is equipped with a pullback connection $\sigma^*\nabla$. Then, ∇ is a B_2-connection if and only if $\sigma^*\nabla$ has curvature of Hodge type $(1,1)$.

■

There exists an analogue of direct and inverse twistor transform as well.

Theorem 7.14: For any B_2-connection (B, ∇), consider the corresponding holomorphic vector bundle

$$(\sigma^*B, (\sigma^*\nabla)^{0,1}).$$

The restriction of $(\sigma^*B, (\sigma^*\nabla)^{0,1})$ to a line $\sigma^{-1}(m) \cong \mathbb{C}P^1$ is a trivial vector bundle, for any point $m \in M$. Denote by $\mathcal{C}_0$ the category of holomorphic vector bundles C on $\operatorname{Tw}(M)$, such that the restriction of C to $\sigma^{-1}(m)$ is trivial, for all $m \in M$, and by $\mathcal{A}$ the category of B_2-bundles (not necessarily Hermitian). Consider the functor

$$(\sigma^*\bullet)^{0,1} : \mathcal{A} \longrightarrow \mathcal{C}_0$$

constructed above. Then it is an equivalence of categories.

Proof: It is easy to modify the proof of the direct and inverse twistor transform theorem from [KV] to work in quaternionic-Kähler situation. ■

We will not use Theorem 7.14, except for its consequence, which was proven in [N1].

Corollary 7.15: Consider the functor

$$(\sigma^*\bullet)^{0,1} : \mathcal{A} \longrightarrow \mathcal{C}_0$$

constructed in Theorem 7.14. Then $(\sigma^*\bullet)^{0,1}$ gives an injection κ from the set of equivalence classes of Hermitian B_2-connections over M to the set of equivalence classes of holomorphic connections over $\mathrm{Tw}(M)$.
■

Let M be a quaternionic-Kähler manifold. The space $\mathrm{Tw}(M)$ has a natural Kähler metric g, such that the standard map $\sigma: \mathrm{Tw}(M) \longrightarrow M$ is a Riemannian submersion, and the restriction of g to the fibers $\sigma^{-1}(m)$ of σ is a metric of constant curvature on $\sigma^{-1}(m) = \mathbb{C}P^1$ ([Sal], [Bes]).

Example 7.16: In the case $M = \mathbb{H}P^n$, we have $\mathrm{Tw}(M) = \mathbb{C}P^{2n+1}$, and the Kähler metric g is proportional to the Fubini-Study metric on $\mathbb{C}P^{2n+1}$.

Theorem 7.17: (T. Nitta) Let M be a quaternionic-Kähler manifold of positive scalar curvature, $\mathrm{Tw}(M)$ its twistor space, equipped with a natural Kähler structure, and B a Hermitian B_2-bundle on M. Consider the pull-back σ^*B, equipped with a Hermitian connection. Then σ^*B is a Yang-Mills bundle on $\mathrm{Tw}(M)$, and $\deg c_1(\sigma^*B) = 0$.
Proof: [N2]. ■

Let κ be the map considered in Corollary 7.15. Assume that M is a compact manifold. In [N2], T. Nitta defined the moduli space of Hermitian B_2-bundles. By Uhlenbeck-Yau theorem, Yang-Mills bundles are polystable. Then the map κ provides an embedding from the moduli of non-decomposable Hermitian B_2-bundles to the moduli $\mathcal{M}$ of stable bundles on $\mathrm{Tw}(M)$. The image of κ is a totally real subvariety in $\mathcal{M}$ ([N2]).

7.4 Hyperkähler manifolds with special $\mathbb{H}^*$-action and quaternionic-Kähler manifolds of positive scalar curvature

Further on, we shall need the following definition.

Definition 7.18: An almost hypercomplex manifold is a smooth manifold M with an action of quaternion algebra in its tangent bundle For each $L \in \mathbb{H}$, $L^2 = -1$, L gives an almost complex structure on M. The manifold M is caled **hypercomplex** if the almost complex structure L is integrable, for all possible choices $L \in \mathbb{H}$.

The twistor space for a hypercomplex manifold is defined in the same way as for hyperkähler manifolds. It is also a complex manifold ([K]). The formalism of direct and inverse twistor transform can be repeated for hypercomplex manifolds verbatim.

Let $\mathbb{H}^*$ be the group of non-zero quaternions. Consider an embedding $SU(2) \hookrightarrow \mathbb{H}^*$. Clearly, every quaternion $h \in \mathbb{H}^*$ can be uniquely represented as $h = |h| \cdot g_h$, where $g_h \in SU(2) \subset \mathbb{H}^*$. This gives a natural decomposition $\mathbb{H}^* = SU(2) \times \mathbb{R}^{>0}$. Recall that $SU(2)$ acts naturally on the set of induced complex structures on a hyperkähler manifold.

Definition 7.19: Let M be a hyperkähler manifold equipped with a free smooth action ρ of the group $\mathbb{H}^* = SU(2) \times \mathbb{R}^{>0}$. The action ρ is called **special** if the following conditions hold.

(i) The subgroup $SU(2) \subset \mathbb{H}^*$ acts on M by isometries.

(ii) For $\lambda \in \mathbb{R}^{>0}$, the corresponding action $\rho(\lambda) : M \longrightarrow M$ is compatible with the hyperholomorphic structure (which is a fancy way of saying that $\rho(\lambda)$ is holomorphic with respect to any of induced complex structures).

(iii) Consider the smooth $\mathbb{H}^*$-action $\rho_e : \mathbb{H}^* \times \operatorname{End}(TM) \longrightarrow \operatorname{End}(TM)$ induced on $\operatorname{End}(TM)$ by ρ. For any $x \in M$ and any induced complex structure I, the restriction $I\Big|_x$ can be considered as a point in the total space of $\operatorname{End}(TM)$. Then, for all induced complex structures I, all $g \in SU(2) \subset \mathbb{H}^*$, and all $x \in M$, the map $\rho_e(g)$ maps $I\Big|_x$ to $g(I)\Big|_{\rho_e(g)(x)}$.

Speaking informally, this can be stated as "$\mathbb{H}^*$-action interchanges the induced complex structures".

(iv) Consider the automorphism of S^2T^*M induced by $\rho(\lambda)$, where $\lambda \in \mathbb{R}^{>0}$. Then $\rho(\lambda)$ maps the Riemannian metric tensor $s \in S^2T^*M$ to $\lambda^2 s$.

Example 7.20: Consider the flat hyperkähler manifold $M_{\mathrm{fl}} = \mathbb{H}^n \backslash 0$. There is a natural action of $\mathbb{H}^*$ on $\mathbb{H}^n \backslash 0$. This gives a special action of $\mathbb{H}^*$ on M_{fl}.

The case of a flat manifold $M_{fl} = \mathbb{H}^n \backslash 0$ is the only case where we apply the results of this section. However, the general statements are just as difficult to prove, and much easier to comprehend.

Definition 7.21: Let M be a hyperkähler manifold with a special action ρ of $\mathbb{H}^*$. Assume that $\rho(-1)$ acts non-trivially on M. Then $M/\rho(\pm 1)$ is also a hyperkähler manifold with a special action of $\mathbb{H}^*$. We say that the manifolds (M, ρ) and $(M/\rho(\pm 1), \rho)$ are **hyperkähler manifolds with special action of $\mathbb{H}^*$ which are special equivalent**. Denote by H_{sp} the category of hyperkähler manifolds with a special action of $\mathbb{H}^*$ defined up to special equivalence.

A. Swann ([Sw]) developed an equivalence between the category of quaternionic-Kähler manifolds of positive scalar curvature and the category H_{sp}. The purpose of this Subsection is to give an exposition of Swann's formalism.

Let Q be a quaternionic-Kähler manifold. The restricted holonomy group of Q is $Sp(n) \cdot Sp(1)$, that is, $(Sp(n) \times Sp(1))/\{\pm 1\}$. Consider the principal bundle $\mathcal{G}$ with the fiber $Sp(1)/\{\pm 1\}$, corresponding to the subgroup

$$Sp(1)/\{\pm 1\} \subset (Sp(n) \times Sp(1))/\{\pm 1\}.$$

of the holonomy. There is a natural $Sp(1)/\{\pm 1\}$-action on the space $\mathbb{H}^*/\{\pm 1\}$. Let

$$\mathcal{U}(Q) := \mathcal{G} \times_{Sp(1)/\{\pm 1\}} \mathbb{H}^*/\{\pm 1\}.$$

Clearly, $\mathcal{U}(Q)$ is fibered over Q, with fibers which are isomorphic to $\mathbb{H}^*/\{\pm 1\}$. We are going to show that the manifold $\mathcal{U}(Q)$ is equipped with a natural hypercomplex structure.

There is a natural smooth decomposition $\mathcal{U}(Q) \cong \mathcal{G} \times \mathbb{R}^{>0}$ which comes from the isomorphism $\mathbb{H}^* \cong Sp(1) \times \mathbb{R}^{>0}$.

Consider the standard 4-dimensional bundle W on Q. Let $x \in Q$ be a point. The fiber $W\big|_q$ is isomorphic to $\mathbb{H}$, in a non-canonical way. The choices of isomorphism $W\big|_q \cong \mathbb{H}$ are called **quaternion frames in** q. The set of quaternion frames gives a fibration over Q, with a fiber $\operatorname{Aut}(\mathbb{H}) \cong Sp(1)/\{\pm 1\}$. Clearly, this fibration coincides with the principal bundle $\mathcal{G}$ constructed above. Since $\mathcal{U}(Q) \cong \mathcal{G} \times \mathbb{R}^{>0}$, a choice of $u \in \mathcal{U}(Q)\big|_q$ determines an isomorphism $W\big|_q \cong \mathbb{H}$.

Let (q, u) be the point of $\mathcal{U}(Q)$, with $q \in Q$, $u \in \mathcal{U}(Q)\Big|_q$. The natural connection in $\mathcal{U}(Q)$ gives a decomposition

$$T_{(q,u)}U(Q) = T_u\left(\mathcal{U}(Q)\Big|_q\right) \oplus T_qQ.$$

The space $\mathcal{U}(Q)\Big|_q \cong \mathbb{H}^*/\{\pm 1\}$ is equipped with a natural hypercomplex structure. This gives a quaternion action on $T_u\left(\mathcal{U}(Q)\Big|_q\right)$ The choice of $u \in \mathcal{U}(Q)\Big|_q$ determines a quaternion action on T_qQ, as we have seen above. We obtain that the total space of $\mathcal{U}(Q)$ is an almost hypercomplex manifold.

Proposition 7.22: (A. Swann) Let Q be a quaternionic-Kähler manifold. Consider the manifold $\mathcal{U}(Q)$ constructed as above, and equipped with a quaternion algebra action in its tangent space. Then $\mathcal{U}(Q)$ is a hypercomplex manifold.

Proof: Clearly, the manifold $\mathcal{U}(Q)$ is equipped with a $\mathbb{H}^*$-action, which is related with the almost hypercomplex structure as prescribed by Definition 7.19 (ii)-(iii). Pick an induced complex structure $I \in \mathbb{H}$. This gives an algebra embedding $\mathbb{C} \longrightarrow \mathbb{H}$. Consider the corresponding $\mathbb{C}^*$-action ρ_I on an almost complex manifold $(\mathcal{U}(Q), I)$. This $\mathbb{C}^*$-action is compatible with the almost complex structure. The quotient $\mathcal{U}(Q)/\rho(I)$ is an almost complex manifold, which is naturally isomorphic to the twistor space $\operatorname{Tw}(Q)$. Let L^* be a complex vector bundle of all $(1,0)$-vectors $v \in T(\operatorname{Tw}(Q))$ tangent to the fibers of the standard projection $\sigma: \operatorname{Tw}(Q) \longrightarrow Q$, and L be the dual vector bundle. Denote by $Tot_{\neq 0}(L)$ the complement $\operatorname{Tot}(L)\backslash N$, where $N = \operatorname{Tw}(Q) \subset \operatorname{Tot}(L)$ is the zero section of L. Using the natural connection in L, we obtain an almost complex structure on $\operatorname{Tot}(L)$.

Consider the natural projection $\varphi: Tot_{\neq 0}(L) \longrightarrow Q$. The fibers $\varphi^{-1}(q)$ of φ are identified with the space of non-zero vectors in the total space of the cotangent bundle $T^*\sigma^{-1}(q) \cong T^*(\mathbb{C}P^1)$. This space is naturally isomorphic to

$$\mathcal{G}\Big|_q \times \mathbb{R}^{>0} = \mathcal{U}(Q)\Big|_q \cong \mathbb{H}^*/\{\pm 1\}.$$

This gives a canonical isomorphism of almost complex manifolds

$$(\mathcal{U}(Q), I) \longrightarrow Tot_{\neq 0}(L).$$

Therefore, to prove that $(\mathcal{U}(Q), I)$ is a complex manifold, it suffices to show that the natural almost complex structure on $Tot_{\neq 0}(L) \subset \operatorname{Tot}(L)$ is integrable. Consider the natural connection ∇_L on L. To prove that $\operatorname{Tot}(L)$ is a complex manifold, it suffices to show that ∇_L is a holomorphic connection. The bunlde L is known under the name of **holomorphic contact bundle**, and it is known to be holomorphic ([Sal], [Bes]). ∎

Remark 7.23: The result of Proposition 7.22 is well known. We have given its proof because we shall need the natural identification $Tot_{\neq 0}(L) \cong \mathcal{U}(Q)$ further on in this paper.

Theorem 7.24: Let Q be a quaternionic-Kähler manifold of positive scalar curvature, and $\mathcal{U}(Q)$ the hypercomplex manifold constructed above. Then $\mathcal{U}(Q)$ admits a unique (up to a scaling) hyperkähler metric compatible with the hypercomplex structure.

Proof: [Sw]. ∎

Consider the action of $\mathbb{H}^*$ on $\mathcal{U}(M)$ defined in the proof of Proposition 7.22. This action satisfies the conditions (ii) and (iii) of Definition 7.19. The conditions (i) and (iv) of Definition 7.19 are easy to check (see [Sw] for details). This gives a functor from the category $\mathcal{C}$ of quaternionic-Kähler manifolds of positive scalar curvature to the category H_{sp} of Definition 7.21.

Theorem 7.25: The functor $Q \longrightarrow \mathcal{U}(Q)$ from $\mathcal{C}$ to H_{sp} is an equivalence of categories.

Proof: [Sw]. ∎

The inverse functor from H_{sp} to C is constructed by taking a quotient of M by the action of $\mathbb{H}^*$. Using the technique of quaternionic-Kähler reduction anf hyperkähler potentials ([Sw]), one can equip the quotient $M/\mathbb{H}^*$ with a natural quaternionic-Kähler structure.

8 $\mathbb{C}^*$-equivariant twistor spaces

In Section 7, we gave an exposition of the twistor transform, B_2-bundles and Swann's formalism. In the present Section, we give a synthesis of these theories, obtaining a construction with should be thought of as Swann's formalism for vector bundles.

Consider the equivalence of categories $Q \longrightarrow \mathcal{U}(Q)$ constructed in Theorem 7.25 (we call this equivalence "Swann's formalism"). We show that B_2-bundles on Q are in functorial bijective correspondence with $\mathbb{C}^*$-equivariant holomorphic bundles on $\operatorname{Tw}(\mathcal{U}(Q))$ (Theorem 8.5).

In Subsection 8.4, this equivalence is applied to the vector bundle $\pi^*(F)$ of Theorem 6.1. We use it to construct a canonical Yang-Mills connection on $\pi^*(F)\Big|_C$, where C is a special fiber of $\pi : \widetilde{M} \longrightarrow (M, I)$ (see Theorem 6.1 for details and notation). This implies that the holomorphic bundle $\pi^*(F)\Big|_C$ is polystable (Theorem 8.15).

8.1 B_2-bundles on quaternionic-Kähler manifolds and $\mathbb{C}^*$-equivariant holomorphic bundles over twistor spaces

For the duration of this Subsection, we fix a hyperkähler manifold M, equipped with a special $\mathbb{H}^*$-action ρ, and the corresponding quaternionic-Kähler manifold $Q = M/\mathbb{H}^*$. Denote the natural quotient map by $\varphi : M \longrightarrow Q$.

Lemma 8.1: Let ω be a 2-form over Q, and $\varphi^*\omega$ its pullback to M. Then the following conditions are equivalent

(i) ω is a B_2-form

(ii) $\varphi^*\omega$ is of Hodge type $(1, 1)$ with respect to some induced complex structure I on M

(iii) $\varphi^*\omega$ is $SU(2)$-invariant.

Proof: Let I be an induced complex structure on M. As we have shown in the proof of Proposition 7.22, the complex manifold (M, I) is identified with an open subset of the total space $\operatorname{Tot}(L)$ of a holomorphic line bundle L over $\operatorname{Tw}(Q)$. The map φ is represented as a composition of the projections $h : \operatorname{Tot}(L) \longrightarrow \operatorname{Tw}(Q)$ and $\sigma_Q : \operatorname{Tw}(Q) \longrightarrow Q$. Since the map h is smooth and holomorphic, the form $\varphi^*\omega$ is of Hodge type $(1, 1)$ if and only if $\sigma_Q^*\omega$ is of type $(1, 1)$. By Claim 7.13 (i), this happens if and only if ω is a B_2-form. This proves an equivalence (i) $\Leftrightarrow$ (ii). Since the choice of I is arbitrary, the pullback $\varphi^*\omega$ of a B_2-form is of Hodge type $(1, 1)$ with respect to all induced complex structures. By Lemma 2.6, this proves the implication (i) $\Rightarrow$ (iii). The implication (iii) $\Rightarrow$ (ii) is clear. ∎

Proposition 8.2: Let (B, ∇) be a complex vector bundle with connection over Q, and $(\varphi^* B, \varphi^* \nabla)$ its pullback to M. Then the following conditions are equivalent

(i) (B, ∇) is a B_2-form

(ii) The curvature of $(\varphi^* B, \varphi^* \nabla)$ is of Hodge type $(1,1)$ with respect to some induced complex structure I on M

(iii) The bundle $(\varphi^* B, \varphi^* \nabla)$ is autodual

Proof: Follows from Lemma 8.1 applied to $\omega = \nabla^2$. ∎

For any point $I \in \mathbb{C}P^1$, consider the corresponding algebra embedding $\mathbb{C} \stackrel{c_I}{\hookrightarrow} \mathbb{H}$. Let ρ_I be the action of $\mathbb{C}^*$ on (M, I) obtained as a restriction of ρ to $c_I(\mathbb{C}^*) \subset \mathbb{H}^*$. Clearly from Definition 7.19 (ii), ρ_I acts on (M, I) by holomorphic automorphisms.

Consider $\operatorname{Tw}(M)$ as a union

$$\operatorname{Tw}(M) = \bigcup_{I \in \mathbb{C}P^1} \pi^{-1}(I), \quad \pi^{-1}(I) = (M, I)$$

Gluing $\rho(I)$ together, we obtain a smooth $\mathbb{C}^*$-action $\rho_{\mathbb{C}}$ on $\operatorname{Tw}(M)$.

Claim 8.3: Consider the action $\rho_{\mathbb{C}} : \mathbb{C}^* \times \operatorname{Tw}(M) \longrightarrow \operatorname{Tw}(M)$ constructed above. Then $\rho_{\mathbb{C}}$ is holomorphic.

Proof: It is obvious from construction that $\rho_{\mathbb{C}}$ is compatible with the complex structure on $\operatorname{Tw}(M)$. ∎

Example 8.4: Let $M = \mathbb{H}^n \backslash 0$. Since $\operatorname{Tw}(\mathbb{H}^n)$ is canonically isomorphic to a total space of the bundle $\mathcal{O}(1)^n$ over $\mathbb{C}P^1$, the twistor space $\operatorname{Tw}(M)$ is $\operatorname{Tot}(\mathcal{O}(1)^n)$ without zero section. The group $\mathbb{C}^*$ acts on $\operatorname{Tot}(\mathcal{O}(1)^n)$ by dilatation, and the restriction of this action to $\operatorname{Tw}(M)$ coincides with $\rho_{\mathbb{C}}$.

Consider the map $\sigma : \operatorname{Tw}(M) \longrightarrow M$. Let (B, ∇) be a B_2-bundle over Q. Since the bundle $(\varphi^* B, \varphi^* \nabla)$ is autodual, the curvature of $\sigma^* \varphi^* \nabla$ has type $(1, 1)$. Let $(\sigma^* \varphi^* B, (\sigma^* \varphi^* \nabla)^{0,1})$ be the holomorphic bundle obtained from $(\varphi^* B, \varphi^* \nabla)$ by twistor transform. Clearly, this bundle is $\mathbb{C}^*$-equivariant, with respect to the natural $\mathbb{C}^*$-action on $\operatorname{Tw}(M)$. It turns out that any $\mathbb{C}^*$-equivariant bundle $\mathcal{F}$ on $\operatorname{Tw}(M)$ can be obtained this way, assuming that $\mathcal{F}$ is compatible with twistor transform.

Theorem 8.5: In the above assumptions, let $\mathcal{C}_{B_2}$ be the category of of B_2-bundles on Q, and $\mathcal{C}_{\mathrm{Tw},\mathbb{C}^*}$ the category of $\mathbb{C}^*$-equivariant holomorphic bundles on $\mathrm{Tw}(M)$ which are compatible with the twistor transform. Consider the functor

$$(\sigma^*\varphi^*)^{0,1} : \mathcal{C}_{B_2} \longrightarrow \mathcal{C}_{\mathrm{Tw},\mathbb{C}^*},$$

$(B, \nabla) \longrightarrow (\sigma^*\varphi^* B, (\sigma^*\varphi^*\nabla)^{0,1})$, constructed above. Then $(\sigma^*\varphi^*)^{0,1}$ establishes an equivalence of categories.

We prove Theorem 8.5 in Subsection 8.3.

Remark 8.6: Let Q be an arbitrary quaternionic-Kähler manifold, and $M = \mathcal{U}(Q)$ the corresponding fibration. Then M is hypercomplex, and its twistor space is equipped with a natural holomorphic action of $\mathbb{C}^*$. This gives necessary ingredients needed to state Theorem 8.5 for Q with negative scalar curvature. The proof which we give for Q with positive scalar curvature will in fact work for all quaternionic-Kähler manifolds.

Question 8.7: What happens with this construction when Q is a hyperkähler manifold?

In this paper, we need Theorem 8.5 only in the case $Q = \mathbb{H}P^n$, $M = \mathbb{H}^n\backslash 0$, but the general proof is just as difficult.

8.2 $\mathbb{C}^*$-equivariant bundles and twistor transform

Let M be a hyperkähler manifold, and $\mathrm{Tw}(M)$ its twistor space. Recall that $\mathrm{Tw}(M) = \mathbb{C}P^1 \times M$ is equipped with a canonical anticomplex involution ι, which acts as identity on M and as central symmetry $I \longrightarrow -I$ on $\mathbb{C}P^1 = S^2$.

Proposition 8.8: Let M be a hyperkähler manifold, and $\mathrm{Tw}(M)$ its twistor space. Assume that $\mathrm{Tw}(M)$ is equipped with a free holomorphic action $\rho(z) : \mathrm{Tw}(M) \longrightarrow \mathrm{Tw}(M)$ of $\mathbb{C}^*$, acting along the fibers of $\pi : \mathrm{Tw}(M) \longrightarrow \mathbb{C}P^1$. Assume, moreover, that $\iota \circ \rho(z) = \rho(\overline{z}) \circ \iota$, where ι is the natural anticomplex involution of $\mathrm{Tw}(M)$.[1] Let $\mathcal{F}$ be a $\mathbb{C}^*$-equivariant holomorphic vector bundle on $\mathrm{Tw}(M)$. Assume that $\mathcal{F}$ is compatible with

[1]These assumptions are automatically satisfied when M is equipped with a special $\mathbb{H}^*$-action, and $\rho(z)$ is the corresponding $\mathbb{C}^*$-action on $\mathrm{Tw}(M)$.

the twistor transform. Let $\nabla_{\mathcal{F}}$ be the natural connection on $\mathcal{F}$ (Remark 7.5). Then $\nabla_{\mathcal{F}}$ is flat along the leaves of ρ.

Proof: First of all, let us recall the construction of the natural connection $\nabla_{\mathcal{F}}$. Let $\mathcal{F}$ be an arbitrary bundle compatible with the twistor transform. We construct $\nabla_{\mathcal{F}}$ in terms of the isomorphism $\Psi_{1,2}$ defined in Lemma 6.7.

Consider an induced complex structure I. Let F_I be the restriction of $\mathcal{F}$ to $(M,I) = \pi^{-1}(I) \subset \operatorname{Tw}(M)$. Consider the evaluation map

$$p_I : \operatorname{Lin}(M) \longrightarrow (M,I)$$

(Subsection 6.1). In a similar way we define the holomorphic vector bundle F_{-I} on $(M,-I)$ and the map $p_{-I} : \operatorname{Lin}(M) \longrightarrow (M,-I)$. Denote by F_1, F_{-1} the sheaves $p_I^*(F_I)$, $p_{-I}^*(F_{-I})$. In Lemma 6.7, we constructed an isomorphism $\Psi_{1,-1} : F_1 \longrightarrow F_{-1}$.

Let us identify $\operatorname{Lin}(M)$ with $(M,I) \times (M,I)$ (this identification is naturally defined in a neighbourhood of $\operatorname{Hor} \subset \operatorname{Lin}(M)$ – see Proposition 6.4). Then the maps p_I, p_{-I} became projections to the relevant components. Let

$$\overline{\partial} : F_1 \longrightarrow F_1 \otimes p_I^*\Omega^1(M,-I),$$
$$\partial : F_{-1} \longrightarrow F_{-1} \otimes p_{-I}^*\Omega^1(M,I),$$

be the sheaf maps obtained as pullbacks of de Rham differentials (the tensor product is taken in the category of coherent sheaves over $\operatorname{Lin}(M)$). Twisting ∂ by an isomorphism $\Psi_{1,-1} : F_1 \longrightarrow F_{-1}$, we obtain a map

$$\partial^{\Psi} : F_1 \longrightarrow F_1 \otimes p_I^*\Omega^1(M,I).$$

Adding $\overline{\partial}$ and ∂^{Ψ}, we obtain

$$\nabla : F_1 \longrightarrow F_1 \otimes \left(p_I^*\Omega^1(M,I) \oplus p_I^*\Omega^1(M,-I) \right).$$

Clearly, ∇ satisfies the Leibniz rule. Moreover, the sheaf $p_I^*\Omega^1(M,I) \oplus p_I^*\Omega^1(M,-I)$ is naturally isomorphic to the sheaf of differentials over

$$\operatorname{Lin}(M) = (M,I) \times (M,-I).$$

Therefore, ∇ can be considered as a connection in F_1, or as a real analytic connection in a real analytic complex vector bundle underlying F_I. From the definition of $\nabla_{\mathcal{F}}$ ([KV]), it is clear that $\nabla_{\mathcal{F}}\big|_{(M,I)}$ equals ∇.

Return to the proof of Proposition 8.8. Consider a $\mathbb{C}^*$-action $\rho_I(z)$ on (M, I), $(M, -I)$ induced from the natural embeddings $(M, I) \hookrightarrow \operatorname{Tw}(M)$, $(M, -I) \hookrightarrow \operatorname{Tw}(M)$. Then F_I is a $\mathbb{C}^*$-equivariant bundle. Since $\iota \circ \rho(z) = \rho(\bar{z}) \circ \iota$, the identification $\operatorname{Lin}(M) = (M, I) \times (M, I)$ is compatible with $\mathbb{C}^*$-action. Let $\mathbf{r} = \frac{d}{dr}$ be the holomorphic vector field on (M, I) corresponding to the $\mathbb{C}^*$-action. To prove Proposition 8.8, we have to show that the operator

$$[\nabla_{\mathbf{r}}, \nabla_{\bar{\mathbf{r}}}] : F_I \longrightarrow F_I \otimes \Lambda^{1,1}(M, I)$$

vanishes.

Consider the equivariant structure operator

$$\rho(z)^F : \rho_I(z)^* F_I \longrightarrow F_I.$$

Let U be a $\mathbb{C}^*$-invariant Stein subset of (M, I). Consider $\rho(z)^F$ an an endomorphism of the space of global holomorphic sections $\Gamma_U(F_I)$. Let

$$D_r(f) := \lim_{\varepsilon \to 0} \frac{\rho_I(1+\varepsilon)}{\varepsilon},$$

for $f \in \Gamma_U(F_I)$. Clearly, D_r is a well defined sheaf endomorphism of F_I, satisfying

$$D_r(\alpha \cdot f) = \frac{d}{dr}\alpha \cdot f + \alpha \cdot D_r(f),$$

for all $\alpha \in \mathcal{O}_{(M,I)}$. We say that a holomorphic section f of F_I is $\mathbb{C}^*$-**invariant** if $D_r(f) = 0$. Clearly, the $\mathcal{O}_{(M,I)}$-sheaf F_I is generated by $\mathbb{C}^*$-invariant sections. Therefore, it suffices to check the equality

$$[\nabla_{\mathbf{r}}, \nabla_{\bar{\mathbf{r}}}](f) = 0$$

for holomorphic $\mathbb{C}^*$-invariant $f \in F_I$.

Since f is holomorphic, we have $\nabla_{\bar{\mathbf{r}}} f = 0$. Thus,

$$[\nabla_{\mathbf{r}}, \nabla_{\bar{\mathbf{r}}}](f) = \nabla_{\bar{\mathbf{r}}} \nabla_{\mathbf{r}}(f).$$

We obtain that Proposition 8.8 is implied by the following lemma.

Lemma 8.9: In the above assumptions, let f be a $\mathbb{C}^*$-invariant section of F_I. Then $\nabla_{\mathbf{r}}(f) = 0$.

Proof: Return to the notation we used in the beginning of the proof of Proposition 8.8. Then, $\nabla(f) = \bar{\partial}(f) + \partial^{\Psi}(f)$. Since f is holomorphic,

$\overline{\partial}(f) = 0$, so we need to show that $\partial^\Psi(f)(\mathbf{r}) = 0$. By definition of ∂^Ψ, this is equivalent to proving that

$$\partial\Psi_{1,-1}(f)(\mathbf{r}) = 0.$$

Consider the $\mathbb{C}^*$-action on $\mathrm{Lin}(M)$ which is induced by the $\mathbb{C}^*$-action on $\mathrm{Tw}(M)$. Since the maps p_I, p_{-I} are compatible with the $\mathbb{C}^*$-action, the sheaves F_1, F_{-1} are $\mathbb{C}^*$-equivariant. We can repeat the construction of the operator D_r for the sheaf F_{-I}. This allows one to speak of holomorphic $\mathbb{C}^*$-invariant sections of F_{-I}. Pick a $\mathbb{C}^*$-invariant Stein subset $U \subset (M, -I)$. Since the statement of Lemma 8.9 is local, we may assume that $M = U$. Let $g_1, ..., g_n$ be a set of $\mathbb{C}^*$-invariant sections of F_I which generated F_I. Then, the sections $p_{-I}^*(g_1), ..., p_{-I}^*(g_n)$ generate F_{-1}. Consider the section $\Psi_{1,-1}(f)$ of F_{-1}. Clearly, $\Psi_{1,-1}$ commutes with the natural $\mathbb{C}^*$-action. Therefore, the section $\Psi_{1,-1}(f)$ is $\mathbb{C}^*$-invariant, and can be written as

$$\Psi_{1,-1}(f) = \sum \alpha_i p_{-I}^*(g_i),$$

where the functions α_i are $\mathbb{C}^*$-invariant. By definition of ∂ we have

$$\partial\left(\sum \alpha_i p_{-I}^*(g_i)\right) = \sum \partial(\alpha_i p_{-I}^*(g_i)) + \sum \alpha_i \partial(p_{-I}^*(g_i)).$$

On the other hand, g_i is a holomorphic section of F_{-I}, so $\partial p_{-I}^*(g_i) = 0$. We obtain

$$\partial\left(\sum \alpha_i p_{-I}^* \cdot (g_i)\right) = \sum \partial\alpha_i p_{-I}^*(g_i).$$

Thus,

$$\partial\Psi_{1,-1}(f)(\mathbf{r}) = \sum \frac{\partial\alpha_i}{\partial r} p_{-I}^*(g_i),$$

but since the functions α_i are $\mathbb{C}^*$-invariant, their derivatives along $\mathbf{r}$ vanish. We obtain $\partial\Psi_{1,-1}(f)(\mathbf{r}) = 0$. This proves Lemma 8.9. Proposition 8.8 is proven. ∎

8.3 Twistor transform and the $\mathbb{H}^*$-action

For the duration of this Subsection, we fix a hyperkähler manifold M, equipped with a special $\mathbb{H}^*$-action ρ, and the corresponding quaternionic-Kähler manifold $Q = M/\mathbb{H}^*$. Denote the natural quotient map by $\varphi : M \longrightarrow Q$. Clearly, Theorem 8.5 is an immediate consequence of the following theorem.

Theorem 8.10: Let $\mathcal{F}$ be a $\mathbb{C}^*$-equivariant holomorphic bundle over $\operatorname{Tw}(M)$, which is compatible with the twistor transform. Consider the natural connection $\nabla_{\mathcal{F}}$ on $\mathcal{F}$. Then $\nabla_{\mathcal{F}}$ is flat along the leaves of $\mathbb{H}^*$-action.

Proof: The leaves of $\mathbb{H}^*$-action are parametrized by the points of $q \in Q$. Consider such a leaf $M_q := \varphi^{-1}(q) \subset M$. Clearly, M_q is a hyperkähler submanifold in M, equipped with a special action of $\mathbb{H}^*$. Moreover, the restriction of $\mathcal{F}$ to $\operatorname{Tw}(M_q) \subset \operatorname{Tw}(M)$ satisfies assumptions of Theorem 8.10. To prove that $\nabla_{\mathcal{F}}$ is flat along the leaves of $\mathbb{H}^*$-action, we have to show that $\mathcal{F}\Big|_{\operatorname{Tw}(M_q)}$ is flat, for all q. Therefore, it suffices to prove Theorem 8.10 for $\dim_{\mathbb{H}} M = 1$.

Lemma 8.11: We work in notation and assumptions of Theorem 8.10. Assume that $\dim_{\mathbb{H}} M = 1$. Then the connection $\nabla_{\mathcal{F}}$ is flat.

Proof: Let I be an induced complex structure, and $F_I := F\Big|_{(M,I)}$ the corresponding holomorphic bundle on (M, I). Denote by z_I the vector field corresponding to the $\mathbb{C}^*$-action ρ_I on (M, I). By definition, the connection $\nabla\Big|_{F_I}$ has $SU(2)$-invariant curvature Θ_I. On the other hand, $\Theta_I(z_I, \overline{z}_I) = 0$ by Proposition 8.8. Since $\nabla_{\mathcal{F}} = \sigma^*\nabla$ is a pullback of an autodual connection ∇ on M, its curvature is a pullback of Θ_I. In particular, $\Theta = \Theta_I$ is independent from the choice of induced complex structure I. We obtain that $\Theta(z_I, \overline{z}_I) = 0$ for all induced complex structures I on M.

Now Lemma 8.11 is implied by the following linear-algebraic claim.

Claim 8.12: Let M be a hyperkähler manifold equipped with a special $\mathbb{H}^*$-action, $\dim_{\mathbb{H}} M = 1$. Consider the vectors z_I, $\overline{z}_I$ defined above. Let Θ be a smooth $SU(2)$-invariant 2-form, such that for all induced complex structures, I, we have $\Theta(z_I, \overline{z}_I) = 0$. Then $\Theta = 0$.

Proof: The proof of Claim 8.12 is an elementary calculation. Fix a point $m_0 \in M$. Consider the flat hyperkähler manifold $\mathbb{H}\backslash 0$, equipped with a natural special action of $\mathbb{H}^*$. From the definition of a special action, it is clear that the map ρ defines a covering $\mathbb{H}\backslash 0 \longrightarrow M$, $h \longrightarrow \rho(h)m_0$ of hyperkähler manifolds, and this covering is compatible with the special action. Therefore, the hyperkähler manifold M is flat, and the $\mathbb{H}^*$-action is linear in the flat coordinates.

Let

$$\Lambda^2(M) = \Lambda^+(M) \oplus \Lambda^-(M)$$

be the standard decomposition of $\Lambda^2(M)$ according to the eigenvalues of the Hodge $*$ operator. Consider the natural Hermitian metric on $\Lambda^2(M)$. Then $\Lambda^-(M)$ is the bundle of $SU(2)$-invariant 2-forms (see, e. g., [V1]), and $\Lambda^+(M)$ is its orthogonal complement. Consider the corresponding orthogonal projection $\Pi : \Lambda^2(M) \longrightarrow \Lambda^-(M)$. Denote by $dz_I \wedge d\bar{z}_I$ the differential form which is dual to the bivector $z_I \wedge \bar{z}_I$. Let $R \subset \Lambda^-(M)$ be the $C^\infty(M)$-subsheaf of $\Lambda^-(M)$ generated by $\Pi(dz_I \wedge d\bar{z}_I)$, for all induced complex structures I on M. Clearly, $\Theta \in \Lambda^-(M)$ and Θ is orthogonal to $R \subset \Lambda^-(M)$. Therefore, to prove that $\Theta = 0$ it suffices to show that $R = \Lambda^-(M)$. Since M is covered by $\mathbb{H}\backslash 0$, we may prove $R = \Lambda^-(M)$ in assumption $M = \mathbb{H}\backslash 0$.

Let γ be the real vector field corresponding to dilatations of $M = \mathbb{H}\backslash 0$, and $d\gamma$ the dual 1-form. Clearly,

$$dz_I \wedge d\bar{z}_I = 2\sqrt{-1}\, d\gamma \wedge I(d\gamma).$$

Averaging $d\gamma \wedge I(d\gamma)$ by $SU(2)$, we obtain

$$\Pi(dz_I \wedge d\bar{z}_I) = \sqrt{-1}\left(d\gamma \wedge I(d\gamma) - J(d\gamma) \wedge K(d\gamma)\right)$$

where I, J, K is the standard triple of generators for quaternion algebra. Similarly,

$$\Pi(dz_J \wedge d\bar{z}_J) = \sqrt{-1}\left(d\gamma \wedge J(d\gamma) + K(d\gamma) \wedge I(d\gamma)\right)$$

and

$$\Pi(dz_K \wedge d\bar{z}_K) = \sqrt{-1}\left(d\gamma \wedge K(d\gamma) + I(d\gamma) \wedge J(d\gamma)\right)$$

Thus, $\Pi(R)$ is a 3-dimensional sub-bundle of $\Lambda^-(M)$. Since $\dim \Lambda^-(M) = 3$, we have $\Pi(R) = \Lambda^-(M)$. This proves Claim 8.12. Lemma 8.11 and Theorem 8.10 is proven. ∎

8.4 Hyperholomorphic sheaves and $\mathbb{C}^*$-equivariant bundles over $M_{\mathfrak{n}}$

Let M be a hyperkähler manifold, I an induced complex structure and F a reflexive sheaf over (M, I), equipped with a hyperholomorphic connection. Assume that F has an isolated singularity in $x \in M$. Consider the sheaf $\mathcal{F}$ on $\mathrm{Tw}(M)$ corresponding to $\mathcal{F}$ as in the proof of Proposition 3.17. Let $s_x \subset \mathrm{Tw}(M)$ be the horizontal twistor line corresponding to x, and $\mathfrak{m}$ its ideal. Consider the associated graded sheaf of $\mathfrak{m}$. Denote by Tw^{gr} the spectre of this associated graded sheaf. Clearly, Tw^{gr} is naturally isomorphic to $\operatorname{Tw}(T_xM)$, where T_xM is the flat hyperkähler manifold corresponding to the space T_xM with induced quaternion action. Consider the natural $\mathbb{H}^*$-action on T_xM. This provides the hyperkähler manifold $T_xM\backslash 0$ with a special $\mathbb{H}^*$-action.

Let $s_0 \subset \operatorname{Tw}^{gr}$ be the horizontal twistor line corresponding to s_x. The space $\operatorname{Tw}^{gr}\backslash s_0$ is equipped with a holomorphic $\mathbb{C}^*$-action (Claim 8.3). Denote by $\mathcal{F}^{gr}$ the sheaf on Tw^{gr} associated with $\mathcal{F}$. Clearly, $\mathcal{F}^{gr}$ is $\mathbb{C}^*$-equivariant. In order to be able to apply Theorem 8.5 and Theorem 8.10 to $\mathcal{F}^{gr}\Big|_{\operatorname{Tw}^{gr}\backslash s_0}$, we need only to show that $\mathcal{F}^{gr}$ is compatible with twistor transform.

Proposition 8.13: Let M be a hyperkähler manifold, I an induced complex structure and F a reflexive sheaf over (M, I), equipped with a hyperholomorphic connection. Assume that $\mathcal{F}$ has an isolated singularity in $x \in M$. Let $\mathcal{F}^{gr}$ be the $\mathbb{C}^*$-equivariant bundle on $\operatorname{Tw}^{gr}\backslash s_0$ constructed above. Then

(i) the bundle $\mathcal{F}^{gr}$ is compatible with twistor transform.

(ii) Moreover, the natural connection $\nabla_{\mathcal{F}^{gr}}$ (Remark 7.5) is Hermitian.

Proof: The argument is clear, but cumbersome, and essentially hinges on taking associate graded quotients everywhere and checking that all equations remain true. We give a simplified version of the proof, which omits some details and notation.

Consider the bundle $\mathcal{F}\Big|_{M\backslash s_x}$. This bundle is compatible with twistor transform, and therefore, is equipped with a natural connection $\nabla_{\mathcal{F}}$. This connection is constructed using the isomorphism $\Psi_{1,-1} : F_1 \longrightarrow F_{-1}$ (see the proof of Proposition 8.8). We apply the same consideration to $\mathcal{F}^{gr}\Big|_{(T_xM,I)}$, and show that the resulting connection $\nabla_{\mathcal{F}^{gr}}$ is hyperholomorphic. This implies that $\mathcal{F}^{gr}$ admits a $(1,1)$-connection which is a pullback of some connection on $\mathcal{F}^{gr}\Big|_{(T_xM,I)}$. This argument is used to prove that $\mathcal{F}^{gr}$ is compatible with the twistor transform.

We use the notation introduced in the proof of Proposition 8.8. Let Lin^{gr} be the space of twistor maps in Tw^{gr}. Consider the maps $p^{gr}_{\pm I} : \operatorname{Lin}^{gr} \longrightarrow (T_xM, \pm I)$ and the sheaves $F^{gr}_{\pm 1} := (p^{gr}_{\pm I})^* \mathcal{F}^{gr}_{\pm I}$ obtained in the same way as the maps $p_{\pm I}$ and the sheaves $F_{\pm 1}$ from the corresponding associated graded objects. Taking the associated graded of $\Psi_{1,-1}$ gives an isomorphism $\Psi^{gr}_{1,-1} : F^{gr}_1 \longrightarrow F^{gr}_{-1}$. Using the same construction as in the proof of Proposition 8.8, we obtain a connection operator

$$\overline{\partial}^{gr} + \partial^{\Psi^{gr}} = \nabla^{gr}_I : F^{gr}_1 \longrightarrow F^{gr}_1 \otimes \Big((p^{gr}_{-I})^* \Omega^1(T_x M, I) \oplus (p^{gr}_I)^* \Omega^1(T_x M, -I) \Big).$$

Since $(\overline{\partial}^{gr})^2 = (\partial^{\Psi^{gr}})^2 = 0$, the curvature of ∇^{gr}_I has Hodge type $(1,1)$ with respect to I. To prove that ∇^{gr}_I is hyperholomorphic, we need to show that the curvature of ∇^{gr}_I has type $(1,1)$ with respect to every induced complex structure. Starting from another induced complex structure J, we obtain a connection ∇^{gr}_J, with the curvature of type $(1,1)$ with respect to J. To prove that ∇^{gr}_J is hyperholomorphic it remains to show that $\nabla^{gr}_J = \nabla^{gr}_I$.

Let ∇_I, ∇_J be the corresponding operators on F_1. From the construction, it is clear that ∇^{gr}_I, ∇^{gr}_J are obtained from ∇_I, ∇_J by taking the associated graded quotients. On the other hand, $\nabla_I = \nabla_J$. Therefore, the connections ∇^{gr}_I and ∇^{gr}_J are equal. We proved that the bundle $\mathcal{F}^{gr}\Big|_{\mathrm{Tw}^{gr}\backslash s_0}$ is compatible with the twistor transform. To prove Proposition 8.13, it remains to show that the natural connection on $\mathcal{F}^{gr}$ is Hermitian.

The bundle $\mathcal{F}\Big|_{\mathrm{Tw}(M\backslash x_0)}$ is by definition Hermitian. Consider the corresponding isomorphism $\mathcal{F} \longrightarrow (\iota^* \overline{\mathcal{F}})^*$ (Proposition 7.7). Taking an associate graded map, we obtain an isomorphism

$$\mathcal{F}^{gr} \tilde{\rightarrow} (\iota^* \overline{\mathcal{F}^{gr}})^*.$$

This gives a non-degenerate semilinear form h^{gr} on $\mathcal{F}^{gr}$. It remains only to show that h^{gr} is pseudo-Hermitian (i. e. satisfies $h(x,y) = \overline{h(y,x)}$) and positive definite.

Let $M^{gr}_{\mathbb{C}}$ be a complexification of $M^{gr} = T_x M$, $M^{gr}_{\mathbb{C}} = \mathrm{Lin}(M^{gr})$. Consider the corresponding complex vector bundle $\mathcal{F}^{gr}_{\mathbb{C}}$ over $M^{gr}_{\mathbb{C}}$ underlying $\mathcal{F}^{gr}$. The metric h^{gr} can be considered as a semilinear form $\mathcal{F}^{gr}_{\mathbb{C}} \times \mathcal{F}^{gr}_{\mathbb{C}} \longrightarrow \mathcal{O}_{M^{gr}_{\mathbb{C}}}$. This semilinear form is obtained from the corresponding form h on $\mathcal{F}$ by taking the associate graded quotients. Since h is Hermitian, the form h^{gr} is pseudo-Hermitian. To prove that h^{gr} is positive semidefinite, we need to show that for all $f \in \mathcal{F}^{gr}_{\mathbb{C}}$, the function $h^{gr}(f, \overline{f})$ belongs to $\mathcal{O}_{M^{gr}_{\mathbb{C}}} \cdot \overline{\mathcal{O}}_{M^{gr}_{\mathbb{C}}}$, where $\mathcal{O}_{M^{gr}_{\mathbb{C}}} \cdot \overline{\mathcal{O}}_{M^{gr}_{\mathbb{C}}}$ denotes the $\mathbb{R}^{>0}$-semigroup of $\mathcal{O}_{M^{gr}_{\mathbb{C}}}$ generated by $x \cdot \overline{x}$, for all $x \in \mathcal{O}_{M^{gr}_{\mathbb{C}}}$. A similar property for h holds, because h is positive definite. Clearly, taking associated graded quotient of the semigroup $\mathcal{O}_{M_{\mathbb{C}}} \cdot \overline{\mathcal{O}}_{M_{\mathbb{C}}}$, we obtain $\mathcal{O}_{M^{gr}_{\mathbb{C}}} \cdot \overline{\mathcal{O}}_{M^{gr}_{\mathbb{C}}}$. Thus,

$$h^{gr}(f, \overline{f}) \in \left(\mathcal{O}_{M_{\mathbb{C}}} \cdot \overline{\mathcal{O}}_{M_{\mathbb{C}}}\right)^{gr} = \mathcal{O}_{M^{gr}_{\mathbb{C}}} \cdot \overline{\mathcal{O}}_{M^{gr}_{\mathbb{C}}}$$

This proves that h^{gr} is positive semidefinite. Since h^{gr} is non-degenerate, this form in positive definite. Proposition 8.13 is proven. ∎

Remark 8.14: Return to the notations of Theorem 6.1. Consider the bundle $\pi^* F\big|_C$, where $C = \mathbb{P}T_x M$ is the blow-up divisor. Clearly, this bundle corresponds to the graded sheaf $F_I^{gr} = \mathcal{F}^{gr}\big|_{(M,I)}$ on $(T_x M, I)$. By Proposition 8.13 (see also Theorem 8.10), the bundle $\pi^* F\big|_C$ is equipped with a natural $\mathbb{H}^*$-invariant connection and Hermitian structure.[2] The sheaf $\pi^* F\big|_{\widetilde{M}\backslash C}$ is a hyperholomorphic bundle over $\widetilde{M}\backslash C \cong M\backslash x_0$. Therefore, $\pi^* F\big|_{\widetilde{M}\backslash C}$ is equipped with a natural metric and a hyperholomorphic connection. It is expected that the natural connection and metric on $\pi^* F\big|_{\widetilde{M}\backslash C}$ can be extended to $\pi^* F$, and the rectriction of the resulting connection and metric to $\pi^* F\big|_C$ coincides with that given by Proposition 8.13 and Theorem 8.10. This will give an alternative proof of Proposition 8.13 (ii), because a continuous extension of a positive definite Hermitian metric is a positive semidefinite Hermitian metric.

8.5 Hyperholomorphic sheaves and stable bundles on $\mathbb{C}P^{2n+1}$

The purpose of the current Section was to prove the following result, which is a consequence of Proposition 8.13 and Theorem 8.10.

Theorem 8.15: Let M be a hyperkähler manifold, I an induced complex structure and F a reflexive sheaf on (M, I) admitting a hyperholomorphic connection. Assume that F has an isolated singularity in $x \in M$, and is locally trivial outside of x. Let $\pi : \widetilde{M} \longrightarrow (M, I)$ be the blow-up of (M, I) in x. Consider the holomorphic vector bundle $\pi^* F$ on $\widetilde{M}$ (Theorem 6.1). Let $C \subset (M, I)$ be the blow-up divisor, $C = \mathbb{P}T_x M$. Then the holomorphic bundle $\pi^* F\big|_C$ admits a natural Hermitian connection ∇ which is flat along the leaves of the natural $\mathbb{H}^*$-action on $\mathbb{P}T_x M$. Moreover, the connection ∇ is Yang-Mills, with respect to the Fubini-Study metric on $C = \mathbb{P}T_x M$, the degree $\deg c_1 \left(\pi^* F\big|_C\right)$ vanishes, and the holomorphic vector bundle $\pi^* F\big|_C$ is polystable.

---

[2] As usually, coherent sheaves over projective variety X correspond to finitely generated graded modules over the graded ring $\oplus \Gamma(\mathcal{O}_X(i))$.

Proof: By definition, coherent sheaves on $C = \mathbb{P}T_xM$ correspond bijectively to $\mathbb{C}^*$-equivariant sheaves on $T_xM\backslash 0$. Let F^{gr} be the associated graded sheaf of F (Subsection 8.4). Consider F^{gr} as a bundle on $T_xM\backslash 0$. In the notation of Proposition 8.13, $F^{gr} = \mathcal{F}^{gr}\Big|_{(M,I)}$. By Proposition 8.13, the sheaf $\mathcal{F}^{gr}$ is $\mathbb{C}^*$-equivariant and compatible with the twistor transform. According to the Swann's formalism for bundles (Theorem 8.10), the bundle $F^{gr}\Big|_{T_xM\backslash 0}$ is equipped with a natural Hermitian connection $\nabla_{F^{gr}}$ which is flat along the leaves of $\mathbb{H}^*$-action. Let $(B, \nabla_{\mathbb{H}})$ be the corresponding B_2-bundle on

$$\mathbb{P}_{\mathbb{H}}T_xM := \left(T_xM\backslash 0\right)/\mathbb{H}^* \cong \mathbb{H}P^n.$$

Then $\pi^*F\Big|_C$ is a holomorphic bundle over the corresponding twistor space $C = \operatorname{Tw}(\mathbb{P}_{\mathbb{H}}T_xM)$, obtained as a pullback of $(B, \nabla_{\mathbb{H}})$ as in Claim 7.13. The natural Kähler metric on the twistor space $C = \operatorname{Tw}(\mathbb{P}_{\mathbb{H}}T_xM)$ is the Fubini-Study metric (Example 7.16). By Theorem 7.17, the bundle $\pi^*F\Big|_C$ is Yang-Mills and has $\deg c_1\left(\pi^*F\Big|_C\right) = 0$. Finally, by Uhlenbeck-Yau theorem (Theorem 2.24), the bundle $\pi^*F\Big|_C$ is polystable. ∎

9 Moduli spaces of hyperholomorphic sheaves and bundles

9.1 Deformation of hyperholomorphic sheaves with isolated singularities

The following theorem is an elementary consequence of Theorem 8.15. The proof uses well known results on stability and reflexization (see, for instance, [OSS]). The main idea of the proof is the following. Given a family of hyperholomorphic sheaves with an isolated singularity, we blow-up this singularity and restrict the obtained family to a blow-up divisor. We obtain a family of coherent sheaves $\mathfrak{V}_s$, $s \in S$ over $\mathbb{C}P^{2n+1}$, with fibers semistable of slope zero. Assume that for all $s \in S$, $s \neq s_0$, the sheaf $\mathfrak{V}_s$ is trivial. Then the family $\mathfrak{V}$ is also trivial, up to a reflexization.

We use the following property of reflexive sheaves.

Definition 9.1: Let X be a complex manifold, and F a torsion-free coherent sheaf. We say that F is **normal** if for all open subvarieties $U \subset X$, and all closed subvarieties $Y \subset U$ of codimension 2, the restriction

$$\Gamma_U(F) \longrightarrow \Gamma_{U\backslash Y}(F)$$

is an isomorphism.

Lemma 9.2: Let X be a complex manifold, and F a torsion-free coherent sheaf. Then F is reflexive if and only if F is normal.

Proof: [OSS], Lemma 1.1.12. ∎

Theorem 9.3: Let M be a hyperkähler manifol, I an induced complex structure, S a complex variety and $\mathfrak{F}$ a family of coherent sheaves over $(M, I) \times S$. Consider the sheaf $F_{s_0} := \mathfrak{F}\big|_{(M,I)\times\{s_0\}}$. Assume that the sheaf F_{s_0} is equipped with a filtration ξ. Let F_i, $i = 1, ..., m$ denote the associated graded components of ξ, and F_i^{**} denote their reflexizations. Assume that $\mathfrak{F}$ is locally trivial outside of $(x_0, s_0) \in (M, I) \times S$. Assume, moreover, that all sheaves F_i^{**}, $i = 1, ..., m$ admit a hyperholomorphic connection. Then the reflexization $\mathfrak{F}^{**}$ is locally trivial.

Proof: Clearly, it suffices to prove Theorem 9.3 for $\mathfrak{F}$ reflexive. Let $\tilde{X}$ be the blow-up of $(M, I) \times S$ in $\{x_0\} \times S$, and $\tilde{\mathfrak{F}}$ the pullback of $\mathfrak{F}$ to $\tilde{X}$. Clearly, $\tilde{X} = \widetilde{M} \times S$, where $\widetilde{M}$ is a blow-up of (M, I) in x_0. Denote by $C \subset \widetilde{M}$ the blow-up divisor of $\widetilde{M}$. Taking S sufficiently small, we may assume that the bundle $\mathfrak{F}\big|_{\{x_0\}\times(S\backslash\{s_0\})}$ is trivial. Thus, the bundle $\tilde{\mathfrak{F}}\big|_{(C\times S)\backslash(C\times\{s_0\})}$, which is a pullback of $\mathfrak{F}\big|_{\{x_0\}\times(S\backslash\{s_0\})}$ under the natural projection $(C \times S)\backslash(C \times \{s_0\}) \longrightarrow \{x_0\} \times (S\backslash\{s_0\}$ is trivial. To prove that $\mathfrak{F}$ is locally trivial, we have to show that $\tilde{\mathfrak{F}}$ is locally trivial, and that the restriction of $\tilde{\mathfrak{F}}$ to $C \times S$ is trivial along the fibers of the natural projection $C \times S \longrightarrow S$. Clearly, to show that $\tilde{\mathfrak{F}}$ is locally trivial we need only to prove that the fiber $\tilde{\mathfrak{F}}\big|_z$ has constant dimension for all $z \in C \times S$. Thus, $\tilde{\mathfrak{F}}$ is locally trivial if and only if $\tilde{\mathfrak{F}}\big|_{C\times S}$ is locally trivial. This sheaf is reflexive, since it corresponds to an associate graded sheaf of a reflexive sheaf, in the sense of Footnote to Remark 8.14. It is non-singular in codimension 2, because all reflexive sheaves are non-singular in codimension 2 ([OSS], Ch. II, Lemma 1.1.10). By Theorem 8.15, the sheaf $\tilde{\mathfrak{F}}\big|_{C\times\{s\}}$ is semistable of slope zero. Theorem 9.3 is implied by the following lemma, applied to the sheaf $\tilde{\mathfrak{F}}\big|_{C\times S}$.

Lemma 9.4: Let C be a complex projective space, S a complex variety and $\mathfrak{F}$ a torsion-free sheaf over $C \times S$. Consider an open set $U \stackrel{j}{\hookrightarrow} C \times S$, which is a complement of $C \times \{s_0\} \subset C \times S$. Assume that the sheaf $\mathfrak{F}\big|_U$ is trivial: $\mathfrak{F}\big|_U \cong \mathcal{O}_U^n$. Assume, moreover, that $\mathfrak{F}$ is non-singular in codimension 2, the sheaf $\left(\mathfrak{F}\big|_{C\times\{s_0\}}\right)^{**}$ is semistable of slope zero and

$$\operatorname{rk}\mathfrak{F} = \operatorname{rk}\mathfrak{F}\big|_{C\times\{s_0\}}.$$

Then the reflexization $\mathfrak{F}^{**}$ of $\mathfrak{F}$ is a trivial bundle.

Proof: Using induction, it suffices to prove Lemma 9.4 assuming that it is proven for all $\mathfrak{F}'$ with $\operatorname{rk}\mathfrak{F}' < \operatorname{rk}\mathfrak{F}$. We may also assume that S is Stein, smooth and 1-dimensional.

Step 1: *We construct an exact sequence*

$$0 \longrightarrow \mathfrak{F}_2 \longrightarrow \mathfrak{F} \longrightarrow \operatorname{im} p_{O_1} \longrightarrow 0$$

of sheaves of positive rank, which, as we prove in Step 3, satisfy assumptions of Lemma 9.4.

Consider the pushforward sheaf $j_*\mathcal{O}_U^n$. From the definition of j_*, we obtain a canonical map

$$\mathfrak{F} \longrightarrow j_*\mathcal{O}_U^n, \tag{9.1}$$

and the kernel of this map is a torsion subsheaf in $\mathfrak{F}$.

Let f be a coordinate function on S, which vanishes in $s_0 \in S$. Clearly,

$$j_*\mathcal{O}_U^n \cong \mathcal{O}_{C\times S}^n\left[\frac{1}{f}\right].$$

On the other hand, the sheaf $\mathcal{O}_{C\times S}\left[\frac{1}{f}\right]$ is a direct limit of the following diagram:

$$\mathcal{O}_{C\times S}^n \xrightarrow{\cdot f} \mathcal{O}_{C\times S}^n \xrightarrow{\cdot f} \mathcal{O}_{C\times S}^n \xrightarrow{\cdot f} \ldots,$$

where $\cdot f$ is the injection given by the multiplication by f. Thus, the map (9.1) gives an embedding

$$\mathfrak{F} \stackrel{p}{\hookrightarrow} \mathcal{O}_{C\times S}^n,$$

which is identity outside of (x_0, s_0). Multiplying p by $\frac{1}{f}$ if necessary, we may assume that the restriction $p\big|_{C\times\{s_0\}}$ is non-trivial. Thus, p gives a map

$$\mathfrak{F}\big|_{C\times\{s_0\}} \longrightarrow \mathcal{O}^n_{C\times\{s_0\}}. \tag{9.2}$$

with image of positive rank. Since both sides of (9.2) are semistable of slope zero, and $\mathcal{O}^n_{C\times\{s_0\}}$ is polystable, the map (9.2) satisfies the following conditions. (see [OSS], Ch. II, Lemma 1.2.8 for details).

> Let $F_1 := \operatorname{im} p\big|_{C\times\{s_0\}}$, and $F_2 := \ker p\big|_{C\times\{s_0\}}$. Then the reflexization of F_1 is a trivial bundle $\mathcal{O}^k_{C\times\{s_0\}}$, and p maps F_1 to the direct summand $O_1' = \mathcal{O}^k_{C\times\{s_0\}} \subset \mathcal{O}^n_{C\times\{s_0\}}$.

Let $O_1 = \mathcal{O}^k_{C\times S} \subset \mathcal{O}^n_{C\times S}$ be the corresponding free subsheaf of $\mathcal{O}^n_{C\times S}$. Consider the natural projection π_{O_1} of $\mathcal{O}^n_{C\times S}$ to O_1. Let p_{O_1} be the composition of p and π_{O_1}, $\mathfrak{F}_1$ the image of p_{O_1}, and $\mathfrak{F}_2$ the kernel of p_{O_1}.

Step 2: *We show that the sheaves $\mathfrak{F}_2$ and $\mathfrak{F}_1$ and non-singular in codimension 2.*

Consider the exact sequence

$$Tor^1(\mathcal{O}_{C\times\{s_0\}}, \mathfrak{F}_1) \longrightarrow \mathfrak{F}_2\big|_{C\times\{s_0\}} \longrightarrow \mathfrak{F}\big|_{C\times\{s_0\}} \longrightarrow \mathfrak{F}_1\big|_{C\times\{s_0\}} \longrightarrow 0$$

obtained by tensoring the sequence

$$0 \longrightarrow \mathfrak{F}_2 \longrightarrow \mathfrak{F} \longrightarrow \mathfrak{F}_1 \longrightarrow 0$$

with $\mathcal{O}_{C\times\{s_0\}}$. From this sequence, we obtain an isomorphism $\mathfrak{F}_1\big|_{C\times\{s_0\}} \cong F_1$.

A torsion-free coherent sheaf over a smooth manifold is non-singular in codimension 1 ([OSS], Ch. II, Corollary 1.1.8).

Since $\mathfrak{F}$ is non-singular in codimension 2, the restriction $\mathfrak{F}\big|_{C\times\{s_0\}}$ is non-singular in codimension 1. Therefore, the torsion of $\mathfrak{F}\big|_{C\times\{s_0\}}$ has support of codimension at least 2 in $C \times \{s_0\}$. Since the sheaf F_2 is a subsheaf of $\mathfrak{F}\big|_{C\times\{s_0\}}$, its torsion has support of codimension at least 2. Therefore, the singular set of F_2 has codimension at least 2 in $C \times \{s_0\}$. The rank of F_2 is by definition equal to $n - k$.

Since F_1 has rank k, the singular set of $\mathfrak{F}_1$ coincides with the singular set of F_1. Since the restriction $\mathfrak{F}_1\Big|_{C\times\{s_0\}} = F_1$, is a subsheaf of a trivial bundle of dimension k on $C \times \{s_0\}$, it is torsion-free. Therefore, the singularities of $\mathfrak{F}_1$ have codimension at least 2 in $C \times \{s_0\}$.

We obtain that the support of $Tor^1(\mathcal{O}_{C\times\{s_0\}}, \mathfrak{F}_1)$ has codimension at least 2 in $C \times \{s_0\}$. Since the quotient sheaf

$$\mathfrak{F}_2\Big|_{C\times\{s_0\}} \Big/ Tor^1(\mathcal{O}_{C\times\{s_0\}}, \mathfrak{F}_1) \cong F_2 \tag{9.3}$$

is isomorphic to the sheaf F_2, this quotient is non-singular in codimension 1. Since we proved that F_2 is non-singular in codimension 1, the sheaf $\mathfrak{F}_2\Big|_{C\times\{s_0\}}$ is also non-singular in codimension 1, and its rank is equal to the rank of F_2.

Let R be the union of singular sets of the sheaves $\mathfrak{F}_2$, $\mathfrak{F}$, $\mathfrak{F}_1$. Clearly, R is contained in $C \times \{s_0\}$, and R coincides with the set of all $x \in C \times \{s_0\}$ where the dimension of the fiber of the sheaves $\mathfrak{F}_2$, $\mathfrak{F}$, $\mathfrak{F}_1$ is not equal to $n-k$, n, k. We have seen that the restrictions of $\mathfrak{F}_2$, $\mathfrak{F}_1$ to $C \times \{s_0\}$ have ranks $n-k$, k. Therefore, the singular sets of $\mathfrak{F}_2$, $\mathfrak{F}_1$ coincide with the singular sets of $\mathfrak{F}_2\Big|_{C\times\{s_0\}}$, $\mathfrak{F}_1\Big|_{C\times\{s_0\}}$. We have shown that these singular sets have codimension at least 2 in $C \times \{s_0\}$. On the other hand, $\mathfrak{F}$ is non-singular in codimension 2, by the conditions of Lemma 9.4. Therefore, R has codimension at least 3 in $C \times S$.

Step 3: *We check the assumptions of Lemma 9.4 applied to the sheaves* $\mathfrak{F}_2$, $\mathfrak{F}_1$.

Since the singular set of $\mathfrak{F}_1$ has codimension 2 in $C \times \{s_0\}$, the $\mathcal{O}_{C\times\{s_0\}}$-sheaf $Tor^1(\mathcal{O}_{C\times\{s_0\}}, \mathfrak{F}_1)$ is a torsion sheaf with support of codimension 2 in $C \times \{s_0\}$. By (9.3), the reflexization of $\mathfrak{F}_2\Big|_{C\times\{s_0\}}$ coincides with the reflexization of F_2. Thus, the sheaf $\left(\mathfrak{F}_2\Big|_{C\times\{s_0\}}\right)^{**}$ is semistable. On the other hand, outside of $C \times \{s_0\}$, the sheaf $\mathfrak{F}_2$ is a trivial bundle. Thus, $\mathfrak{F}_2$ satisfies assumptions of Lemma 9.4. Similarly, the sheaf $\mathfrak{F}_1$ is non-singular in codimension 2, its restriction to $C \times \{s_0\}$ has trivial reflexization, and it is free outside of $C \times \{s_0\}$.

Step 4: *We apply induction and prove Lemma 9.4.*

By induction assumption, the reflexization of $\mathfrak{F}_2$ is isomorphic to a trivial bundle $\mathcal{O}^{n-k}_{C\times S}$, and reflexization of $\mathfrak{F}_1$ is $\mathcal{O}^k_{C\times S}$. We obtain an exact sequence

$$0 \longrightarrow \mathfrak{F}_2 \longrightarrow \mathfrak{F} \longrightarrow \mathfrak{F}_1 \longrightarrow 0, \tag{9.4}$$

where the sheaves $\mathfrak{F}_2$ and $\mathfrak{F}_1$ have trivial reflexizations.

Let $V := C \times S \backslash R$. Restricting the exact sequence (9.4) to V, we obtain an exact sequence

$$0 \longrightarrow \mathcal{O}^{n-k}_V \stackrel{a}{\longrightarrow} \mathfrak{F}\Big|_V \stackrel{b}{\longrightarrow} \mathcal{O}^k_V \longrightarrow 0. \tag{9.5}$$

Since V is a complement of a codimension-3 complex subvariety in a smooth Stein domain, the first cohomology of a trivial sheaf on V vanish. Therefore, the sequence (9.5) splits, and the sheaf $\mathfrak{F}\Big|_V$ is a trivial bundle. Consider the pushforward $\zeta_*\mathfrak{F}\Big|_V$, where $\zeta: V \longrightarrow C\times S$ is the standard map. Then $\zeta_*\mathfrak{F}\Big|_V$ is a reflexization of $\mathfrak{F}$ (a pushforward of a reflexive sheaf over a subvariety of codimension 2 or more is reflexive – see Lemma 9.2). On the other hand, since the sheaf $\mathfrak{F}\Big|_V$ is a trivial bundle, its push-forward over a subvariety of codimension at least 2 is also a trivial bundle over $C \times S$. We proved that the sheaf $\mathfrak{F}^{**} = \zeta_*\mathfrak{F}\Big|_V$ is a trivial bundle over $C \times S$. The push-forward $\zeta_*\mathfrak{F}\Big|_V$ coincides with reflexization of $\mathfrak{F}$, by Lemma 9.2. This proves Lemma 9.4 and Theorem 9.3. ■

9.2 The Maruyama moduli space of coherent sheaves

This Subsection is a compilation of results of Gieseker and Maruyama on the moduli of coherent sheaves over projective manifolds. We follow [OSS], [Ma2].

To study the moduli spaces of holomorphic bundles and coherent sheaves, we consider the following definition of stability.

Definition 9.5: (Gieseker–Maruyama stability) ([Gi], [OSS]) Let X be a projective variety, $\mathcal{O}(1)$ the standard line bundle and F a torsion-free coherent sheaf. The sheaf F is called **Gieseker–Maruyama stable** (resp. Gieseker–Maruyama semistable) if for all coherent subsheaves $E \subset F$ with $0 < \operatorname{rk} E < \operatorname{rk} F$, we have

$$p_F(k) < p_E(k) \quad (\text{resp.,} \;\; p_F(k) \leqslant p_E(k))$$

for all sufficiently large numbers $k \in \mathbb{Z}$. Here

$$p_F(k) = \frac{\dim \Gamma_X(F \otimes \mathcal{O}(k))}{\operatorname{rk} F}.$$

Clearly, Gieseker–Maruyama stability is weaker than the Mumford-Takemoto stability. Every Gieseker–Maruyama semistable sheaf F has a so-called Jordan-Hölder filtration $F_0 \subset F_1 \subset ... \subset F$ with Gieseker–Maruyama stable successive quotients F_i/F_{i-1}. The corresponding associated graded sheaf

$$\oplus F_i/F_{i-1}$$

is independent from a choice of a filtration. It is called **the associate graded quotient of the Jordan-Hölder filtration on** F.

Definition 9.6: Let F, G be Gieseker–Maruyama semistable sheaves on X. Then F, G are called **S-equivalent** if the corresponding associate graded quotients $\oplus F_i/F_{i-1}$, $\oplus G_i/G_{i-1}$ are isomorphic.

Definition 9.7: Let X be a complex manifold, F a torsion-free sheaf on X, and Y a complex variety. Consider a sheaf $\mathcal{F}$ on $X \times Y$ which is flat over Y. Assume that for some point $s_0 \in Y$, the sheaf $\mathcal{F}\Big|_{X \times \{s_0\}}$ is isomorphic to F. Then $\mathcal{F}$ is called **a deformation of F parametrized by Y**. We say that a sheaf F' on X is **deformationally equivalent** to F if for some $s \in Y$, the restriction $\mathcal{F}\Big|_{X \times \{s\}}$ is isomorphic to F'. Slightly less formally, such sheaves are called **deformations of** F. If F' is a (semi-)stable bundle, it is called **a (semi-) stable bundle deformation of** F.

Remark 9.8: Clearly, the Chern classes of deformationally equivalent sheaves are equal.

Definition 9.9: Let X be a complex manifold, and F a torsion-free sheaf on X, and $\mathcal{M}_{mar}$ a complex variety. We say that $\mathcal{M}_{mar}$ is a **coarse moduli space of deformations of** F if the following conditions hold.

(i) The points of $s \in \mathcal{M}_{mar}$ are in bijective correspondence with S-equivalence classes of coherent sheaves F_s which are deformationally equivalent to F.

(ii) For any flat deformation $\mathcal{F}$ of F parametrized by Y, there exists a unique morphism $\varphi : Y \longrightarrow \mathcal{M}_{mar}$ such that for all $s \in Y$, the restriction $\mathcal{F}\Big|_{X\times\{s\}}$ is S-equivalent to the sheaf $F_{\varphi(s)}$ corresponding to $\varphi(s) \in \mathcal{M}_{mar}$.

Clearly, the coarse moduli space is unique. By Remark 9.8, the Chern classes of F_s are equal for all $s \in \mathcal{M}_{mar}$.

It is clear how to define other kinds of moduli spaces. For instance, replacing the word *sheaf* by the word *bundle* throughout Definition 9.9, we obtain a definition of **the coarse moduli space of semistable bundle deformations of** F. Further on, we shall usually omit the word "coarse" and say "moduli space" instead.

Theorem 9.10: (Maruyama) Let X be a projective manifold and F a coherent sheaf over X. Then the Maruyama moduli space $\mathcal{M}_{mar}$ of deformations of F exists and is compact.

Proof: See, e. g., [Ma2]. ∎

9.3 Moduli of hyperholomorphic sheaves and C-restricted comples structures

Usually, the moduli space of semistable bundle deformations of a bundle F is not compact. To compactify this moduli space, Maruyama adds points corresponding to the deformations of F which are singular (these deformations can be non-reflexive and can have singular reflexizations). Using the desingularization theorems for hyperholomorphic sheaves, we were able to obtain Theorem 9.3, which states (roughly speaking) that a deformation of a semistable hyperholomorphic bundle is again a semistable bunlde, assuming that all its singularities are isolated. In Section 5, we showed that under certain conditions, a deformation of a hyperholomorphic sheaf is again hyperholomorphic (Theorem 5.14). This makes it possible to prove that a deformation of a semistable hyperholomorphic bundle is locally trivial.

In [V5], we have shown that a Hilbert scheme of a K3 surface has no non-trivial trianalytic subvarieties, for a general hyperkähler structure.

Theorem 9.11: Let M be a compact hyperkähler manifold without non-trivial trianalytic subvarieties, $\dim_{\mathbb{H}} \geqslant 2$, and I an induced complex structure. Consider a hyperholomorphic bundle F on (M, I) (Definition 3.11). Assume that I is a C-restricted complex structure, $C = \deg_I c_2(F)$. Let $\mathcal{M}$ be the moduli space of semistable bundle deformations of F over (M, I). Then $\mathcal{M}$ is compact.

Proof: The complex structure I is by definition algebraic, with unique polarization. This makes it possible to speak of Gieseker–Maruyama stability on (M, I). Denote by $\mathcal{M}_{mar}$ the Maruyama moduli of deformations of F. Then $\mathcal{M}$ is naturally an open subset of $\mathcal{M}_{mar}$. Let $s \in \mathcal{M}_{mar}$ be an arbitrary point and F_s the corresponding coherent sheaf on (M, I), defined up to S-equivalence. According to Remark 9.8, the Chern classes of F and F_s are equal. Thus, by Theorem 5.14, the sheaf F_s is hyperholomorphic. Therefore, F_s admits a filtration with hyperholomorphic stable quotient sheaves F_i, $i = 1, ..., m$. By Claim 3.16, the singular set S of F_s is trianalytic. Since M has no proper trianalytic subvarieties, S is a collection of points. We obtain that F_s has isolated singularities. Let $\mathfrak{F}$ be a family of deformations of F, parametrized by Y. The points $y \in Y$ correspond to deformations F_y of F_s. Assume that for all $y \in Y$, $y \neq s$, the sheaf F_y is a bundle. Since $\mathcal{M}$ is open in $\mathcal{M}_{mar}$, such a deformation always exists.

The sheaf F_s has isolated singularities and admits a filtration with hyperholomorphic stable quotient sheaves. This implies that the family $\mathfrak{F}$ satisfies the conditions of Theorem 9.3. By Theorem 9.3, the reflexization $\mathfrak{F}^{**}$ is locally trivial. To prove that $\mathcal{M} = \mathcal{M}_{mar}$, we have to show that for all $s \in \mathcal{M}_{mar}$, the corresponding coherent sheaf F_s is locally trivial. Therefore, to finish the proof of Theorem 9.11, it remains to prove the following algebro-geometric claim.

Claim 9.12: Let X be a compact complex manifold, $\dim_{\mathbb{C}} X > 2$, and $\mathfrak{F}$ a torsion-free coherent sheaf over $X \times Y$ which is flat over Y. Assume that the reflexization of $\mathfrak{F}$ is locally trivial, $\mathfrak{F}$ has isolated singularities, and for some point $s \in Y$, the restriction of $\mathfrak{F}$ to the complement $(X \times Y) \backslash (X \times \{s\})$ is locally trivial. Then the reflexization $\left(\mathfrak{F}\big|_X \times \{s\}\right)^{**}$ is locally trivial.

Remark 9.13: We say that a kernel of a map from a bundle to an Artinian sheaf is **a bundle with holes**. In slightly more intuitive terms, Claim 9.12 states that a flat deformation of a bundle with holes is again a bundle with holes, and cannot be smooth, assuming that $\dim_{\mathbb{C}} X > 2$.

Proof of Claim 9.12: Claim 9.12 is well known. Here we give a sketch of a proof. Consider a coherent sheaf $F_s = \mathfrak{F}\big|_{X\times\{s\}}$, and an exact sequence

$$0 \longrightarrow F_s \longrightarrow F_s^{**} \longrightarrow k \longrightarrow 0,$$

where k is an Artinian sheaf. By definition, the sheaf F_s^{**} is locally trivial. The flat deformations of F_s are infinitesimally classified by $Ext^1(F_s, F_s)$. Replacing F_s by a quasi-isomorphic complex of sheaves $F_s^{**} \longrightarrow k$, we obtain a spectral sequence converging to $Ext^\bullet(F_s, F_s)$. In the E_2-term of this sequence, we observe the group

$$Ext^1(F_s^{**}, F_s^{**}) \oplus Ext^1(k, k) \oplus Ext^2(k, F_s^{**}) \oplus Ext^0(F_s^{**}, k).$$

which is responsible for $Ext^1(F_s, F_s)$. The term $Ext^1(F_s^{**}, F_s^{**})$ is responsible for deformations of the bundle F_s^{**}, the term $Ext^0(F_s^{**}, k)$ for the deformations of the map $F_s^{**} \longrightarrow k$, and the term $Ext^1(k, k)$ for the deformations of the Artinian sheaf k. Thus, the term $Ext^2(k, F_s^{**})$ is responsible for the deformations of F_s which change the dimension of the cokernel of the embedding $F_s \longrightarrow F_s^{**}$. We obtain that whenever $Ext^2(k, F_s^{**}) = 0$, all deformations of F_s are singular. On the other hand, $Ext^2(k, F_s^{**}) = 0$, because the i-th Ext from the skyscraper to a free sheaf on a manifold of dimension more than i vanishes (this is a basic result of Grothendieck's duality, [H-Gro]). ∎

10 New examples of hyperkähler manifolds

10.1 Twistor paths

This Subsection contains an exposition and further elaboration of the results of [V3-bis] concerning the twistor curves in the moduli space of complex structures on a complex manifold of hyperkähler type.

Let M be a compact manifold admitting a hyperkähler structure. In Definition 5.8, we defined the coarse, marked moduli space of complex structures on M, denoted by *Comp*. For the duration of this section, we fix a compact simple hyperkähler manifold M, and its moduli *Comp*.

Further on, we shall need the following fact.

Claim 10.1: Let M be a hyperkähler manifold, I an induced complex structure of general type, and B a holomorphic vector bundle over (M, I). Then B is stable if an only if B is simple.[1]

[1]Simple sheaves are coherent sheaves which have no proper subsheaves

Proof: By Lemma 2.26, for all $\omega \in Pic(M, I)$, we have $\deg_I(\omega) = 0$. Therefore, every subsheaf of B is destabilising. ∎

Remark 10.2: In assumptions of Claim 10.1, all stable bundles are hyperholomorphic (Theorem 2.27). Therefore, Claim 10.1 implies that B is hyperholomorphic if it is simple.

In Subsection 5.2, we have shown that every hyperkähler structure $\mathcal{H}$ corresponds to a holomorphic embedding

$$\kappa(\mathcal{H}) : \mathbb{C}P^1 \longrightarrow Comp, \quad L \longrightarrow (M, L).$$

Definition 10.3: A projective line $C \subset Comp$ is called **a twistor curve** if $C = \kappa(\mathcal{H})$ for some hyperkähler structure $\mathcal{H}$ on M.

The following theorem was proven in [V3-bis].

Theorem 10.4: ([V3-bis], Theorem 3.1) Let $I_1, I_2 \in Comp$. Then there exist a sequence of intersecting twistor curves which connect I_1 with I_2. ∎

Definition 10.5: Let $P_0, ..., P_n \subset Comp$ be a sequence of twistor curves, supplied with an intersection point $x_{i+1} \in P_i \cap P_{i+1}$ for each i. We say that $\gamma = P_0, ..., P_n, x_1, ..., x_n$ is a **twistor path**. Let $I, I' \in Comp$. We say that γ is **a twistor path connecting** I **to** I' if $I \in P_0$ and $I' \in P_n$. The lines P_i are called **the edges**, and the points x_i **the vertices** of a twistor path.

Recall that in Definition 2.13, we defined induced complex structures which are generic with respect to a hyperkähler structure.

Given a twistor curve P, the corresponding hyperkähler structure $\mathcal{H}$ is unique (Theorem 5.11). We say that a point $x \in P$ is **of general type**, or **generic with respect to** P if the corresponding complex structure is generic with respect to $\mathcal{H}$.

Definition 10.6: Let $I, J \in Comp$ and $\gamma = P_0, ..., P_n$ be a twistor path from I to J, which corresponds to the hyperkähler structures $\mathcal{H}_0, ..., \mathcal{H}_n$. We say that γ is **admissible** if all vertices of γ are of general type with respect to the corresponding edges.

Remark 10.7: In [V3-bis], admissible twistor paths were defined slightly differently. In addition to the conditions above, we required that I, J are of general type with respect to $\mathcal{H}_0$, $\mathcal{H}_n$.

Theorem 10.4 proves that every two points I, I' in $Comp$ are connected with a twistor path. Clearly, each twistor path induces a diffeomorphism $\mu_\gamma : (M, I) \longrightarrow (M, I')$. In [V3-bis], Subsection 5.2, we studied algebro-geometrical properties of this diffeomorphism.

Theorem 10.8: Let I, $J \in Comp$, and γ be an admissible twistor path from I to J. Then

(i) There exists a natural isomorphism of tensor cetegories

$$\Phi_\gamma : Bun_I(\mathcal{H}_0) \longrightarrow Bun_J(\mathcal{H}_n),$$

where $Bun_I(\mathcal{H}_0)$, $Bun_J(\mathcal{H}_n)$ are the categories of polystable hyperholomorphic vector bundles on (M, I), (M, J), taken with respect to $\mathcal{H}_0$, $\mathcal{H}_n$ respectively.

(ii) Let $B \in Bun_I(\mathcal{H}_0)$ be a stable hyperholomorphic bundle, and

$$\mathcal{M}_{I,\mathcal{H}_0}(B)$$

the moduli of stable deformations of B, where stability is taken with respect to the Kähler metric induced by $\mathcal{H}_0$. Then Φ_γ maps stable bundles which are deformationally equivalent to B to the stable bundles which are deformationally equivalent to $\Phi_\gamma(B)$. Moreover, obtained this way bijection

$$\Phi_\gamma : \mathcal{M}_{I,\mathcal{H}_0}(B) \longrightarrow \mathcal{M}_{J,\mathcal{H}_n}(\Phi_\gamma(B))$$

induces a real analytic isomorphism of deformation spaces.

Proof: Theorem 10.8 (i) is a consequence of [V3-bis], Corollary 5.1. Here we give a sketch of its proof.

Let I be an induced complex structure of general type. By Claim 10.1, a bundle B over (M, I) is stable if and only if it is simple. Thus, the category $Bun_I(\mathcal{H})$ is independent from the choice of $\mathcal{H}$ (Claim 10.1).

In Theorem 3.27, we constructed the equivalence of categories $\Phi_{I,J}$, which gives the functor Φ_γ for twistor path which consists of a single twistor curve. This proves Theorem 10.8 (i) for $n = 1$. A composition of isomorphisms $\Phi_{I,J} \circ \Phi_{J,J'}$ is well defined, because the category $Bun_I(\mathcal{H})$ is independent from the choice of $\mathcal{H}$. Taking successive compositions of the maps $\Phi_{I,J}$, we obtain an isomorphism Φ_γ. This proves Theorem 10.8 (i).

The variety $\mathcal{M}_{I,\mathcal{H}}(B)$ is singular hyperkähler ([V1]), and the variety $\mathcal{M}_{J,\mathcal{H}}(B)$ is the same singular hyperkähler variety, taken with another induced complex structure. By definition of singular hyperkähler varieties, this implies that $\mathcal{M}_{I,\mathcal{H}}(B)$, $\mathcal{M}_{J,\mathcal{H}}(B)$ are real analytic equivalent, with equivalence provided by $\Phi_{I,J}$. This proves Theorem 10.8 (ii). ∎

For $I \in Comp$, denote by $Pic(M,I)$ the group $H^{1,1}(M,I) \cap H^2(M,\mathbb{Z})$, and by $Pic(I,\mathbb{Q})$ the space $H^{1,1}(M,I) \cap H^2(M,\mathbb{Q}) \subset H^2(M)$. Let $Q \subset H^2(M,\mathbb{Q})$ be a subspace of $H^2(M,\mathbb{Q})$, and

$$Comp_Q := \{I \in Comp \mid Pic(I,\mathbb{Q}) = Q\}.$$

Theorem 10.9: Let $\mathcal{H}$, $\mathcal{H}'$ be hyperkähler structures, and I, I' be complex structures of general type to and induced by $\mathcal{H}$, $\mathcal{H}'$. Assume that $Pic(I,\mathbb{Q}) = Pic(I',\mathbb{Q}) = Q$, and I, I' lie in the same connected component of $Comp_Q$. Then I, I' can be connected by an admissible path.

Proof: This is [V3-bis], Theorem 5.2. ∎

For a general Q, we have no control over the number of connected components of $Comp_Q$ (unless global Torelli theorem is proven), and therefore we cannot directly apply Theorem 10.9 to obtain results from algebraic geometry.[2] However, when $Q = 0$, $Comp_Q$ is clearly connected and dense in $Comp$. This is used to prove the following corollary.

Corollary 10.10: Let I, $I' \in Comp_0$. Then I can be connected to I' by an admissible twistor path.

Proof This is [V3-bis], Corollary 5.2. ∎

Definition 10.11: Let $I \in Comp$ be a complex structure, ω be a Kähler form on (M,I), and $\mathcal{H}$ the corresponding hyperkähler metric, which exists by Calabi-Yau theorem. Then ω is called **a generic polarization** if any of the following conditions hold

(i) For all $a \in Pic(M,I)$, the degree $\deg_\omega(a) \neq 0$, unless $a = 0$.

[2]Exception is a K3 surface, where Torelli holds. For K3, $Comp_Q$ is connected for all $Q \subset H^2(M,\mathbb{Q})$.

(ii) For all $SU(2)$-invariant integer classes $a \in H^2(M, \mathbb{Z})$, we have $a = 0$.

The conditions (i) and (ii) are equivalent by Lemma 2.26.

Claim 10.12: Let $I \in Comp$ be a complex structure, ω be a Kähler form on (M, I), and $\mathcal{H}$ the corresponding hyperkähler structure, which exists by Calabi-Yau theorem. Then ω is generic if and only if for all integer classes $a \in H^{1,1}(M, I)$, the class a is not orthogonal to ω with respect to the Bogomolov-Beauville pairing.

Proof: Clearly, the map $\deg_\omega : H^2(M) \longrightarrow \mathbb{R}$ is equal (up to a scalar multiplier) to the orthogonal projection onto the line $\mathbb{R} \cdot \omega$. Then, Claim 10.12 is equivalent to Definition 10.11, (i). ∎

From Claim 10.12 it is clear that the set of generic polarizations is a complement to a countable union of hyperplanes. Thus, generic polarizations are dense in the Kähler cone of (M, I), for all I.

Claim 10.13: Let $I, J \in Comp$, and a, b be generic polarizations on (M, I). Consider the corresponding hyperkähler structures $\mathcal{H}_0$ and $\mathcal{H}_n$ inducing I and J. Then there exists an admissible twistor path starting from $I, \mathcal{H}_0$ and ending with $\mathcal{H}_n, J$.

Proof: Consider the twistor curves P_0, P_n corresponding to $\mathcal{H}_0$, $\mathcal{H}_n$. Since a, b are generic, the curves P_0, P_n intersect with $Comp_0$. Applying Corollary 10.10, we connect the curves P_0 and P_n by an admissible path. ∎

Putting together Claim 10.13 and Theorem 10.8, we obtain the following result.

Theorem 10.14: Let I, $J \in Comp$ be complex structures, and a, b be generic polarizations on (M, I), (M, J). Then

(i) There exist an isomorphism of tensor cetegories

$$\Phi_\gamma : Bun_I(a) \longrightarrow Bun_J(a),$$

where $Bun_I(a)$, $Bun_J(b)$ are the categories of polystable hyperholomorphic vector bundles on (M, I), (M, J), taken with respect to the hyperkähler structures defined by the Kähler classes a, b as in Theorem 2.4.

(ii) Let $B \in Bun_I(a)$ be a stable hyperholomorphic bundle, and

$$\mathcal{M}_{I,a}(B)$$

the moduli of stable deformations of B, where stability is taken with respect to the polarization a. Then Φ_γ maps stable bundles which are deformationally equivalent to B to the stable bundles which are deformationally equivalent to $\Phi_\gamma(B)$. Moreover, obtained this way bijection

$$\Phi_\gamma : \mathcal{M}_{I,a}(B) \longrightarrow \mathcal{M}_{J,b}(\Phi_\gamma(B))$$

induces a real analytic isomorphism of deformation spaces.

■

Lemma 10.15: In assumptions of Theorem 10.8, let B be a holomorphic tangent bundle of (M, I). Then $\Phi_\gamma(B)$ is a holomorphic tangent bundle of (M, J).

Proof: Clear. ■

Corollary 10.16: Let $I, J \in Comp$ be complex structures, and a, b generic polarizations on (M, I), (M, J). Assume that the moduli of stable deformations $\mathcal{M}_{I,a}(T(M, I))$ of the holomorphic tangent bundle $T^{1,0}(M, I)$ is compact. Then the space $\mathcal{M}_{J,b}(T(M, J))$ is also compact.

Proof: Let γ be the twistor path of Claim 10.13. By Lemma 10.15, $\Phi_\gamma(T(M, I)) = T(M, J)$. Applying Theorem 10.8, we obtain a real analytic equivalence from $\mathcal{M}_{I,a}(T(M, I))$ to $\mathcal{M}_{J,b}(T(M, J))$. ■

10.2 New examples of hyperkähler manifolds

Theorem 10.17: Let M be a compact hyperkähler manifold without nontrivial trianalytic subvarieties, $\dim_{\mathbb{H}} M \geqslant 2$, and I an induced complex structure. Consider a hyperholomorphic bundle F on M (Definition 3.28). Let F_I be the corresponding holomorphic bundle over (M, I). Assume that I is a C-restricted complex structure, $C = \deg_I c_2(F)$. Assume, moreover, that all semistable bundle deformations of F_I are stable.[3] Denote by $\mathcal{M}^I_F$ the moduli of stable bundle deformations of F_I over (M, I). Then

(i) the normalization $\widetilde{\mathcal{M}^I_F}$ is a compact and smooth complex manifold equipped with a natural hyperkähler structure.

[3] This may happen, for instance, when $\operatorname{rk} F = \dim_{\mathbb{C}} M = n$, and the number $c_n(F)$ is prime.

(ii) Moreover, for all induced complex structures J on M, the the variety $\mathcal{M}^J_F$ is compact, and has a smooth normalization $\widetilde{\mathcal{M}}^J_F$, which is also equipped with a natural hyperkähler structure.

(iii) Finally, the hyperkähler manifolds $\widetilde{\mathcal{M}}^J_F$, $\widetilde{\mathcal{M}}^I_F$ are naturally isomorphic.

Proof: The variety $\mathcal{M}^I_F$ is compact by Theorem 9.11. In [V1], it was proven that the space $\mathcal{M}^I_F$ of stable deformations of F is a singular hyperkähler variety (see also [KV] for an explicit construction of the twistor space of $\mathcal{M}^I_F$). Then Theorem 10.17 is a consequence of the Desingularization Theorem for singular hyperkähler varietiess (Theorem 2.16). ∎

The assumptions of Theorem 10.17 are quite restrictive. Using the technique of twistor paths, developed in Subsection 10.1, it is possible to prove a more accessible form of Theorem 10.17.

Let M be a hyperkähler manifold, and I, J induced complex structures. Given an admissible twistor path from I to J, we obtain an equivalence Φ_γ between the category of hyperholomorphic bundles on (M, I) and (M, J).

Theorem 10.18: Let M be a compact simple hyperkähler manifold, $\dim_{\mathbb{H}} M > 1$, and I a complex structure on M. Consider a generic polarization a on (M, I). Let $\mathcal{H}$ be the corresponding hyperkähler structure, and F a hyperholomorphic bundle on (M, I). Fix a hyperkähler structure $\mathcal{H}'$ on M admitting C-restricted complex structures, such that M has no trianalytic subvarieties with respect to $\mathcal{H}'$. Assume that for some C-restricted complex structure J induced by $\mathcal{H}'$, $C = \deg_I c_2(F)$, all admissible twistor paths γ from I to J, and all semistable bundles F' which are deformationally equivalent to $\Phi_\gamma(F)$, the bundle F' is stable. Then the space of stable deformations of F is compact.

Remark 10.19: The space of stable deformations of F is singular hyperkähler ([V1]) and its normalization is smooth and hyperkähler (Theorem 2.16).

Proof of Theorem 10.18: Clearly, F' satisfies assumptions of Theorem 10.17, and the moduli space of its stable deformations is compact. Since Φ_γ induces a homeomorphism of moduli spaces (Theorem 10.8), the space of stable deformations of F is also compact. ∎

Applying Theorem 10.18 to the holomorphic tangent bundle $T(M, I)$, we obtain the following corollary.

Theorem 10.20: Let M be a compact simple hyperkähler manifold, $\dim_{\mathbb{H}}(M) > 1$. Assume that for a generic hyperkähler structure $\mathcal{H}$ on M, this manifold admits no trianalytic subvarieties.[4] Assume, moreover, that for some C-restricted induced complex structure I, all semistable bundle deformations of $T(M, I)$ are stable, for $C > \deg_I c_2(M)$. Then, for all complex structures J on M and all generic polarizations ω on (M, J), the deformation space $\mathcal{M}_{J,\omega}(T(M, J))$ is compact.

Proof: Follows from Theorem 10.18 and Corollary 10.16. ∎

10.3 How to check that we obtained new examples of hyperkähler manifolds?

A. Beauville [Bea] described two families of compact hyperkähler manifolds, one obtained as the Hilbert scheme of points on a K3-surface, another obtained as the Hilbert scheme of a 2-dimensional torus factorized by the free torus action.

Conjecture 10.21: There exist compact simple hyperkähler manifolds which are not isomorphic to deformations of these two fundamental examples.

Here we explain our strategy of a proof of Conjecture 10.21 using results on compactness of the moduli space of hyperholomorphic bundles.

The results of this subsection are still in writing, so all statements below this line should be considered as conjectures. We give an idea of a proof for each result and label it as "proof", but these "proofs" are merely sketches.

First of all, it is possible to prove the following theorem.

Theorem 10.22: Let M be a complex K3 surface without automorphisms. Assume that M is Mumford-Tate generic with respect to some hyperkähler structure. Consider the Hilbert scheme $M^{[n]}$ of points on M, $n > 1$. Pick a hyperkähler structure $\mathcal{H}$ on $M^{[n]}$ which is compatible with

[4] This assumption holds for a Hilbert scheme of points on a K3 surface.

the complex structure. Let B be a hyperholomorphic bundle on $(M^{[n]}, \mathcal{H})$, $\operatorname{rk} B = 2$. Then B is a trivial bundle.

Proof: The proof of Theorem 10.22 is based on the same ideas as the proof of Theorem 2.17. ∎

For a compact complex manifold X of hyperkähler type, denote its coarse, marked moduli space (Definition 5.8) by $Comp(X)$.

Corollary 10.23: Let M be a K3 surface, $I \in Comp(X)$ an arbitrary complex structure on $X = M^{[n]}$, $n > 1$, and a a generic polarization on (X, J). Consider the hyperkähler structure $\mathcal{H}$ which corresponds to (I, a) as in Theorem 2.4. Let B, $\operatorname{rk} B = 2$ be a hyperholomorphic bundle over $(X, \mathcal{H})$. Then B is trivial.

Proof: Follows from Theorem 10.22 and Theorem 10.14. ∎

Corollary 10.24: Let M be a K3 surface, $I \in Comp(X)$ an arbitrary complex structure on $X = M^{[n]}$, $n > 1$, and a a generic polarization on (X, I). Consider the hyperkähler structure $\mathcal{H}$ which corresponds to (J, a) (Theorem 2.4). Let B, $\operatorname{rk} B \leqslant 6$ be a stable hyperholomorphic bundle on $(X, \mathcal{H})$. Assume that the Chern class $c_{\operatorname{rk} B}(B)$ is non-zero. Assume, moreover, that I is C-restricted, $C = \deg_I(c_2(B))$. Let B' be a semistable deformation of B over (X, I). Then B' is stable.

Proof: Consider the Jordan–Hölder serie for B'. Let $Q_1 \oplus Q_2 \oplus ...$ be the associated graded sheaf. By Theorem 5.14, the stable bundles Q_i are hyperholomorphic. Since $c_{\operatorname{rk} B}(B) \neq 0$, we have $c_{\operatorname{rk} Q_i}(Q_i) \neq 0$. Therefore, the bundles Q_i are non-trivial. By Corollary 10.23, $\operatorname{rk} Q_i > 2$. Since all the Chern classes of the bundles Q_i are $SU(2)$-invariant, the odd Chern classes of Q_i vanish (Lemma 2.6). Therefore, $\operatorname{rk} Q_i \geqslant 4$ for all i. Since $\operatorname{rk} B \leqslant 6$, we have $i = 1$ and the bundle B' is stable. ∎

Let M be a K3 surface, $X = M^{[i]}$, $i = 2, 3$ be its second or third Hilbert scheme of points, $I \in Comp(X)$ arbitrary complex structure on X, and a a generic polarization on (X, I). Consider the hyperkähler structure $\mathcal{H}$ which corresponds to J and a by Calabi-Yau theorem (Theorem 2.4). Denote by TX the tangent bundle of X, considered as a hyperholomorphic bundle. Let $\operatorname{Def}(TX)$ denote the hyperkähler desingularization of the moduli of stable deformations of TX. By Theorem 10.14, the real analytic subvariety underlying $\operatorname{Def}(TX)$ is independent from the choice of I. Therefore, its dimension is also independent from the choice of I. The dimension of the deformation space $\mathrm{Def}(TX)$ can be estimated by a direct computation, for X a Hilbert scheme. We obtain that $\dim \mathrm{Def}(TX) > 40$.

Claim 10.25: In these assumptions, the space $\mathrm{Def}(TX)$ is a compact hyperkähler manifold.

Proof: By Corollary 10.24, all semistable bundle deformations of TX are stable. Then Claim 10.25 is implied by Theorem 10.20. ∎

Clearly, deforming the complex structure on X, we obtain a deformation of complex structures on $\mathrm{Def}(TX)$. This gives a map

$$Comp(X) \longrightarrow Comp(\mathrm{Def}(TX)). \tag{10.1}$$

It is easy to check that the map (10.1) is complex analytic, and maps twistor curves to twistor curves.

Claim 10.26: Let X, Y be hyperkähler manifolds, and

$$\varphi: Comp(X) \longrightarrow Comp(Y)$$

be a holomorphic map of corresponding moduli spaces which maps twistor curves to twistor curves. Then φ is locally an embedding.

Proof: An elementary argument using the period maps, in the spirit of Subsection 5.2. ∎

The following result, along with Theorem 10.22, is the major stumbling block on the way to proving Conjecture 10.21. The other results of this Subsection are elementary or routinely proven, but the complete proof of Theorem 10.22 and Theorem 10.27 seems to be difficult.

Theorem 10.27: Let X be a simple hyperkähler manifold without proper trianalytic subvarieties, B a hyperholomorphic bundle over X, and I an induced complex structure. Denote the corresponding holomorphic bundle over (X, I) by B_I. Assume that the space $\mathcal{M}$ of stable bundle deformations of B is compact. Let $\mathrm{Def}(B)$ be the hyperkähler desingularization of $\mathcal{M}$. Then $\mathrm{Def}(M)$ is a simple hyperkähler manifold.

Proof: Given a decomposition $\mathrm{Def}(M) = M_1 \times M_2$, we obtain a parallel 2-form on Ω_1 on $\mathrm{Def}(B)$, which is a pullback of the holomorphic symplectic form on M_1. Consider the space $\mathcal{A}$ of connections on B, which is an infinitely-dimensional complex analytic Banach manifold. Then Ω_1 corresponds to a holomorphic 2-form $\tilde{\Omega}_1$ on $\mathcal{A}$. Since Ω_1 is parallel with respect to the natural connection on $\operatorname{Def}(B)$, the form $\tilde{\Omega}_1$ is also a parallel 2-form on the tangent space to $\mathcal{A}$, which is identified with $\Omega^1(X, \operatorname{End}(B))$. It is possible to prove that this 2-form is obtained as

$$A, B \longrightarrow \int_Y \Theta\left(A\Big|_Y, B\Big|_Y\right) \operatorname{Vol}(Y)$$

where

$$\Theta : \Omega^1(Y, \operatorname{End}(B)) \times \Omega^1(Y, \operatorname{End}(B)) \longrightarrow \mathcal{O}_Y$$

is a certain holomorphic pairing on the bundle $\Omega^1(Y, \operatorname{End}(B))$, and Y is a trianalytic subvariety of X. Since X has no trianalytic subvarieties, $\tilde{\Omega}_1$ is obtained from a $\mathcal{O}_X$-linear pairing

$$\Omega^1(X, \operatorname{End}(B)) \times \Omega^1(X, \operatorname{End}(B)) \longrightarrow \mathcal{O}_X.$$

Using stability of B, it is possible to show that such a pairing is unique, and thus, Ω_1 coincides with the holomorphic symplectic form on $\operatorname{Def}(B)$. Therefore, $\operatorname{Def}(B) = M_1$, and this manifold is simple. ∎

Return to the deformations of tangent bundles on $X = M^{[i]}$, $i = 2, 3$. Recall that the second Betti number of a Hilbert scheme of points on a K3 surface is equal to 23, and that of the generalized Kummer variety is 7 ([Bea]). Consider the map (10.1). By Theorem 10.27, the manifold $\operatorname{Def}(TX)$ is simple. By Bogomolov's theorem (Theorem 5.9), we have

$$\dim \mathit{Comp}(\operatorname{Def}(TX)) = \dim H^2(\operatorname{Def}(TX)) - 2.$$

Therefore, either $\dim H^2(\operatorname{Def}(TX)) > \dim H^2(X) = 23$, or the map (10.1) is etale. In the first case, the second Betti number of $\operatorname{Def}(TX)$ is bigger than that of known simple hyperkähler manifolds, and thus, $\operatorname{Def}(TX)$ is a new example of a simple hyperkähler manifold; this proves Conjecture 10.21. Therefore, to prove Conjecture 10.21, we may assume that $\dim H^2(\operatorname{Def}(TX)) = 23$, the map (10.1) is etale, and $\operatorname{Def}(TX)$ is a deformation of a Hilbert scheme of points on a K3 surface.

Consider the universal bundle $\tilde{B}$ over $X \times \operatorname{Def}(TX)$. Restricting $\tilde{B}$ to $\{x\} \times \operatorname{Def}(TX)$, we obtain a bundle B on $\operatorname{Def}(TX)$. Let $\operatorname{Def}(B)$ be the hyperkähler desingularization of the moduli space of stable deformations of B. Clearly, the manifold $\operatorname{Def}(B)$ is independent from the choice of $x \in X$.

Taking the generic hyperkähler structure on X, we may assume that the hyperkähler structure $\mathcal{H}$ on $\mathrm{Def}(TX)$ is also generic. Thus, $(\mathrm{Def}(TX), \mathcal{H})$ admits C-restricted complex structures and has no trianalytic subvarieties. In this situation, Corollary 10.24 implies that the hyperkähler manifold $\mathrm{Def}(B)$ is compact. Applying Claim 10.26 again, we obtain a sequence of maps

$$Comp(X) \longrightarrow Comp(\mathrm{Def}(TX)) \longrightarrow Comp(\mathrm{Def}(B))$$

which are locally closed embeddings. By the same argument as above, we may assume that the composition $Comp(X) \longrightarrow Comp(\mathrm{Def}(B))$ is etale, and the manifold $\mathrm{Def}(B)$ is a deformation of a Hilbert scheme of points on K3. Using Mukai's version of Fourier transform ([O], [BBR]), we obtain an embedding of the corresponding derived categories of coherent sheaves,

$$D(X) \longrightarrow D(\mathrm{Def}(TX)) \longrightarrow D(\mathrm{Def}(B)).$$

Using this approach, it is easy to prove that

$$\dim X \leqslant \dim \mathrm{Def}(TX) \leqslant \dim \mathrm{Def}(B).$$

Let $x \in X$ be an arbitrary point. Consider the complex $C_x \in D(\mathrm{Def}(B))$ of coherent sheaves on $\mathrm{Def}(B)$, obtained as a composition of the Fourier-Mukai transform maps. It is easy to check that the lowest non-trivial cohomology sheaf of C_x is a skyscraper sheaf in a point $F(x) \in \mathrm{Def}(B)$. This gives an embedding

$$F: X \longrightarrow \mathrm{Def}(B).$$

The map F is complex analytic for all induced complex structure. We obtained the following result.

Lemma 10.28: In the above assumptions, the embedding

$$F: X \longrightarrow \mathrm{Def}(B)$$

is compatible with the hyperkähler structure.

■

By Lemma 10.28, the manifold $\mathrm{Def}(B)$ has a trianalytic subvariety $F(X)$, of dimension $0 < \dim F(X) < 40 < \dim \mathrm{Def}(B)$. On the other hand, for a hyperkähler structure on X generic, the corresponding hyperkähler structure on $\mathrm{Def}(B)$ is also generic, so this manifold has no trianalytic subvarieties. We obtained a contradiction. Therefore, either $\mathrm{Def}(TX)$ or $\mathrm{Def}(B)$ is a new example of a simple hyperkähler manifold. This proves Conjecture 10.21.

Acknowledegments: I am grateful to V. Batyrev, A. Beilinson, P. Deligne, D. Gaitsgory, D. Kaledin, D. Kazhdan, M. Koncevich and T. Pantev for valuable discussions. My gratitude to D. Kaledin, who explained to me the results of [Sw]. This paper uses many ideas of our joint work on direct and inverse twistor transform ([KV]).

Bibliography

[BS] Bando, S., Siu, Y.-T, *Stable sheaves and Einstein-Hermitian metrics*, In: Geometry and Analysis on Complex Manifolds, Festschrift for Professor S. Kobayashi's 60th Birthday, ed. T. Mabuchi, J. Noguchi, T. Ochiai, World Scientific, 1994, pp. 39-50.

[BBR] Bartocci, C., Bruzzo, U., Hernandez Ruiperez, D., *A Fourier-Mukai transform for stable bundles on K3 surfaces*, J. Reine Angew. Math. **486**, pp. 1-16 (1997).

[Bea] Beauville, A. *Varietes Kählеriennes dont la première classe de Chern est nulle*. J. Diff. Geom. **18**, pp. 755-782 (1983).

[Bes] Besse, A., *Einstein Manifolds*, Springer-Verlag, New York (1987)

[Bo] Bogomolov, F.A., *Hamiltonian Kähler manifolds*, (English. Russian original) Sov. Math., Dokl. **19**, 1462-1465 (1978); translation from Dokl. Akad. Nauk SSSR **243**, 1101-1104 (1978).

[C] Calabi, E., *Metriques kähleriennes et fibrès holomorphes*, Ann. Ecol. Norm. Sup. **12** (1979), 269-294.

[D] Deligne, P., *Categories tannakiennes*, The Grothendieck Festschrift, Collect. Artic. in Honor of the 60th Birthday of A. Grothendieck. Vol. II, Prog. Math. **87**, 111-195 (1990).

[Do] Douady, A., *Le probleme des modules pour les sous-espaces analitiques compactes d'un espace analitique donné*, Ann. Inst. Fourier, **16** (1966), vol. 1, pp. 1-95.

[O'G] O'Grady, K. G., *Desingularized moduli spaces of sheaves on a K3*, alg-geom/9708009, 41 pages, preliminary version.

[Gi] Gieseker, D. *On the moduli of vector bundles on an algebraic surface*, Ann. Math., II. Ser. **106**, 45-60 (1977).

[GH] Griffiths, Ph., Harris, J., *Principles of Algebraic Geometry*, Wiley-Intersience, New York, 1978.

[GMT] Guaraldo, F., Macri, P., Tancredi, A., *Topics on real analytic spaces*, Advanced lectures in mathematics, Braunschweig: F. Vieweg, 1986.

[F] Fujiki, A., *On primitively symplectic compact Kähler V-manifolds of dimension four*, in Classification of algebraic and analytic manifolds, (Katata, 1982), 71–250, Progr. Math., 39, Birkhäuser Boston, 1983.

[H-Gro] Hartshorne, R., Local cohomology; a seminar given by A. Grothendieck, Harvard University, Fall, 1961. // Lecture notes in mathematics, vol. 41, Springer-Verlag, 1967.

[HKLR] N.J. Hitchin, A. Karlhede, U. Lindström, M. Roček, *Hyperkähler metrics and supersymmetry*, Comm. Math. Phys (1987).

[H] Huybrechts, D., Birational symplectic manifolds and their deformations, alg-geom/9601015.

[K] Kaledin, D., *Integrability of the twistor space for a hypercomplex manifold*, alg-geom/9612016, 9 pages, Latex2e.

[KV] Kaledin, D., Verbitsky, M., *Non-Hermitian Yang-Mills connections*, // alg-geom 9606019 (1996), 48 pages, LaTeX 2e.

[Ko] Kobayashi S., *Differential geometry of complex vector bundles*, // Princeton University Press, 1987.

[Ma1] Maruyama, Masaki *On boundedness of families of torsion free sheaves*, J. Math. Kyoto Univ. **21**, 673-701 (1981)

[Ma2] Maruyama, Masaki, *Construction of moduli spaces of stable sheaves via Simpson's idea*, Lect. Notes Pure Appl. Math. **179**, 147-187 (1996).

[LY] Li, Jun, and Yau, S.-T., *Hermitian Yang-Mills connections on non-Kahler manifolds*, in "Mathematical aspects of string theory" (S.T. Yau ed.), World Scientific Publ., London, 1987, pp. 560-573.

[N1] Nitta, Takashi, *Vector bundles over quaternionic Kähler manifolds*, Tohoku Math. J., II. Ser. 40, No.3, 425-440 (1988).

[N2] Nitta, Takashi, *Moduli spaces of Yang-Mills connections over quaternionic Kähler manifolds*, Osaka J. Math. **26**, No.2, 287-298 (1989).

[OSS] Christian Okonek, Michael Schneider, Heinz Spindler, *Vector bundles on complex projective spaces*. Progress in mathematics, vol. 3, Birkhauser, 1980.

[O] Orlov, D. *Equivalences of derived categories and K3 surfaces*, alg-geom electronic preprint 9606006 (1996).

[Sal] Salamon, S., *Quaternionic Kähler manifolds*, Inv. Math. **67** (1982), 143-171.

[Sw] Swann, A., *HyperKähler and quaternionic Kähler geometry*, Math. Ann. **289**, No. 3, 421-450 (1991).

[UY] Uhlenbeck K., Yau S. T., *On the existence of Hermitian Yang-Mills connections in stable vector bundles*, // Comm. on Pure and Appl. Math., **39**, p. S257-S293 (1986).

[V0] M. Verbitsky, *On the action of the Lie algebra* $\mathfrak{so}(5)$ *on the cohomology of a hyperkähler manifold*, Func. Anal. and Appl. **24** (1990), 70-71.

[V1] Verbitsky M., *Hyperholomorphic bundles over a hyperkähler manifold*, alg-geom electronic preprint 9307008 (1993), 43 pages, LaTeX, also published in: Journ. of Alg. Geom., **5** no. 4 (1996) pp. 633-669.

[V2] Verbitsky M., *Hyperkähler and holomorphic symplectic geometry I*, Journ. of Alg. Geom., **5** no. 3 (1996) pp. 401-415. A different version appeared as alg-geom/9307009

[V2(II)] Verbitsky M., *Hyperkähler embeddings and holomorphic symplectic geometry II*, alg-geom electronic preprint 9403006 (1994), 14 pages, LaTeX, also published in: GAFA **5** no. 1 (1995), 92-104.

[V3] Verbitsky, M., *Cohomology of compact hyperkähler manifolds*. alg-geom electronic preprint 9501001, 89 pages, LaTeX.

[V3-bis] Verbitsky, M., *Cohomology of compact hyperkähler manifolds and its applications*, alg-geom electronic preprint 9511009, 12 pages, LaTeX, also published in: GAFA vol. 6 (4) pp. 601-612 (1996).

[V4] Verbitsky M., *Deformations of trianalytic subvarieties of hyperkähler manifolds*, alg-geom electronic preprint 9610010 (1996), 51 pages, LaTeX2e.

[V5] Verbitsky M., *Trianalytic subvarieties of the Hilbert scheme of points on a K3 surface*, alg-geom/9705004.

[V-a] Verbitsky M., *Algebraic structures on hyperkähler manifolds*, alg-geom electronic preprint 9609011, also published in: Math. Res. Lett., **3**, 763-767 (1996).

[V-d] Verbitsky M., *Desingularization of singular hyperkähler varieties I*, electronic preprint 9611015 (1996), 13 pages, LaTeX 2e, also published in: Math. Res. Lett. **4** (1997), no. 2-3, pp. 259–271.

[V-d2] Verbitsky M., *Desingularization of singular hyperkähler varieties II*, alg-geom/9612013 (1996), 15 pages, LaTeX 2e.

[V-d3] Verbitsky M., *Hypercomplex Varieties*, alg-geom/9703016 (1997), 40 pages, LaTeX (to appear in Comm. in An. and Geom.)

[Y] Yau, S. T., *On the Ricci curvature of a compact Kähler manifold and the complex Monge-Ampère equation I.* // Comm. on Pure and Appl. Math. 31, 339-411 (1978).

E-mail: `verbit@thelema.dnttm.rssi.ru`, `verbit@ihes.fr`

Part II. Hyperkähler structures on total spaces of holomorphic cotangent bundles

D. Kaledin

Introduction

A hyperkähler manifold is by definition a Riemannian manifold M equipped with two anti-commuting almost complex structures I, J parallel with respect to the Levi-Civita connection. Hyperkähler manifolds were introduced by E. Calabi in [C]. Since then they have become the topic of much research. We refer the reader to [B] and to [HKLR] for excellent overviews of the subject.

Let M be a hyperkähler manifold. The almost complex structures I and J generate an action of the quaternion algebra $\mathbb{H}$ in the tangent bundle $\Theta(M)$ to the manifold M. This action is parallel with respect to the Levi-Civita connection. Every quaternion $h \in \mathbb{H}$ with $h^2 = -1$, in particular, the product $K = IJ \in \mathbb{H}$, defines by means of the $\mathbb{H}$-action an almost complex structure M_h on M. This almost complex structure is also parallel, hence integrable and Kähler. Thus every hyperkähler manifold M is canonically Kähler, and in many different ways. For the sake of convenience, we will consider M as a Kähler manifold by means of the complex structure M_I, unless indicated otherwise.

One of the basic facts about hyperkähler manifolds is that the Kähler manifold M_I underlying a hyperkähler manifold M is canonically holomorphically symplectic. To see this, let ω_J, ω_K be the Kähler forms for the complex structures M_J, M_K on the manifold M, and consider the 2-form $\Omega = \omega_J + \sqrt{-1}\omega_K$ on M. It is easy to check that the form Ω is of Hodge type $(2,0)$ for the complex structure M_I on M. Since it is obviously non-degenerate and closed, it is holomorphic, and the Kähler manifold M_I equipped with the form Ω is a holomorphically symplectic manifold.

It is natural to ask whether every holomorphically symplectic manifold $\langle M, \Omega\rangle$ underlies a hyperkähler structure on M, and if so, then how many such hyperkähler structures are there. Note that if such a hyperkähler structure exists, it is completely defined by the Kähler metric h on M. Indeed, the Kähler forms ω_J and ω_K are by definition the real and imaginary parts of the form Ω, and the forms ω_J and ω_K together with the metric define the complex structures J and K on M and, consequently, the whole $\mathbb{H}$-action in the tangent bundle $\Theta(M)$. For the sake of simplicity, we will call a metric h on a holomorphically symplectic manifold $\langle M, \Omega\rangle$ hyperkähler if the Riemannian manifold $\langle M, h\rangle$ with the quaternionic action associated to the pair $\langle \Omega, h\rangle$ is a hyperkähler manifold.

It is known (see, e.g., [Beau]) that if the holomorphically symplectic manifold M is compact, for example, if M is a $K3$-surface, then every Kähler class in $H^{1,1}(M)$ contains a unique hyperkähler metric. This is, in fact, a consequence of the famous Calabi-Yau Theorem, which provides the canonical Ricci-flat metric on M with the given cohomology class. This Ricci-flat metric turns out to be hyperkähler. Thus in the compact case holomorphically symplectic and hyperkähler manifolds are essentially the same.

The situation is completely different in the general case. For example, all holomorphically symplectic structures on the formal neighborhood of the origin $0 \in \mathbb{C}^{2n}$ in the $2n$-dimensional complex vector space $\mathbb{C}^{2n}$ are isomorphic by the appropriate version of the Darboux Theorem. On the other hand, hyperkähler structures on this formal neighborhood form an infinite-dimensional family (see, e.g., [HKLR], where there is a construction of a smaller, but still infinite-dimensional family of hyperkähler metrics defined on the whole $\mathbb{C}^{2n}$). Thus, to obtain meaningful results, it seems necessary to restrict our attention to holomorphically symplectic manifolds belonging to some special class.

The simplest class of non-compact holomorphically symplectic manifolds is formed by total spaces T^*M to the cotangent bundle to complex manifolds M. In fact, the first examples of hyperkähler manifolds given by Calabi in [C] were of this type, with M being a Kähler manifold of constant holomorphic sectional curvature (for example, a complex projective space). It has been conjectured for some time that every total space T^*M of the cotangent bundle to a Kähler manifold admits a hyperkähler structure. The goal of this paper is to prove that this is indeed the case, if one agrees to consider only an open neighborhood $U \subset T^*M$ of the zero section $M \subset T^*M$. Our main result is the following.

Theorem 1 *Let M be a complex manifold equipped with a Kähler metric. The metric on M extends to a hyperkähler metric h defined in the formal neighborhood of the zero section $M \subset T^*M$ in the total space T^*M to the holomorphic cotangent bundle to M. The extended metric h is invariant under the action of the group $U(1)$ on T^*M given by dilatations along the fibers of the canonical projection $\rho : T^*M \to M$. Moreover, every other $U(1)$-invariant hyperkähler metric on the holomorphically symplectic manifold T^*M becomes equal to h after a holomorphic symplectic $U(1)$-equivariant automorphism of T^*M over M. Finally, if the Kähler metric on M is real-analytic, then the formal hyperkähler metric h converges to a real-analytic metric in an open neighborhood $U \subset T^*M$ of the zero section $M \subset T^*M$.*

Many of the examples of hyperkähler metrics obtained by Theorem 1 are already known. (See, e.g., [Kr1], [Kr2], [N], [H2], [BG], [DS].) In these examples M is usually a generalized flag manifold or a homogeneous space of some kind. On the other hand, very little is known for manifolds of general type. In particular, it seems that even for curves of genus $g \geq 2$ Theorem 1 is new.

We would like to stress the importance of the $U(1)$-invariance condition on the metric in the formulation of Theorem 1. This condition for a total space T^*M of a cotangent bundle is equivalent to a more general compatibility condition between a $U(1)$-action and a hyperkähler structure on a smooth manifold introduced by N.J. Hitchin (see, e.g., [H2]). Thus Theorem 1 can be also regarded as answering a question of Hitchin's in [H2], namely, whether every Kähler manifold can be embedded as the submanifold of $U(1)$-fixed points in a $U(1)$-equivariant hyperkähler manifold. On the other hand, it is this $U(1)$-invariance that guarantees the uniqueness of the metric h claimed in Theorem 1.

We also prove a version of Theorem 1 "without the metrics". The Kähler metric on M in this theorem is replaced with a holomorphic connection ∇ on the cotangent bundle to M without torsion and $(2,0)$-curvature. We call such connections *Kählerian.* The total space of the cotangent bundle T^*M is replaced with the total space $\overline{T}M$ of the complex-conjugate to the tangent bundle to M. (Note that *a priori* there is no complex structure on $\overline{T}M$, but the $U(1)$-action by dilatations on this space is well-defined.) The analog of the notion of a hyperkähler manifold "without the metric" is the notion if a hypercomplex manifold (see, e.g., [B]). We define a version of Hitchin's condition on the $U(1)$-action for hypercomplex manifolds and prove the following.

Theorem 2 *Let M be a complex manifold, and let $\overline{T}M$ be the total space of the complex-conjugate to the tangent bundle to M equipped with an action of the group $U(1)$ by dilatation along the fibers of the projection $\overline{T}M \to M$. There exists a natural bijection between the set of all Kählerian connections on the cotangent bundle to M and the set of all isomorphism classes of $U(1)$-equivariant hypercomplex structures on the formal neighborhood of the zero section $M \subset \overline{T}M$ in $\overline{T}M$ such that the projection $p : \overline{T}M \to M$ is holomorphic. If the Kählerian connection on M is real-analytic, the corresponding hypercomplex structure is defined in an open neighborhood $U \subset \overline{T}M$ of the zero section.*

Our main technical tool in this paper is the relation between $U(1)$-equivariant hyperkähler manifolds and the theory of $\mathbb{R}$-Hodge structures discovered by P. Deligne and C. Simpson (see [D2], [S1]). To emphasize this relation, we use the name *Hodge manifolds* for the hypercomplex manifolds equipped with a compatible $U(1)$-action.

It must be noted that many examples of hyperkähler manifolds equipped with a compatible $U(1)$-action are already known. Such are, for example, many of the manifolds constructred by the so-called hyperkähler reduction from flat hyperkähler spaces (see [H2] and [HKLR]). An important class of such manifolds is formed by the so-called quiver varieties, studied by H. Nakajima ([N]). On the other hand, the moduli spaces $\mathcal{M}$ of flat connections on a complex manifold M, studied by Hitchin ([H1]) when M is a curve and by Simpson ([S2]) in the general case, also belong to the class of Hodge manifolds, as Simpson has shown in [S1]. Some parts of our theory, especially the uniqueness statement of Theorem 1, can be applied to these known examples.

We now give a brief outline of the paper. Sections 1-3 are preliminary and included mostly to fix notation and terminology. Most of the facts in these sections are well-known.

- In Section 1 we have collected the necessary facts from linear algebra about quaternionic vector spaces and $\mathbb{R}$-Hodge structures. Everything is standard, with an exception, perhaps, of the notion of *weakly Hodge map*, which we introduce in Subsection 1.4.

- We begin Section 2 with introducing a technical notion of a *Hodge bundle* on a smooth manifold equipped with a $U(1)$-action. This notion will be heavily used throughout the paper. Then we switch to our main subject, namely, various differential-geometric objects related to a quaternion action in the tangent bundle. The rest of Section 2 deals with almost quaternionic manifolds and the compatibility conditions between an almost quaternionic structure and a $U(1)$-action on a smooth manifold M.

- In Section 3 we describe various integrability conditions on an almost quaternionic structure. In particular, we recall the definition of a hypercomplex manifold and introduce $U(1)$-equivariant hypercomplex manifolds under the name of ***Hodge manifolds.*** We then rewrite the definition of a Hodge manifold in the more convenient language of Hodge bundles, to be used throughout the rest of the paper. Finally, in Subsection 3.3 we discuss metrics on hypercomplex and Hodge manifolds. We recall the definition of a hyperkähler manifold and define the notion of a *polarization* of a Hodge manifold. A polarized Hodge manifold is the same as a hyperkähler manifold equipped with a $U(1)$-action compatible with the hyperkähler structure of the sense of Hitchin, [H2].

- The main part of the paper begins in Section 4. We start with arbitrary Hodge manifolds and prove that in a neighborhood of the subset $M^{U(1)}$ of "regular" $U(1)$-fixed points

- termsregular fixed point every such manifold M is canonically isomorphic to an open neighborhood of the zero section in a total space $\overline{T}M^{U(1)}$ of the tangent bundle to the fixed point set. A fixed point $m \in M$ is "regular" if the group $U(1)$ acts on the tangent space $T_m M$ with weights 0 and 1. We call this canonical isomorphism *the linearization* of the regular Hodge manifold.

  The linearization construction can be considered as a hyperkähler analog of the Darboux-Weinstein Theorem in the symplectic geometry. Apart from the cotangent bundles, it can be applied to the Hitchin-Simpson moduli space $\mathcal{M}$ of flat connections on a Kähler manifold X. The regular points in this case correspond to stable flat connections such that the underlying holomorphic bundle is also stable. The linearization construction provides a canonical embedding of the subspace $\mathcal{M}^{reg} \subset \mathcal{M}$ of regular points into the total space $T^*\mathcal{M}_0$ of the cotangent bundle to the space $\mathcal{M}_0$ of stable holomorphic bundles on X. The unicity statement of Theorem 1 guarantees that the hyperkähler metric on $\mathcal{M}^{reg}$ provided by the Simpson theory is the same as the canonical metric constructed in Theorem 1.

- Starting with Section 5, we deal only with total spaces $\overline{T}M$ of the complex-conjugate to tangent bundles to complex manifolds M. In Section 5 we describe Hodge manifolds structures on $\overline{T}M$ in terms of certain connections on the locally trivial fibration $\overline{T}M \to M$. "Connection" here is understood as a choice of the subbundle of horizontal vectors, regardless of the vector bundle structure on the fibration $\overline{T}M \to M$. We establish a correspondence between Hodge manifold structures on $\overline{T}M$ and connections on $\overline{T}M$ over M of certain type, which we call *Hodge connection.*

- In Section 6 we restrict our attention to the formal neighborhood of the zero section $M \subset \overline{T}M$. We introduce the appropriate "formal" versions of all the definitions and then establish a correspondence between formal Hodge connections on $\overline{T}M$ over M and certain differential operators on the manifold M itself, which we call *extended connections*

- *termsextended connection.* We also introduce a certain canonical algebra bundle $\mathcal{B}^{\bullet}(M, \mathbb{C})$ on the complex manifold M, which we call *the Weil algebra* of the manifold M. Extended connections give rise to natural derivations of the Weil algebra.

- Before we can proceed with classification of extended connections

- termsextended connection on the manifold M and therefore of regular Hodge manifolds, we need to derive some linear-algebraic facts on the Weil algebra $\mathcal{B}^{\bullet}(M, \mathbb{C})$. This is the subject of Section 7. We begin with introducing a certain version of the de Rham complex of a smooth complex manifold, which we call *the total de Rham complex.* Then we combine it the material of Section 6 to define the so-called *total Weil algebra* of the manifold M and establish some of its properties. Section 7 is the most technical part of the whole paper. The reader is advised to skip reading it until needed.

- Section 8 is the main section of the paper. In this section we prove, using the technical results of Section 7, that extended

- termsextended connection connections on M are in a natural one-to-one correspondence with Kählerian connections on the cotangent bundle to M (Theorem 8.1). This proves the formal part of Theorem 2.
- In Section 9 we deal with polarizations. After some preliminaries, we use Theorem 2 to deduce the formal part of Theorem 1 (see Theorem 9.1).
- Finally, in Section 10 we study the convergence of our formal series and prove Theorem 1 and Theorem 2 in the real-analytic case.
- We have also attached a brief Appendix, where we sketch a more conceptual approach to some of the linear algebra used in the paper, in particular, to our notion of a weakly Hodge map. This approach also allows us to describe a simple and conceptual proof of Proposition 7.1, the most technical of the facts proved in Section 7. The Appendix is mostly based on results of Deligne and Simpson ([D2], [S1]).

Acknowledgment. I would like to thank A. Beilinson, D. Kazhdan, A. Levin, L. Posicelsky, A. Shen and A. Tyurin for stimulating discussions. I am especially grateful to my friends Misha Verbitsky and Tony Pantev for innumerable discussions, constant interest and help, without which this paper most probably would not have been written. I would also like to mention here how much I have benefited from a course on moduli spaces and Hodge theory given by Carlos Simpson at MIT in the Fall of 1993. On a different note, I would like to express my dearest gratitude to Julie Lynch, formerly at International Press in Cambridge, and also to the George Soros's Foundation and to CRDF for providing me with a source of income during the preparation of this paper.

1. Preliminary facts from linear algebra

1.1. Quaternionic vector spaces

1.1.1. Throughout the paper denote by $\mathbb{H}$ the $\mathbb{R}$-algebra of quaternions.

Definition. A *quaternionic vector space* is a finite-dimensional left module over the algebra $\mathbb{H}$.

Let V be a quaternionic vector space. Every algebra embedding $I : \mathbb{C} \to \mathbb{H}$ defines by restriction an action of $\mathbb{C}$ on V. Denote the corresponding complex vector space by V_I.

Fix once and for all an algebra embedding $I : \mathbb{C} \to \mathbb{H}$ and call the complex structure V_I *the preferred complex structure* on V.

1.1.2. Let the group $\mathbb{C}^*$ act on the algebra $\mathbb{H}$ by conjugation by $I(\mathbb{C}^*)$. Since $I(\mathbb{R}^*) \subset I(\mathbb{C}^*)$ lies in the center of the algebra $\mathbb{H}$, this action factors through the map $N : \mathbb{C}^* \to \mathbb{C}^*/\mathbb{R}^* \cong U(1)$ from $\mathbb{C}^*$ to the one-dimensional unitary group defined by $N(a) = a^{-1}\overline{a}$. Call this action *the standard action* of $U(1)$ on $\mathbb{H}$.

The standard action commutes with the multiplication and leaves invariant the subalgebra $I(\mathbb{C})$. Therefore it extends to an action of the complex algebraic group $\mathbb{C}^* \supset U(1)$ on the algebra $\mathbb{H}$ considered as a right complex vector space over $I(\mathbb{C})$. Call this action *the standard action* of $\mathbb{C}^*$ on $\mathbb{H}$.

1.1.3. Definition. An *equivariant* quaternionic vector space is a quaternionic vector space V equipped with an action of the group $U(1)$ such that the action map

$$\mathbb{H} \otimes_{\mathbb{R}} V \to V$$

is $U(1)$-equivariant.

The $U(1)$-action on V extends to an action of $\mathbb{C}^*$ on the complex vector space V_I. The action map $\mathbb{H} \otimes_{\mathbb{R}} V \to V$ factors through a map

$$\mathrm{Mult} : \mathbb{H} \otimes_{\mathbb{C}} V_I \to V_I$$

of complex vector spaces. This map is $\mathbb{C}^*$-equivariant if and only if V is an equivariant quaternionic vector space.

1.1.4. The category of complex algebraic representations V of the group $\mathbb{C}^*$ is equivalent to the category of graded vector spaces $V = \oplus V^{\bullet}$. We will say that a representation W is *of weight i* if $W = W^i$, that is, if an element $z \in \mathbb{C}^*$ acts on W by multiplication by z^k. For every representation W we will denote by $W(k)$ the representation corresponding to the grading

$$W(k)^i = W^{k+i}.$$

1.1.5. The algebra $\mathbb{H}$ considered as a complex vector space by means of right multiplication by $I(\mathbb{C})$ decomposes $\mathbb{H} = I(\mathbb{C}) \oplus \overline{\mathbb{C}}$ with respect to the standard $\mathbb{C}^*$-action. The first summand is of weight 0, and the second is of weight 1. This decomposition is compatible with the left $I(\mathbb{C})$-actions as well and induces for every complex vector space W a decomposition

$$\mathbb{H} \otimes_{\mathbb{C}} W = W \oplus \overline{\mathbb{C}} \otimes_{\mathbb{C}} W.$$

If W is equipped with an $\mathbb{C}^*$-action, the second summand is canonically isomorphic to $\overline{W}(1)$, where $\overline{W}$ is the vector space complex-conjugate to W.

1.1.6. Let V be an equivariant quaternionic vector space. The action map

$$\mathrm{Mult} : \mathbb{H} \otimes_{\mathbb{C}} V_I \cong V_I \oplus \overline{\mathbb{C}} \otimes V_I \to V_I$$

decomposes Mult $= \mathsf{id} \oplus j$ for a certain map $j : \overline{V_I}(1) \to V_I$. The map j satisfies $j \circ \overline{j} = -\mathsf{id}$, and we obviously have the following.

Lemma. *The correspondence $V \longmapsto \langle V_I, j\rangle$ is an equivalence of categories between the category of equivariant quaternionic vector spaces and the category of pairs $\langle W, j\rangle$ of a graded complex vector space W and a map $j : \overline{W^{\bullet}} \to W^{1-\bullet}$ satisfying $j \circ \overline{j} = -\mathsf{id}$.*

1.1.7. We will also need a particular class of equivariant quaternionic vector spaces which we will call regular.

Definition. An equivariant quaternionic vector space V is called *regular* if every irreducible $\mathbb{C}^*$-subrepresentation of V_I is either trivial or of weight 1.

Lemma. *Let V be a regular $\mathbb{C}^*$-equivariant quaternionic vector space and let $V_I^0 \subset V_I$ be the subspace of $\mathbb{C}^*$-invariant vectors. Then the action map*

$$V_I^0 \oplus \overline{V_I^0} \cong \mathbb{H} \otimes_{\mathbb{C}} V_I^0 \to V_I$$

is invertible.

Proof. Let $V_I^1 \subset V_I$ be the weight 1 subspace with respect to the gm-action. Since V is regular, $V_I = V_I^0 \oplus V_I^1$. On the other hand, $j : \overline{V_I} \to V_I$ interchanges V_I^0 and V_I^1. Therefore $V_I^1 \cong \overline{V_I^0}$ and we are done. □

Thus every regular equivariant quaternionic vector space is a sum of several copies of the algebra $\mathbb{H}$ itself. The corresponding Hodge structure has Hodge numbers $h^{1,0} = h^{0,1}$, $h^{p,q} = 0$ otherwise.

1.2. The complementary complex structure

1.2.1. Let $J : \mathbb{C} \to \mathbb{H}$ be another algebra embedding. Say that embeddings I and J are *complementary* if

$$J(\sqrt{-1})I(\sqrt{-1}) = -I(\sqrt{-1})J(\sqrt{-1}).$$

Let V be an equivariant quaternionic vector space. The standard $U(1)$-action on $\mathbb{H}$ induces an action of $U(1)$ on the set of all algebra embeddings $\mathbb{C} \to \mathbb{H}$. On the subset of embeddings complementary to I this action is transitive. Therefore for every two embeddings $J_1, J_2 : \mathbb{C} \to \mathbb{H}$ complementary to I the complex vector spaces V_{J_1} and V_{J_2} are canonically isomorphic. We will from now on choose for convenience an algebra embedding $J : \mathbb{C} \to \mathbb{H}$ complementary to I and speak of *the complementary complex structure* V_J on V; however, nothing depends on this choice.

1.2.2. For every equivariant quaternionic vector space V the complementary embedding $J : \mathbb{C} \to \mathbb{H}$ induces an isomorphism

$$J \otimes \mathsf{id} : \mathbb{C} \otimes_{\mathbb{R}} V \to \mathbb{H} \otimes_{I(\mathbb{C})} V_I$$

of complex vector spaces. Let $\mathrm{Mult} : \mathbb{H} \otimes_{I(\mathbb{C})} V_I \to V_I$, $\mathrm{Mult} : \mathbb{C} \otimes_{\mathbb{R}} V \to V_J$ be the action maps. Then there exists a unique isomorphism $H : V_J \to V_I$ of complex vector spaces such that the diagram

$$\begin{array}{ccc}
\mathbb{C} \otimes_{\mathbb{R}} V & \xrightarrow{J \otimes \mathsf{id}} & \mathbb{H} \otimes_{I(\mathbb{C})} V_I \\
{\scriptstyle \mathrm{Mult}}\downarrow & & {\scriptstyle \mathrm{Mult}}\downarrow \\
V_J & \xrightarrow{H} & V_I
\end{array}$$

is commutative. Call the map $H : V_J \to V_I$ *the standard isomorphism* between the complementary and the preferred complex structures on the equivariant quaternionic vector space V.

1.2.3. Note that both V_I and V_J are canonically isomorphic to V as real vector spaces; therefore the map $H : V_J \to V_I$ is in fact an automorphism of the real vector space V. Up to a constant this automorphism is given by the action of the element $I(\sqrt{-1}) + J(\sqrt{-1}) \in \mathbb{H}$ on the $\mathbb{H}$-module V.

1.3. $\mathbb{R}$-Hodge structures

1.3.1. Recall that *a pure $\mathbb{R}$-Hodge structure W of weight i* is a pair of a graded complex vector space $W = \oplus_{p+q=i} W^{p,q}$ and a *real structure* map $\overline{} : W^{p,q} \to W^{q,p}$ satisfying $\overline{} \circ \overline{} = \mathsf{id}$. The bigrading $W^{p,q}$ is called *the Hodge type bigrading.* The dimensions $h^{p,q} = \dim_{\mathbb{C}} W^{p,q}$ are called *the Hodge numbers* of the pure $\mathbb{R}$-Hodge structure W. Maps between pure Hodge structures are by definition maps of their underlying complex vector spaces compatible with the bigrading and the real structure maps.

1.3.2. Recall also that for every k the *Hodge-Tate* pure $\mathbb{R}$-Hodge structure $\mathbb{R}(k)$ of weight $-2k$ is by definition the 1-dimensional complex vector space with complex conjugation equal to $(-1)^k$ times the usual one, and with Hodge bigrading

$$\mathbb{R}(k)^{p,q} = \begin{cases} \mathbb{R}(k), & p = q = -k, \\ 0, & \text{otherwise.} \end{cases}$$

For a pure $\mathbb{R}$-Hodge structure V denote, as usual, by $V(k)$ the tensor product $V(k) = V \otimes \mathbb{R}(k)$.

1.3.3. We will also need special notation for another common $\mathbb{R}$-Hodge structure, which we now introduce. Note that for every complex V be a complex vector space the complex vector space $V \otimes_{\mathbb{R}} \mathbb{C}$ carries a canonical $\mathbb{R}$-Hodge structure of weight 1 with Hodge bigrading given by

$$V^{1,0} = V \subset V \otimes_{\mathbb{R}} \mathbb{C} \qquad V^{0,1} = \overline{V} \otimes_{\mathbb{R}} \mathbb{C}.$$

In particular, $\mathbb{C} \otimes_{\mathbb{R}} \mathbb{C}$ carries a natural $\mathbb{R}$-Hodge structure of weight 1 with Hodge numbers $h^{1,0} = h^{0,1} = 1$. Denote this Hodge structure by $\mathcal{W}_1$.

1.3.4. Let $\langle W, \bar{\ } \rangle$ be a pure Hodge structure, and denote by $W_{\mathbb{R}} \subset W$ the $\mathbb{R}$-vector subspace of elements preserved by $\bar{\ }$. Define the *Weil operator* $C : W \to W$ by

$$C = (\sqrt{-1})^{p-q} : W^{p,q} \to W^{q,p}.$$

The operator $C : W \to W$ preserves the $\mathbb{R}$-Hodge structure, in particular, the subspace $W_{\mathbb{R}} \subset W$. On pure $\mathbb{R}$-Hodge structures of weight 0 the Weil operator C corresponds to the action of $-1 \in U(1) \subset \mathbb{C}^*$ in the corresponding representation.

1.3.5. For a pure Hodge structure W of weight i let

$$j = C \circ \bar{\ } : \overline{W^{\bullet,1-\bullet}} \to W^{1-\bullet,\bullet}.$$

If i is odd, in particular, if $i = 1$, then $j \circ \bar{j} = -\mathsf{id}$. Together with Lemma 1.1.6 this gives the following.

Lemma. *The category of equivariant quaternionic vector spaces is equivalent to the category of pure* $\mathbb{R}$*-Hodge structures of weight* 1.

1.3.6. Let V be an equivariant quaternionic vector space, and let $\langle W, \bar{\ } \rangle$ be $\mathbb{R}$-Hodge structure of weight 1 corresponding to V under the equivalence of Lemma 1.3.5. By definition the complex vector space W is canonically isomorphic to the complex vector space V_I with the preferred complex structure on V. It will be more convenient for us to identify W with the complementary complex vector space V_J by means of the standard isomorphism $H : V_J \to V_I$. The multiplication map

$$\mathrm{Mult} : V_I \otimes_{\mathbb{R}} \mathbb{C} \cong V \otimes_{\mathbb{C}} \mathbb{H} \to V_J$$

is then a map of $\mathbb{R}$-Hodge structures. The complex conjugation $\bar{\ } : W \to \overline{W}$ is given by

$$\bar{\ } = C \circ H \circ j \circ H^{-1} = C \circ i : V_J \to \overline{V_J}, \tag{1.1}$$

where $i : V_J \to \overline{V_J}$ is the action of the element $I(\sqrt{-1}) \subset \mathbb{H}$.

1.4. Weakly Hodge maps

1.4.1. Recall that the category of pure $\mathbb{R}$-Hodge structures of weight i is equivalent to the category of pairs $\langle V, F^\bullet \rangle$ of a real vector space V and a decreasing filtration $F^\bullet$ on the complex vector space $V_\mathbb{C} = V \otimes_\mathbb{R} \mathbb{C}$ satisfying

$$V_\mathbb{C} = \bigoplus_{p+q=i} F^p V_\mathbb{C} \cap \overline{F^q V_\mathbb{C}}.$$

The filtration $F^\bullet$ is called *the Hodge filtration.* The Hodge type bigrading and the Hodge filtration are related by $V^{p,q} = F^p V_\mathbb{C} \cap \overline{F^q V_\mathbb{C}}$ and $F^p V_\mathbb{C} = \oplus_{k \geq p} V^{k,i-k}$.

1.4.2. Let $\langle V, F^\bullet \rangle$ and $\langle W, F^\bullet \rangle$ be pure $\mathbb{R}$-Hodge structures of weights n and m respectively. Usually maps of pure Hodge structures are required to preserve the weights, so that $\operatorname{Hom}(V, W) = 0$ unless $n = m$. In this paper we will need the following weaker notion of maps between pure $\mathbb{R}$-Hodge structures.

Definition. An $\mathbb{R}$-linear map $f : V \to W$ is said to be *weakly Hodge* if it preserves the Hodge filtrations.

Equivalently, the complexified map $f : V_\mathbb{C} \to W_\mathbb{C}$ must decompose

$$f = \sum_{0 \leq p \leq m-n} f^{p,m-n-p}, \tag{1.2}$$

where the map $f^{p,m-n-p} : V_\mathbb{C} \to W_\mathbb{C}$ is of Hodge type $(p, m-n-p)$. Note that this condition is indeed weaker than the usual definition of a map of Hodge structures: a weakly Hodge map $f : V \to W$ can be non-trivial if m is strictly greater than n. If $m < n$, then f must be trivial, and if $m = n$, then weakly Hodge maps from V to W are the same as usual maps of $\mathbb{R}$-Hodge structures.

1.4.3. We will denote by $\mathcal{WH}odge$ the category of pure $\mathbb{R}$-Hodge structures of arbitrary weight with weakly Hodge maps as morphisms. For every integer n let $\mathcal{WH}odge_n$ be the full subcategory in $\mathcal{WH}odge$ consisting of pure $\mathbb{R}$-Hodge structures of weight n, and let $\mathcal{WH}odge_{\geq n}$ be the full subcategory of $\mathbb{R}$-Hodge structures of weight not less than n. Since weakly Hodge maps between $\mathbb{R}$-Hodge structures of the same weight are the same as usual maps of $\mathbb{R}$-Hodge structures, the category $\mathcal{WH}odge_n$ is the usual category of pure $\mathbb{R}$-Hodge structures of weight n.

1.4.4. Let $\mathcal{W}_1 = \mathbb{C} \otimes_\mathbb{R} \mathbb{C}$ be the pure Hodge structure of weight 1 with Hodge numbers $h^{1,0} = h^{0,1} = 1$, as in 1.3.3. The diagonal embedding $\mathbb{C} \to \mathbb{C} \otimes_\mathbb{R} \mathbb{C}$ considered as a map $w_1 : \mathbb{R} \to \mathcal{W}_1$ from the trivial pure $\mathbb{R}$-Hodge structure $\mathbb{R}$ of weight 0 to $\mathcal{W}_1$ is obviously weakly Hodge. It decomposes $w_1 = w_1^{1,0} + w_1^{0,1}$ as in (1.2), and the components $w_1^{1,0} : \mathbb{C} \to \mathcal{W}_1^{1,0}$ and $w_1^{0,1} : \mathbb{C} \to \mathcal{W}_1^{0,1}$ are isomorphisms of complex vectors spaces. Moreover, for every pure $\mathbb{R}$-Hodge structure V of weight n the map w_1 induces a weakly Hodge map $w_1 : V \to \mathcal{W}_1 \otimes V$, and the components $w_1^{1,0} : V_{\mathbb{C}} \to V_{\mathbb{C}} \otimes \mathcal{W}_1^{1,0}$ and $w_1^{0,1} : V_{\mathbb{C}} \to V_{\mathbb{C}} \otimes \mathcal{W}_1^{0,1}$ are again isomorphisms of complex vector spaces.

More generally, for every $k \geq 0$ let $\mathcal{W}_k = S^k\mathcal{W}_1$ be the k-th symmetric power of the Hodge structure $\mathcal{W}_1$. The space $\mathcal{W}_k$ is a pure $\mathbb{R}$-Hodge structure of weight k, with Hodge numbers $h^{k,0} = h^{k-1,1} = \ldots = h^{0,k} = 1$. Let $w_k : \mathbb{R} \to \mathcal{W}_k$ be the k-th symmetric power of the map $w_1 : \mathbb{R} \to \mathcal{W}_1$. For every pure $\mathbb{R}$-Hodge structure V of weight n the map w_k induces a weakly Hodge map $w_k : V \to \mathcal{W}_k \otimes V$, and the components

$$w_k^{p,q} : V_{\mathbb{C}} \to V_{\mathbb{C}} \otimes \mathcal{W}_k^{p,k-p}, \qquad 0 \leq p \leq k$$

are isomorphisms of complex vector spaces.

1.4.5. The map w_k is a universal weakly Hodge map from a pure $\mathbb{R}$-Hodge structures V of weight n to a pure $\mathbb{R}$-Hodge structure of weight $n+k$. More precisely, every weakly Hodge map $f : V \to V'$ from V to a pure $\mathbb{R}$-Hodge structure V' of weight $n+k$ factors uniquely through $w_k : V \to \mathcal{W}_k \otimes V$ by means of a map $f' : \mathcal{W}_k \otimes V \to V'$ preserving the pure $\mathbb{R}$-Hodge structures. Indeed, $V_{\mathbb{C}} \otimes \mathcal{W}_k = \bigoplus_{0 \leq p \leq k} V_{\mathbb{C}} \otimes \mathcal{W}_k^{p,k-p}$, and the maps $w_k^{p,k-p} : V_{\mathbb{C}} \to V_{\mathbb{C}} \otimes \mathcal{W}_k^{p,k-p}$ are invertible. Hence to obtain the desired factorization it is necessary and sufficient to set

$$f' = f^{p,k-p} \circ \left(w_k^{p,k-p}\right)^{-1} : V_{\mathbb{C}} \otimes \mathcal{W}_k^{p,k-p} \to V_{\mathbb{C}} \to V'_{\mathbb{C}},$$

where $f = \sum_{0 \leq p \leq k} f^{p,k-p}$ is the Hodge type decomposition (1.2).

1.4.6. It will be convenient to reformulate the universal properties of the maps w_k as follows. By definition $\mathcal{W}_k = S^k\mathcal{W}_1$, therefore the dual $\mathbb{R}$-Hodge structures equal $\mathcal{W}_k^* = S^k\mathcal{W}_1^*$, and for every $n, k \geq 0$ we have a canonical projection $\mathsf{can} : \mathcal{W}_n^* \otimes \mathcal{W}_k^* \to \mathcal{W}_{n+k}^*$. For every pure $\mathbb{R}$-Hodge structure V of weight $k \geq 0$ let $\Gamma(V) = V \otimes \mathcal{W}_k^*$.

Lemma. *The correspondence $V \mapsto \Gamma(V)$ extends to a functor*

$$\Gamma : \mathcal{W}\mathcal{H}odge_{\geq 0} \to \mathcal{W}\mathcal{H}odge_0$$

adjoint on the right to the canonical embedding $\mathcal{W}\mathcal{H}odge_0 \hookrightarrow \mathcal{W}\mathcal{H}odge_{\geq 0}$.

Proof. Consider a weakly Hodge map $f : V_n \to V_{n+k}$ from $\mathbb{R}$-Hodge structure V_n of weight n to a pure $\mathbb{R}$-Hodge structure V_{n+k} of weight $n+k$. By the universal property the map f factors through the canonical map $w_k : V_n \to V_n \otimes \mathcal{W}_k$ by means of a map $f_k : V_n \otimes \mathcal{W}_k \to V_{n+k}$. Let $f_k' : V_k \to V_{n+k} \otimes \mathcal{W}_k^*$ be the adjoint map, and let

$$\Gamma(f) = \mathsf{can} \circ f_k' : \Gamma(V_n) = V_n \otimes \mathcal{W}_k^* \to V_{n+k} \otimes \mathcal{W}_n^* \otimes \mathcal{W}_k^* \to$$
$$\to \Gamma(V_{n+k}) = V \otimes \mathcal{W}_{n+k}^*.$$

This defines the desired functor $\Gamma : \mathcal{WH}odge_{\geq 0} \to \mathcal{WH}odge_0$. The adjointness is obvious. □

Remark. See Appendix for a more geometric description of the functor $\Gamma : \mathcal{WH}odge_{\geq 0} \to \mathcal{WH}odge_0$.

1.4.7. The complex vector space $\mathcal{W}_1 = \mathbb{C} \otimes_{\mathbb{R}} \mathbb{C}$ is equipped with a canonical skew-symmetric trace pairing $\mathcal{W}_1 \otimes_{\mathbb{C}} \mathcal{W}_1 \to \mathbb{C}$. Let $\gamma : \mathbb{C} \to \mathcal{W}_1^*$ be the map dual to $w_1 : \mathbb{R} \to \mathcal{W}_1$ under this pairing. The map γ is not weakly Hodge, but it decomposes $\gamma = \gamma^{-1,0} + \gamma^{0,-1}$ with respect to the Hodge type bigrading. Denote $\gamma_l = \gamma^{-1,0}$, $\gamma_r = \gamma^{0,-1}$. For every $0 \leq p \leq k$ the symmetric powers of the maps γ_l, γ_r give canonical complex-linear embeddings

$$\gamma_l, \gamma_r : \mathcal{W}_p^* \to \mathcal{W}_k^*.$$

1.4.8. The map γ_l if of Hodge type $(p-k, 0)$, while γ_r is of Hodge type $(0, p-k)$, and the maps γ_l, γ_r are complex conjugate to each other. Moreover, they are each compatible with the natural maps $\mathsf{can} : \mathcal{W}_p^* \otimes \mathcal{W}_q^* \to \mathcal{W}_{p+q}^*$ in the sense that $\mathsf{can} \circ (\gamma_l \otimes \gamma_l) = \gamma_l \circ \mathsf{can}$ and $\mathsf{can} \circ (\gamma_r \otimes \gamma_r) = \gamma_r \circ \mathsf{can}$. For every p, q, k such that $p + q \geq k$ we have a short exact sequence

$$0 \longrightarrow \mathcal{W}_{p+q-k} \xrightarrow{\gamma_r - \gamma_l} \mathcal{W}_p^* \oplus \mathcal{W}_q^* \xrightarrow{\gamma_l + \gamma_l} \mathcal{W}_k^* \longrightarrow 0 \tag{1.3}$$

of complex vector spaces. We will need this exact sequence in 7.1.7.

1.4.9. The functor Γ is, in general, not a tensor functor. However, the canonical maps $\mathsf{can} : \mathcal{W}_n^* \otimes \mathcal{W}_k^* \to \mathcal{W}_{n+k}^*$ define for every two pure $\mathbb{R}$-Hodge structures V_1, V_2 of non-negative weight a surjective map

$$\Gamma(V_1) \otimes \Gamma(V_2) \to \Gamma(V_1 \otimes V_2).$$

These maps are functorial in V_1 and V_2 and commute with the associativity and the commutativity morphisms. Moreover, for every algebra $\mathcal{A}$ in the tensor category $\mathcal{WH}odge_{\geq 0}$ they turn $\Gamma(\mathcal{A})$ into an algebra in $\mathcal{WH}odge_0$.

1.5. Polarizations

1.5.1. Consider a quaternionic vector space V, and let h be a Euclidean metric on V.

Definition. The metric h is called *Quaternionic-Hermitian* if for any algebra embedding $I : \mathbb{C} \to \mathbb{H}$ the metric h is the real part of an Hermitian metric on the complex vector space V_I.

Equivalently, a metric is Quaternionic-Hermitian if it is invariant under the action of the group $SU(2) \subset \mathbb{H}$ of unitary quaternions.

1.5.2. Assume that the quaternionic vector space V is equivariant. Say that a metric h on V is *Hermitian-Hodge* if it is Quaternionic-Hermitian and, in addition, invariant under the $U(1)$-action on V.

Let V_I be the vector space V with the preferred complex structure I, and let

$$V_I = \bigoplus V^{\bullet}$$

and $j : \overline{V^{\bullet}} \to V^{1-\bullet}$ be as in Lemma 1.1.6. The metric h is Hermitian-Hodge if and only if

(i) it is the real part of an Hermitian metric on V_I,

(ii) $h(V^p, V^q) = 0$ whenever $p \neq q$, and

(iii) $h(j(a), b) = -h(a, j(b))$ for every $a, b \in V$.

1.5.3. Recall that a *polarization* S on a pure $\mathbb{R}$-Hodge structure W of weight i is a bilinear form $S : W \otimes W \to \mathbb{R}(-i)$ which is a morphism of pure Hodge structures and satisfies

$$S(a, b) = (-1)^i S(b, a)$$
$$S(a, Ca) > 0$$

for every $a, b \in W$. (Here $C : W \to W$ is the Weil operator.)

1.5.4. Let V be an equivariant quaternionic vector space equipped with an Euclidean metric h, and let $\langle W, -\rangle$ be the pure $\mathbb{R}$-Hodge structure of weight 1 associated to V by Lemma 1.3.5. Recall that $W = V_J$ as a complex vector space. Assume that h extends to an Hermitian metric h_J on V_J, and let $S : W \otimes W \to \mathbb{R}(-1)$ be the bilinear form defined by

$$S(a, b) = h(a, C\bar{b}), \qquad a, b \in W_{\mathbb{R}} \subset W.$$

The form S is a polarization if and only if the metric h is Hermitian-Hodge. This gives a one-to-one correspondence between the set of Hermitian-Hodge metrics on V and the set of polarizations on the Hodge structure W.

1.5.5. Let W^* be the Hodge structure of weight -1 dual to W. The sets of polarizations on W and on W^* are, of course, in a natural one-to-one correspondence. It will be more convenient for us to identify the set of Hermitian-Hodge metrics on V with the set of polarizations on W^* rather then on W.

Assume that the metric h on the equivariant quaternionic vector space V is Hermitian-Hodge, and let $S \in \Lambda^2(W) \subset \Lambda^2(V \otimes \mathbb{C})$ be the corresponding polarization. Extend h to an Hermitian metric h_I on the complex vector space V with the preferred complex structure V_I, and let

$$\omega_I \in V_I \otimes \overline{V_I} \subset \Lambda^2(V \otimes_{\mathbb{R}} \mathbb{C})$$

be the imaginary part of the corresponding Hermitian metric on the dual space V_I^*. Let $i : V_J \to \overline{V_J}$ the action of the element $I(\sqrt{-1}) \in \mathbb{H}$. By (1.1) we have

$$\omega_I(a, b) = h(a, i(b)) = h(a, C\bar{b}) = S(a, b)$$

for every $a, b \in V_J \subset V \otimes \mathbb{C}$. Since ω_I is real, and $V \otimes \mathbb{C} = V_J \oplus \overline{V_J}$, the 2-forms ω_I and S are related by

$$\omega_I = \frac{1}{2}(S + \nu(S)), \tag{1.4}$$

where $\nu : V \otimes \mathbb{C} \to \overline{V \otimes \mathbb{C}}$ is the usual complex conjugation.

2. Hodge bundles and quaternionic manifolds

2.1. Hodge bundles

2.1.1. Throughout the rest of the paper, our main tool in studying hyperkähler structures on smooth manifolds will be the equivalence between equivariant quaternionic vector spaces and pure $\mathbb{R}$-Hodge structures of weight 1, established in Lemma 1.1.6. In order to use it, we will need to generalize this equivalence to the case of vector bundles over a smooth manifold M, rather than just vector spaces. We will also need to consider manifolds equipped with a smooth action of the group $U(1)$, and we would like our generalization to take this $U(1)$-action into account.

Such a generalization requires, among other things, an appropriate notion of a vector bundle equipped with a pure $\mathbb{R}$-Hodge structure. We introduce and study one version of such a notion in this section, under the name of a Hodge bundle (see Definition 2.1.2).

2.1.2. Let M be a smooth manifold equipped with a smooth $U(1)$-action (or *a $U(1)$-manifold*, to simplify the terminology), and let $\iota : M \to M$ be the action of the element $-1 \in U(1) \subset \mathbb{C}^*$.

Definition. An *Hodge bundle of weight k* on M is a pair $\langle \mathcal{E}, \bar{-} \rangle$ of a $U(1)$-equivariant complex vector bundle $\mathcal{E}$ on M and a $U(1)$-equivariant bundle map $\bar{-} : \overline{\iota^*\mathcal{E}}(k) \to \mathcal{E}$ satisfying $\bar{-} \circ \iota^* \bar{-} = \mathrm{id}$.

Hodge bundles of weight k over M form a tensor $\mathbb{R}$-linear additive category, denoted by $\mathcal{WHodge}_k(M)$.

2.1.3. Let $\mathcal{E}$, $\mathcal{F}$ be two Hodge bundles on M of weights m and n. Say that a bundle map, or, more generally, a differential operator $f : \mathcal{E} \to \mathcal{F}$ is *weakly Hodge* if

(i) $f = \overline{\iota^* f}$, and

(ii) there exists a decomposition $f = \sum_{0 \leq n-m} f_i$ with f_i being of degree i with respect to the $U(1)$-equivariant structure. (In particular, $f = 0$ unless $n \geq m$, and we always have $f_k = \overline{\iota^* f_{n-m-k}}$.)

Denote by $\mathcal{WHodge}(M)$ the category of Hodge bundles of arbitrary weight on M, with weakly Hodge bundle maps as morphisms. For every i the category $\mathcal{WHodge}_i(M)$ is a full subcategory in $\mathcal{WHodge}(M)$. Introduce also the category $\mathcal{WHodge}^{\mathcal{D}}(M)$ with the same objects as $\mathcal{WHodge}(M)$ but with weakly Hodge differential operators as morphisms. Both the categories $\mathcal{WHodge}(M)$ and $\mathcal{WHodge}^{\mathcal{D}}(M)$ are additive $\mathbb{R}$-linear tensor categories.

2.1.4. For a weakly Hodge map $f : \mathcal{E} \to \mathcal{F}$ call the canonical decomposition

$$f = \sum_{0 \leq i \leq m-n} f_i$$

the H-type decomposition. For a Hodge bundle $\mathcal{E}$ on M of non-negative weight k let $\Gamma(\mathcal{E}) = \mathcal{E} \otimes \mathcal{W}_k^*$, where $\mathcal{W}_k$ is the canonical pure $\mathbb{R}$-Hodges structure introduced in 1.4.4. The universal properties of the Hodge structures $\mathcal{W}_k$ and Lemma 1.4.6 generalize immediately to Hodge bundles. In particular, Γ extends to a functor $\Gamma : \mathcal{WHodge}_{\geq 0}(M) \to \mathcal{WHodge}_0(M)$ adjoint on the right to the canonical embedding.

2.1.5. If the $U(1)$-action on M is trivial, then Hodge bundles of weight i are the same as real vector bundles $\mathcal{E}$ equipped with a Hodge type bigrading $\mathcal{E} = \oplus_{p+q=i} \mathcal{E}^{p,q}$ on the complexification $\mathcal{E}_{\mathbb{C}} = \mathcal{E} \otimes_{\mathbb{R}} \mathbb{C}$. In particular, if $M = \mathrm{pt}$ is a single point, then $\mathcal{WHodge}(M) \cong \mathcal{WHodge}^{\mathcal{D}}(M)$ is the category of pure $\mathbb{R}$-Hodge structures. Weakly Hodge bundle map are then the same as weakly Hodge maps of $\mathbb{R}$-Hodge structures considered in 1.4.2. (Thus the notion of a Hodge bundle is indeed a generalization of the notion of a pure $\mathbb{R}$-Hodge structure.)

2.1.6. The categories of Hodge bundles are functorial in M, namely, for every smooth map $f : M_1 \to M_2$ of smooth $U(1)$-manifolds M_1, M_2 there exist pull-back functors

$$f^* : \mathcal{WH}odge(M_1) \to \mathcal{WH}odge(M_1)$$
$$f^* : \mathcal{WH}odge^{\mathcal{D}}(M_1) \to \mathcal{WH}odge^{\mathcal{D}}(M_1).$$

In particular, let M be a smooth $U(1)$-manifold and let $\pi : M \to \mathrm{pt}$ be the canonical projection. Then every $\mathbb{R}$-Hodge structure V of weight i defines a constant Hodge bundle π^*V on M, which we denote for simplicity by the same letter V. Thus the trivial bundle $\mathbb{R} = \Lambda^0(M) = \pi^*\mathbb{R}(0)$ has a natural structure of a Hodge bundle of weight 0.

2.1.7. To give a first example of Hodge bundles and weakly Hodge maps, consider a $U(1)$-manifold M equipped with a $U(1)$-invariant almost complex structure M_I. Let $\Lambda^i(M,\mathbb{C}) = \oplus\Lambda^{\bullet,i-\bullet}(M_I)$ be the usual Hodge type decomposition of the bundles $\Lambda^i(M,\mathbb{C})$ of complex valued differential forms on M. The complex vector bundles $\Lambda^{p,q}(M_I)$ are naturally $U(1)$-equivariant. Let

$$- : \Lambda^{p,q}(M_I) \to \iota^*\overline{\Lambda^{q,p}(M_I)}$$

be the usual complex conjugation, and introduce a $U(1)$-equivariant structure on $\Lambda^{\bullet}(M,\mathbb{C})$ by setting

$$\Lambda^i(M,\mathbb{C}) = \bigoplus_{0\leq j\leq i} \Lambda^{j,i-j}(M)(j).$$

The bundle $\Lambda^i(M,\mathbb{C})$ with these $U(1)$-equivariant structure and complex conjugation is a Hodge bundle of weight i on M. The de Rham differential d_M is weakly Hodge, and the H-type decomposition for d_M is in this case the usual Hodge type decomposition $d = \partial + \bar{\partial}$.

2.1.8. Remark. Definition 2.1.2 is somewhat technical. It can be heuristically rephrased as follows. For a complex vector bundle $\mathcal{E}$ on M the space of smooth global section $C^\infty(M,\mathcal{E})$ is a module over the algebra $C^\infty(M,\mathbb{C})$ of smooth $\mathbb{C}$-valued functions on M, and the bundle $\mathcal{E}$ is completely defined by the module $C^\infty(M,\mathcal{E})$. The $U(1)$-action on M induces a representation of $U(1)$ on the algebra $C^\infty(M,\mathbb{C})$. Let $\nu : C^\infty(M,\mathbb{C}) \to C^\infty(M,\mathbb{C})$ be composition of the complex conjugation and the map $\iota^* : C^\infty(M,\mathbb{C}) \to C^\infty(M,\mathbb{C})$. The map ν is an anti-complex involution; together with the $U(1)$-action it defines a pure $\mathbb{R}$-Hodge structure of weight 0 on the algebra $C^\infty(M,\mathbb{C})$.

Giving a weight i Hodge bundle structure on $\mathcal{E}$ is then equivalent to giving a weight i pure $\mathbb{R}$-Hodge structure on the module $C^\infty(M, \mathcal{E})$ such that the multiplication map

$$C^\infty(M, \mathbb{C}) \otimes C^\infty(M, \mathcal{E}) \to C^\infty(M, \mathcal{E})$$

is a map of $\mathbb{R}$-Hodge structures.

2.2. Equivariant quaternionic manifolds

2.2.1. We now turn to our main subject, namely, various differential-geometric structures on smooth manifolds associated to actions of the algebra $\mathbb{H}$ of quaternions.

2.2.2. Definition. A smooth manifold M is called *quaternionic* if it is equipped with a smooth action of the algebra $\mathbb{H}$ on its cotangent bundle $\Lambda^1(M)$.

Let M be a quaternionic manifold. Every algebra embedding $J : \mathbb{C} \to \mathbb{H}$ defines by restriction an almost complex structure on the manifold M. Denote this almost complex structure by M_J.

2.2.3. Assume that the manifold M is equipped with a smooth action of the group $U(1)$, and consider the standard action of $U(1)$ on the vector space $\mathbb{H}$. Call the quaternionic structure and the $U(1)$-action on M *compatible* if the action map

$$\mathbb{H} \otimes_{\mathbb{R}} \Lambda^1(M) \to \Lambda^1(M)$$

is $U(1)$-equivariant.

Equivalently, the quaternionic structure and the $U(1)$-action are compatible if the action preserves the almost complex structure M_I, and the action map

$$\mathbb{H} \otimes_{\mathbb{C}} \Lambda^{1,0}(M_I) \to \Lambda^{1,0}(M_I)$$

is $U(1)$-equivariant.

2.2.4. Definition. A quaternionic manifold M equipped with a compatible smooth $U(1)$-action is called *an equivariant quaternionic manifold.*

For a $U(1)$-equivariant complex vector bundle $\mathcal{E}$ on M denote by $\mathcal{E}(k)$ the bundle $\mathcal{E}$ with $U(1)$-equivariant structure twisted by the 1-dimensional representation of weight k. Lemma 1.1.6 immediately gives the following.

Lemma. *The category of quaternionic manifolds is equivalent to the category of pairs $\langle M_I, j\rangle$ of an almost complex manifold M_I and a $\mathbb{C}$-linear $U(1)$-equivariant smooth map $j : \Lambda^{0,1}(M_I)(1) \to \Lambda^{0,1}(M_I)$ satisfying $j \circ \overline{j} = -\mathsf{id}$.*

2.3. Quaternionic manifolds and Hodge bundles

2.3.1. Let M be a smooth $U(1)$-manifold. Recall that we have introduced in Subsection 2.1 a notion of a Hodge bundle on M. Hodge bundles arise naturally in the study of quaternionic structures on M in the following way. Define a *quaternionic bundle* on M as a real vector bundle $\mathcal{E}$ equipped with a left action of the algebra $\mathbb{H}$, and let $\operatorname{Bun}(M, \mathbb{H})$ be the category of smooth quaternionic vector bundles on the manifold M. Let also $\operatorname{Bun}^{U(1)}(M, \mathbb{H})$ be the category of smooth quaternionic bundles $\mathcal{E}$ on M equipped with a $U(1)$-equivariant structure such that the $\mathbb{H}$-action map $\mathbb{H} \to \mathcal{E}nd\,(\mathcal{E})$ is $U(1)$-equivariant. Lemma 1.3.5 immediately generalizes to give the following.

Lemma. *The category* $\operatorname{Bun}^{U(1)}(M, \mathbb{H})$ *is equivalent to the category of Hodge bundles of weight* 1 *on* M.

2.3.2. Note that if the $U(1)$-manifold M is equipped with an almost complex structure, then the decomposition $\mathbb{H} = \overline{\mathbb{C}} \oplus I(\mathbb{C})$ (see 1.1.5) induces an isomorphism

$$\mathsf{can} : \mathbb{H} \otimes_{I(\mathbb{C})} \Lambda^{0,1} \cong \Lambda^{1,0}(M) \oplus \Lambda^{0,1}(M) \cong \Lambda^1(M, \mathbb{C}).$$

The weight 1 Hodge bundle structure on $\Lambda^1(M, \mathbb{C})$ corresponding to the natural quaternionic structure on $\mathbb{H} \otimes_{I(\mathbb{C})} \Lambda^{0,1}(M)$ is the same as the one considered in 2.1.7.

2.3.3. Assume now that the smooth $U(1)$-manifold M is equipped with a compatible quaternionic structure, and let M_I be the preferred almost complex structure on M. Since M_I is preserved by the $U(1)$-action on M, the complex vector bundle $\Lambda^{0,1}(M_I)$ of $(0,1)$-forms on M_I is naturally $U(1)$-equivariant.

The quaternionic structure on $\Lambda^1(M)$ induces by Lemma 2.3.1 a weight-1 Hodge bundle structure on $\Lambda^{0,1}(M_I)$. The corresponding $U(1)$-action on $\Lambda^{0,1}(M_I)$ is induced by the action on M_I, and the real structure map

$$\overline{} : \Lambda^{1,0}(M_I)(1) \to \Lambda^{0,1}(M_I)$$

is given by $\overline{} = \sqrt{-1}\,(\iota^* \circ j)$. (Here j is induced by quaternionic structure, as in Lemma 2.2.4).

2.3.4. Let M_J be the complementary almost complex structure on the equivariant quaternionic manifold M. Recall that in 1.2 we have defined for every equivariant quaternionic vector space V the standard isomorphism

$H : V_J \to V_I$. This construction can be immediately generalized to give a complex bundle isomorphism

$$H : \Lambda^{0,1}(M_J) \to \Lambda^{0,1}(M_I).$$

Let $P : \Lambda^1(M, \mathbb{C}) \to \Lambda^{0,1}(M_J)$ be the natural projection, and let $\mathrm{Mult} : \mathbb{H} \otimes_{I(\mathbb{C})} \Lambda^{0,1}(M_I) \to \Lambda^{0,1}(M_I)$ be the action map. By definition the diagram

$$\begin{array}{ccc} \Lambda^1(M, \mathbb{C}) & \xrightarrow{\text{can}} & \mathbb{H} \otimes_{I(\mathbb{C})} \Lambda^{0,1}(M_I) \\ {\scriptstyle P}\downarrow & & \downarrow{\scriptstyle \mathrm{Mult}} \\ \Lambda^{0,1}(M_J) & \xrightarrow{H} & \Lambda^{0,1}(M_I) \end{array}$$

is commutative. Since the map Mult is compatible with the Hodge bundle structures, so is the projection P.

Remark. This may seems paradoxical, since the complex conjugation $\bar{\ }$ on $\Lambda^1(M, \mathbb{C})$ does not preserve $\operatorname{Ker} P = \Lambda^{0,1}(M_J)$. However, under our definition of a Hodge bundle the conjugation on $\Lambda^1(M, \mathbb{C})$ is $\iota^* \circ \bar{\ }$ rather than $\bar{\ }$. Both $\bar{\ }$ and ι^* interchange $\Lambda^{1,0}(M_J)$ and $\Lambda^{0,1}(M_J)$.

2.3.5. The standard isomorphism $H : \Lambda^{0,1}(M_J) \to \Lambda^{0,1}(M_I)$ does not commute with the Dolbeult differentials. They are, however, related by means of the Hodge bundle structure on $\Lambda^{0,1}(M_I)$. Namely, we have the following.

Lemma. *The Dolbeult differential $D : \Lambda^0(M, \mathbb{C}) \to \Lambda^{0,1}(M_I)$ for the almost complex structure M_J is weakly Hodgeweakly Hodge map. The $U(1)$-invariant component D_0 in the H-type decomposition $D = D_0 + \overline{D_0}$ of the map D coincides with the Dolbeult differential for the almost complex structure M_I.*

Proof. The differential D is the composition $D = P \circ d_M$ of the de Rham differential $d_M : \Lambda^0(M, \mathbb{C}) \to \Lambda^1(M, \mathbb{C})$ with the canonical projection P. Since both P and d_M are weakly Hodge, so is D. The rest follows from the construction of the standard isomorphism H. □

2.4. Holonomic derivations

2.4.1. Let M be a smooth $U(1)$-manifold. In order to give a Hodge-theoretic description of the set of all equivariant quaternionic structures on M, it is convenient to work not with various complex structures on M, but with associated Dolbeult differentials. To do this, recall the following universal property of the cotangent bundle $\Lambda^1(M)$.

Lemma. *Let M be a smooth manifold, and let $\mathcal{E}$ be a complex vector bundle on M. Every differential operator $\partial : \Lambda^0(M, \mathbb{C}) \to \mathcal{E}$ which is a derivation with respect to the algebra structure on $\Lambda^0(M, \mathbb{C})$ factors uniquely through the de Rham differential $d_M : \Lambda^0(M, \mathbb{C}) \to \Lambda^1(M, \mathbb{C})$ by means of a bundle map $P : \Lambda^1(M, \mathbb{C}) \to \mathcal{E}$.*

2.4.2. We first use this universal property to describe almost complex structures. Let M be a smooth manifold equipped with a complex vector bundle $\mathcal{E}$.

Definition. A derivation $D : \Lambda^0(M, \mathbb{C}) \to \mathcal{E}$ is called *holonomic* if the associated bundle map $P : \Lambda^1(M, \mathbb{C}) \to \mathcal{E}$ induces an isomorphism of the subbundle $\Lambda^1(M, \mathbb{R}) \subset \Lambda^1(M, \mathbb{C})$ of real 1-forms with the real vector bundle underlying $\mathcal{E}$.

By Lemma 2.4.1 the correspondence

$$M_I \mapsto \langle \Lambda^{0,1}(M_I), \bar{\partial} \rangle$$

identifies the set of all almost complex structures M_I on M with the set of all pairs $\langle \mathcal{E}, D \rangle$ of a complex vector bundle $\mathcal{E}$ and a holonomic derivation $D : \Lambda^0(M, \mathbb{C}) \to \mathcal{E}$.

2.4.3. Assume now that the smooth manifold M is equipped with smooth action of the group $U(1)$. Then we have the following.

Lemma. *Let $\mathcal{E}$ be a weight 1 Hodge bundle on the smooth $U(1)$-manifold M, and let*

$$D : \Lambda^0(M, \mathbb{C}) \to \mathcal{E}$$

be a weakly Hodge holonomic derivation. There exists a unique $U(1)$-equivariant quaternionic structure on M such that $\mathcal{E} \cong \Lambda^{0,1}(M_J)$ and D is the Dolbeult differential for the complementary almost complex structures M_J on M.

Proof. Since the derivation M is holonomic, it induces an almost complex structure M_J on M. To construct an almost complex structure M_I complementary to M_J, consider the H-type decomposition $D = D_0 + \overline{D_0}$ of the derivation $D : \Lambda^0(M, \mathbb{C}) \to \mathcal{E}$ (defined in 2.1.4).

The map D_0 is also a derivation. Moreover, it is holonomic. Indeed, by dimension count it is enough to prove that the associated bundle map $P : \Lambda^1(M, \mathbb{R}) \to \mathcal{E}$ is injective. Since the bundle $\Lambda^1(M, \mathbb{R})$ is generated by exact 1-forms, it is enough to prove that any real valued function f on M with $D_0 f = 0$ is constant. However, since D is weakly Hodge,

$$Df = D_0 f + \overline{D_0} f = D_0 f + \overline{D_0 \bar{f}} = D_0 f + \overline{D_0 f},$$

hence $D_0 f = 0$ if and only if $Df = 0$. Since D is holonomic, f is indeed constant.

The derivation D_0, being holonomic, is the Dolbeult differential for an almost complex structure M_I on M. Since D_0 is by definition $U(1)$-equivariant, the almost complex structure M_I is $U(1)$-invariant. Moreover, $\mathcal{E} \cong \Lambda^{0,1}(M_I)$ as $U(1)$-equivariant complex vector bundles. By Lemma 2.3.1 the weight 1 Hodge bundle structure on $\mathcal{E}$ induces an equivariant quaternionic bundle structure on $\mathcal{E}$ and, in turn, a structure of an equivariant quaternionic manifold on M. The almost complex structure M_I coincides by definition with the preferred almost complex structure.

It remains to notice that by Lemma 2.3.5 the Dolbeult differential $\bar{\partial}_J$ for the complementary almost complex structure on M indeed equals $D = D_0 + \overline{D_0}$. □

Together with Lemma 2.3.5, this shows that the set of equivariant quaternionic structures on the $U(1)$-manifold M is naturally bijective to the set of pairs $\langle \mathcal{E}, D \rangle$ of a weight 1 Hodge bundle $\mathcal{E}$ on M and a weakly Hodge holonomic derivation $D : \Lambda^0(M, \mathbb{C}) \to \mathcal{E}$.

3. Hodge manifolds

3.1. Integrability

3.1.1. There exists a notion of integrability for quaternionic manifolds analogous to that for the almost complex ones.

Definition. A quaternionic manifold M is called *hypercomplex* if for two complementary algebra embeddings $I, J : \mathbb{C} \to \mathbb{H}$ the almost complex structures M_I, M_J are integrable.

In fact, if M is hypercomplex, then M_I is integrable for any algebra embedding $I : \mathbb{C} \to \mathbb{H}$. For a proof see, e.g., [K].

3.1.2. When a quaternionic manifold M is equipped with a compatible $U(1)$-action, there exist a natural choice for a pair of almost complex structures on M, namely, the preferred and the complementary one.

Definition. An equivariant quaternionic manifold M is called a *Hodge manifold* if both the preferred and the equivariant almost complex structures M_I, M_J are integrable.

Hodge manifolds are the main object of study in this paper.

3.1.3. There exists a simple Hodge-theoretic description of Hodge manifolds based on Lemma 2.4.3. To give it (see Proposition 3.1), consider an equivariant quaternionic manifold M, and let M_J and M_I be the complementary and the preferred complex structures on M. The weight 1 Hodge bundle structure on $\Lambda^{0,1}(M_J)$ induces a weight i Hodge bundle structure on the bundle $\Lambda^{0,i}(M_J)$ of $(0,i)$-forms on M_J. The standard identification $H : \Lambda^{0,1}(M_J) \to \Lambda^{0,1}(M_I)$ given in 2.3.4 extends uniquely to an algebra isomorphism $H : \Lambda^{0,i}(M_J) \to \Lambda^{0,i}(M_I)$.

Let $D : \Lambda^{0^\bullet}(M_J) \to \Lambda^{0,\bullet+1}(M_J)$ be the Dolbeault differential for the almost complex manifold M_J.

Lemma. *The equivariant quaternionic manifold M is Hodge if and only if the following holds.*

(i) *M_J is integrable, that is, $D \circ D = 0$, and*

(ii) *the differential $D : \Lambda^{0,i}(M_J) \to \Lambda^{0,i+1}(M_J)$ is weakly Hodge for every $i \geq 0$.*

Proof. Assume first that the conditions (i), (ii) hold. Condition (i) means that the complementary almost complex structure M_J is integrable. By (ii) the map D is weakly Hodge.

Let $D = D_0 + \overline{D_0}$ be the H-type decomposition. The map D_0 is an algebra derivation of $\Lambda^{0,\bullet}(M_I)$. Moreover, by Lemma 2.3.5 the map $D_0 : \Lambda^0(M, \mathbb{C}) \to \Lambda^{0,1}(M_J)$ is the Dolbeault differential $\bar\partial_I$ for the preferred almost complex structure M_I on M. (Or, more precisely, is identified with $\bar\partial_I$ under the standard isomorphism H.) But the Dolbeult differential admits at most one extension to a derivation of the algebra $\Lambda^{0,\bullet}(M_J)$. Therefore $D_0 = \bar\partial_I$ everywhere.

The composition $D_0 \circ D_0$ is the $(2,0)$-component in the H-type decomposition of the map $D \circ D$. Since $D \circ D = 0$,

$$D_0 \circ D_0 = \bar\partial_I \circ \bar\partial_I = 0.$$

Therefore the preferred complex structure M_I is also integrable, and the manifold M is indeed Hodge.

Assume now that M is Hodge. The canonical projection $P : \Lambda^1(M, \mathbb{C}) \to \Lambda^{0,1}(M_J)$ extends then to a multiplicative projection

$$P : \Lambda^\bullet(M, \mathbb{C}) \to \Lambda^{0,\bullet}(M_J)$$

from the de Rham complex of the complex manifold M_I to the Dolbeault complex of the complex manifold M_J. The map P is surjective and weakly Hodge, moreover, it commutes with the differentials. Since the de Rham differential preserves the pre-Hodge structures, so does the Dolbeault differential D. □

3.1.4. Lemma 3.1.3 and Lemma 2.4.3 together immediately give the following.

Proposition 3.1 *The category of Hodge manifolds is equivalent to the category of triples $\langle M, \mathcal{E}, D\rangle$ of a smooth $U(1)$-manifold M, a weight 1 Hodge bundle $\mathcal{E}$ on M, and a weakly Hodge algebra derivation*

$$D = D^{\bullet} : \Lambda^{\bullet}(\mathcal{E}) \to \Lambda^{\bullet+1}(\mathcal{E})$$

such that $D \circ D = 0$, and the first component

$$D^0 : \Lambda^0(M, \mathbb{C}) = \Lambda^0(\mathcal{E}) \to \mathcal{E} = \Lambda^1(\mathcal{E})$$

is holonomic in the sense of 2.4.2.

3.2. The de Rham complex of a Hodge manifold

3.2.1. Let M be a Hodge manifold. In this subsection we study in some detail the de Rham complex $\Lambda^{\bullet}(M, \mathbb{C})$ of the manifold M, in order to obtain information necessary for the discussion of metrics on M given in the Subsection 3.3. The reader is advised to skip this subsection until needed.

3.2.2. Let $\Lambda^{0,\bullet}(M_J)$ be the Dolbeault complex for the complementary complex structure M_J on M. By Proposition 3.1 the complex vector bundle $\Lambda^{0,i}(M_J)$ is a Hodge bundle of weight i on M, and the Dolbeult differential $D : \Lambda^{0,\bullet}(M_J) \to \Lambda^{0,\bullet+1}(M_J)$ is weakly Hodge. Therefore D admits an H-type decomposition $D = D_0 + \overline{D_0}$.

3.2.3. Consider the de Rham complex $\Lambda^i(M, \mathbb{C})$ of the smooth manifold M. Let $\Lambda^i(M, \mathbb{C}) = \oplus_{p+q} \Lambda^{p,q}(M_J)$ be the Hodge type decomposition for the complementary complex structure M_J on M, and let $\nu : \Lambda^{p,q}(M_J) \to \overline{\Lambda^{q,p}(M_J)}$ be the usual complex conjugation. Denote also

$$f^{\nu} = \nu \circ f \circ \nu^{-1}$$

for any map $f : \Lambda^{\bullet}(M, \mathbb{C}) \to \Lambda^{\bullet}(M, \mathbb{C})$.

Let $d_M : \Lambda^{\bullet}(M, \mathbb{C}) \to \Lambda^{\bullet+1}(M, \mathbb{C})$ be the de Rham differential, and let $d_M = D + D^{\nu}$ be the Hodge type decomposition for the complementary complex structure M_J on M. Since the Dolbeult differential, in turn, equals $D = D_0 + \overline{D_0}$, we have

$$d_M = D_0 + \overline{D_0} + D_0^{\nu} + \overline{D_0}^{\nu}.$$

3.2.4. Let $\bar{\partial}_I : \Lambda^{\bullet}(M, \mathbb{C}) \to \Lambda^{\bullet+1}(M, \mathbb{C})$ be the Dolbeult differential for the preferred complex structure M_I on M. As shown in the proof of Lemma 3.1.3, the $(0,1)$-component of the differential $\bar{\partial}_I$ with respect to the complex structure M_J equals D_0. Therefore the $(1,0)$-component of the complex-conjugate map $\partial_I = \bar{\partial}_I^\nu$ equals D_0^ν. Since $d_M = \bar{\partial}_I + \partial_I$, we have

$$\bar{\partial}_I = D_0 + \overline{D_0}^\nu$$
$$\partial_I = \overline{D_0} + D_0^\nu$$

3.2.5. The standard isomorphism $H : \Lambda^{0,1}(M_J) \to \Lambda^{0,1}(M_I)$ introduced in 3.1.3 extends uniquely to a bigraded algebra isomorphism $H : \Lambda^{\bullet,\bullet}(M_J) \to \Lambda^{\bullet,\bullet}(M_I)$. By definition of the map H, on $\Lambda^0(M, \mathbb{C})$ we have

$$\begin{aligned} \bar{\partial}_I &= H \circ D_0 \circ H^{-1} \\ \partial_I &= H \circ D_0^\nu \circ H^{-1} \\ d_M = \partial_I + \bar{\partial}_I &= H \circ (D_0 + D_0^\nu) H^{-1}. \end{aligned} \tag{3.1}$$

The right hand side of the last identity is the algebra derivation of the de Rham complex $\Lambda^\bullet(M, \mathbb{C})$. Therefore, by Lemma 2.4.1 it holds not only on $\Lambda^0(M, \mathbb{C})$, but on the whole $\Lambda^\bullet(M, \mathbb{C})$. The Hodge type decomposition for the preferred complex structure M_I then shows that the first two identities also hold on the whole de Rham complex $\Lambda^\bullet(M, \mathbb{C})$.

3.2.6. Let now $\xi = I(\sqrt{-1}) : \Lambda^{0,1}(M_J) \to \Lambda^{1,0}(M_J)$ be the operator corresponding to the preferred almost complex structure M_I on M. Let also $\xi = 0$ on $\Lambda^0(M, \mathbb{C})$ and $\Lambda^{1,0}(M_J)$, and extend ξ to a derivation $\xi : \Lambda^{\bullet,\bullet}(M_J) \to \Lambda^{\bullet-1,\bullet+1}(M_J)$ by the Leibnitz rule. We finish this subsection with the following simple fact.

Lemma. *On $\Lambda^{\bullet,0}(M_J) \subset \Lambda^\bullet(M, \mathbb{C})$ we have*

$$\begin{aligned} \xi \circ D_0 + D_0 \circ \xi &= \overline{D}_0^\nu \\ \xi \circ \overline{D}_0 + \overline{D}_0 \circ \xi &= -D_0^\nu. \end{aligned} \tag{3.2}$$

Proof. It is easy to check that both identities hold on $\Lambda^0(M, \mathbb{C})$. But both sides of these identities are algebra derivations of $\Lambda^{\bullet,0}(M_J)$, and the right hand sides are holonomic in the sense of 2.4.2. Therefore by Lemma 2.4.1 both identities hold on the whole $\Lambda^{\bullet,0}(M_J)$. □

3.3. Polarized Hodge manifolds

3.3.1. Let M be a quaternionic manifold. A Riemannian metric h on M is called *Quaternionic-Hermitian* if for every point $m \in M$ the induced metric h_m on the tangent bundle $T_m M$ is Quaternionic-Hermitian in the sense of 1.5.1.

Definition. A *hyperkähler manifold* is a hypercomplex manifold M equipped with a Quaternionic-Hermitian metric h which is Kähler for both integrable almost complex structures M_I, M_J on M.
Remark. In the usual definition (see, e.g., [B]) the integrability of the almost complex structures M_I, M_J is omitted, since it is automatically implied by the Kähler condition.
3.3.2. Let M be a Hodge manifold equipped with a Riemannian metric h. The metric h is called *Hermitian-Hodge* if it is Quaternionic-Hermitian and, in addition, invariant under the $U(1)$-action on M.
Definition. Say that the manifold M is *polarized* by the Hermitian-Hodge metric h if h is not only Quaternionic-Hermitian, but also hyperkähler.
3.3.3. Let M be a Hodge manifold. By Proposition 3.1 the holomorphic cotangent bundle $\Lambda^{1,0}(M_J)$ for the complementary complex structure M_J on M is a Hodge bundle of weight 1. The holomorphic tangent bundle $\Theta(M_J)$ is therefore a Hodge bundle of weight -1. By 1.5.4 the set of all Hermitian-Hodge metrics h on M is in natural one-to-one correspondence with the set of all polarizations on the Hodge bundle $\Theta(M_J)$.

Since $\theta(M)$ is of odd weight, its polarizations are skew-symmetric as bilinear forms and correspond therefore to smooth $(2,0)$-forms on the complex manifold M_J. A $(2,0)$-form Ω defines a polarization on $\Theta(M_J)$ if and only if $\Omega \in C^\infty(M, \Lambda^{2,0}(M_J))$ considered as a map

$$\Omega : \mathbb{R}(-1) \to \Lambda^{2,0}(M_J)$$

is a map of weight 2 Hodge bundles, and for an arbitrary smooth section $\chi \in C^\infty(M, \Theta(M_J))$ we have

$$\Omega(\chi, \overline{\iota^*(\chi)}) > 0. \tag{3.3}$$

3.3.4. Assume that the Hodge manifold M is equipped with an Hermitian-Hodge metric h. Let $\Omega \in C^\infty(M, \Lambda^{2,0}(M_J))$ be the corresponding polarization , and let $\omega_I \in C^\infty(M, \Lambda^{1,1}(M_I))$ be the $(1,1)$-form on the complex manifold M_I associated to the Hermitian metric h. Either one of the forms Ω, ω_I completely defines the metric h, and by (1.4) we have

$$\Omega + \nu(\Omega) = \omega_I,$$

where $\nu : \Lambda^{\bullet}(M, \mathbb{C}) \to \Lambda^{\bullet}(M, \mathbb{C})$ is the complex conjugation.

Lemma. *The Hermitian-Hodge metric h polarizes M if and only if the corresponding $(2,0)$-form Ω on M_J is holomorphic, that is,*

$$D\Omega = 0,$$

where D is the Dolbeult differential for complementary complex structure M_J.

Proof. Let $\omega_I, \omega_J \in \Lambda^2(M, \mathbb{C})$ be the Kähler forms for the metric h and complex structures M_I, M_J on M. The metric h is hyperkähler, hence polarizes M, if and only if $d_M\omega_I = d_M\omega_J = 0$.

Let $D = D_0 + \overline{D_0}$ be the H-type decomposition and let $H : \Lambda^{\bullet,\bullet}(M_J) \to \Lambda^{\bullet,\bullet}(M_J)$ be the standard algebra identification introduced in 3.2.5. By definition $H(\omega_I) = \omega_J$. Moreover, by (3.1) $H^{-1} \circ d_M \circ H = D_0 + \overline{D_0}^{\nu}$, hence

$$H(d_M\omega_J) = D_0\omega_I + \overline{D_0}^{\nu}\omega_I,$$

and the metric h is hyperkähler if and only if

$$d_M\omega_I = (D_0 + \overline{D_0}^{\nu})\omega_I = 0 \tag{3.4}$$

But $2\omega_I = \Omega + \nu(\Omega)$. Since Ω is of Hodge type $(2,0)$ with respect to the complementary complex structure M_J, (3.4) is equivalent to

$$\overline{D_0}\Omega = D_0\Omega = \overline{D_0}^{\nu}\Omega = D_0^{\nu}\Omega = 0.$$

Moreover, let ξ be as in Lemma 3.2.6. Then $\xi(\Omega) = 0$, and by (3.2) $D_0\Omega = \overline{D}_0\Omega = 0$ implies that $D_0^{\nu}\Omega = \overline{D}_0^{\nu}\Omega = 0$ as well. It remains to notice that since the metric h is Hermitian-Hodge, Ω is of H-type $(1,1)$ as a section of the weight 2 Hodge bundle $\Lambda^{2,0}(M_J)$. Therefore $D\Omega = 0$ implies $\overline{D_0}\Omega = D_0\Omega = 0$, and this proves the lemma. □

Remark. This statement is wrong for general hyperkähler manifolds (everything in the given proof carries over, except for the implication $D\Omega = 0 \Rightarrow D_0\Omega = \overline{D_0}\Omega = \overline{D_0}^{\nu}\Omega = D_0^{\nu}\Omega = 0$, which depends substantially on the $U(1)$-action). To describe general hyperkähler metrics in holomorphic terms, one has to introduce the so-called *twistor spaces* (see, e.g., [HKLR]).

4. Regular Hodge manifolds

4.1. Regular stable points

4.1.1. Let M be a smooth manifold equipped with a smooth $U(1)$-action with differential φ_M (thus φ_M is a smooth vector field on M). Since the group $U(1)$ is compact, the subset $M^{U(1)} \subset M$ of points fixed under $U(1)$ is a smooth submanifold.

Let $m \in M^{U(1)} \subset M$ be a point fixed under $U(1)$. Consider the representation of $U(1)$ on the tangent space T_m to M at m. Call the fixed point m *regular* if every irreducible subrepresentation of T_m is either trivial or isomorphic to the representation on $\mathbb{C}$ given by embedding $U(1) \subset \mathbb{C}^*$. (Here $\mathbb{C}$ is considered as a 2-dimensional real vector space.) Regular fixed points form a union of connected component of the smooth submanifold $M^{U(1)} \subset M$.

4.1.2. Assume that M is equipped with a complex structure preserved by the $U(1)$-action. Call a point $m \in M$ *stable* if for any $t \in \mathbb{R}, t \geq 0$ there exists $\exp(\sqrt{-1}t\varphi_M)m$, and the limit

$$m_0 \in M, m_0 = \lim_{t \to +\infty} \exp(\sqrt{-1}t\varphi_M)m$$

also exists.

4.1.3. For every stable point $m \in M$ the limit m_0 is obviously fixed under $U(1)$. Call a point $m \in M$ *regular stable* if it is stable and the limit $m_0 \in M^{U(1)}$ is a regular fixed point.

Denote by $M^{reg} \subset M$ the subset of all regular stable points. The subset M^{reg} is open in M.

Example. Let Y be a complex manifold with a holomorphic bundle $\mathcal{E}$ and let E be the total space of $\mathcal{E}$. Let $\mathbb{C}^*$ act on M by dilatation along the fibers. Then every point $e \in E$ is regular stable.

4.1.4. Let M be a Hodge manifold. Recall that the $U(1)$-action on M preserves the preferred complex structure M_I.

Definition. A Hodge manifold M is called *regular* if $M_I^{reg} = M_I$.

4.2. Linearization of regular Hodge manifolds

4.2.1. Consider a regular Hodge manifold M. Let $\Delta \subset \mathbb{C}$ be the unit disk equipped with the multiplicative semigroup structure. The group $U(1) \subset \Delta$ is embedded into Δ as the boundary circle.

Lemma. *The action $a : U(1) \times M \to M$ extends uniquely to a holomorphic action $\tilde{a} : \Delta \times M_I \to M_I$. Moreover, for every $x \in \Delta \setminus \{0\}$ the action map $\tilde{a}(x) : M_I \to M_I$ is an open embedding.*

Proof. Since M is regular, the exponential flow $\exp(it\varphi_M)$ of the differential φ_M of the action is defined for all positive $t \in \mathbb{R}$. Therefore $a : U(1) \times M \to M$ extends uniquely to a holomorphic action

$$\tilde{a} : \Delta^* \times M_I \to M_I,$$

where $\Delta^* = \Delta\backslash\{0\}$ is the punctured disk. Moreover, the exponential flow converges as $t \to +\infty$, therefore $\tilde{a}$ extends to $\Delta \times M_I$ continuously. Since this extension is holomorphic on a dense open subset, it is holomorphic everywhere. This proves the first claim.

To prove the second claim, consider the subset $\widetilde{\Delta} \subset \Delta^*$ of points $x \in \Delta$ such that $\widetilde{a}(x)$ is injective and étale. The subset $\widetilde{\Delta}$ is closed under multiplication and contains the unit circle $U(1) \subset \Delta^*$. Therefore to prove that $\widetilde{\Delta} = \Delta^*$, it suffices to prove that $\widetilde{\Delta}$ contains the interval $]0, 1] \subset \Delta^*$.

By definition we have $\widetilde{a}(h) = \exp(-\sqrt{-1}\log h\varphi_M)$ for every $h \in]0, 1] \subset \Delta^*$. Thus we have to prove that if for some $t \in \mathbb{R}, t \geq 0$ and for two points $m_1, m_2 \in M$ we have

$$\exp(\sqrt{-1}t\varphi_M)(m_1) = \exp(\sqrt{-1}t\varphi_M)(m_2),$$

then $m_1 = m_2$. Let m_1, m_2 be such two points and let

$$t = \inf\{t \in \mathbb{R}, t \geq 0, \exp(\sqrt{-1}t\varphi_M)(m_1) = \exp(\sqrt{-1}t\varphi_M)(m_2)\}.$$

If the point $m_0 = \exp(\sqrt{-1}t\varphi_M)(m_1) = \exp(\sqrt{-1}t\varphi_M)(m_2) \in M$ is not $U(1)$-invariant, then it is a regular point for the vector field $\sqrt{-1}\varphi_M$, and by the theory of ordinary differential equations we have $t = 0$ and $m_1 = m_2 = m_0$.

Assume therefore that $m_0 \in M^{U(1)}$ is $U(1)$-invariant. Since the group $U(1)$ is compact, the vector field $\sqrt{-1}\varphi_M$ has only a simple zero at $m_0 \subset M^{U(1)} \subset M$. Therefore $m_0 = \exp(\sqrt{-1}t\varphi_M)m_1$ implies that the point $m_1 \in M$ also is $U(1)$-invariant, and the same is true for the point $m_2 \in M$. But $\widetilde{a}(\exp t)$ acts by identity on $M^{U(1)} \subset M$. Therefore in this case we also have $m_1 = m_2 = m_0$. □

4.2.2. Let $V = M_I^{U(1)} \subset M_I$ be the submanifold of fixed points of the $U(1)$ action. Since the action preserves the complex structure on M_I, the submanifold V is complex.

Lemma. *There exists a unique $U(1)$-invariant holomorphic map*

$$\rho_M : M_I \to V$$

such that $\rho_M|_V = \mathrm{id}$.

Proof. For every point $m \in M$ we must have $\rho_M(m) = \lim_{t\to+\infty} \exp(it\varphi_M)$, which proves uniqueness.

To prove that ρ_M thus defined is indeed holomorphic, notice that the diagram

$$\begin{array}{ccc} M_I & \xrightarrow{0\times\mathrm{id}} & \Delta \times M \\ \rho_M \downarrow & & \downarrow \tilde{a} \\ V & \longrightarrow & M_I \end{array}$$

is commutative. Since the action $\tilde{a} : \Delta \times M_I \to M_I$ is holomorphic, so is the map ρ_M. □

4.2.3. Call the canonical map $\rho_M : M_I \to M_I^{U(1)}$ the *canonical projection* of the regular Hodge manifold M onto the submanifold $V \subset M$ of fixed points.

Lemma. *The canonical projection $\rho_M : M \to M^{U(1)}$ is submersive, that is, for every point $m \in M$ the differential $d\rho_M : T_m M \to T_{\rho(m)} M^{U(1)}$ of the map ρ_M at m is surjective.*

Proof. Since $\rho_M|_{M^{U(1)}} = \mathsf{id}$, the differential $d\rho_M$ is surjective at points $m \in V \subset M$. Therefore it is surjective on an open neighborhood $U \supset V$ of V in M. For any point $m \in M$ there exists a point $x \in \Delta$ such that $x \cdot m \in U$. Since ρ_M is Δ-invariant, this implies that $d\rho_M$ is surjective everywhere on M. □

4.2.4. Let $\Theta(M/V)$ be the relative tangent bundle of the holomorphic map $\rho : M \to V$. Let $\Theta(M)$ and $\Theta(V)$ be the tangent bundles of M and V and consider the canonical exact sequence of complex bundles

$$0 \longrightarrow \Theta(M/V) \longrightarrow \Theta(M) \xrightarrow{d\rho_M} \rho^*\Theta(V) \longrightarrow 0,$$

where $d\rho_M$ is the differential of the projection $\rho_M : M \to V$.

The quaternionic structure on M defines a $\mathbb{C}$-linear map $j : \Theta(M) \to \overline{\Theta}(M)$. Restricting to $\Theta(M/V)$ and composing with $d\rho_M$, we obtain a $\mathbb{C}$-linear map $j : \Theta(M/V) \to \rho^*\overline{\Theta}(V)$.

4.2.5. Let $\overline{T}V$ be the total space of the bundle $\overline{\Theta}(V)$ complex-conjugate to the tangent bundle $\Theta(V)$, and let $\rho : \overline{T}V \to V$ be the projection. Let the group $U(1)$ act on $\overline{T}V$ by dilatation along the fibers of the projection ρ.

Since the canonical projection $\rho_M : M \to V$ is $U(1)$-invariant, the differential φ_M of the $U(1)$-action defines a section

$$\varphi_M \in C^\infty(M, \Theta(V/M)).$$

The section $j(\varphi_M) \in C^\infty(M, \rho_M^*\overline{\Theta}(V))$ defines a map $\mathrm{Lin}_M : M \to \overline{T}V$ such that $\mathrm{Lin}_M \circ \rho = \rho_M : M \to V$. Call the map Lin_M *the linearization* of the regular Hodge manifold M.

Proposition 4.1 *The linearization map* Lin_M *is a* $U(1)$*-equivariant open embedding.*

Proof. The map $j : \Theta(M/V) \to \rho^*\overline{\Theta}(V)$ is of degree 1 with respect to the $U(1)$-action, while the section $\varphi_M \in C^\infty(M, \Theta(M/V))$ is $U(1)$-invariant. Therefore the map Lin_M is $U(1)$-equivariant.

Consider the differential $d\operatorname{Lin}_M : T_m(M) \to T_m(\overline{T}V)$ at a point $m \in V \subset M$. We have

$$d\operatorname{Lin}_M = d\rho_M \oplus d\rho_M \circ j : T_m(M) \to T_m(V) \oplus \overline{T}_m(V)$$

with respect to the decomposition $T_m(\overline{T}V) = T_m(V) \oplus \overline{T}(V)$. The tangent space T_m is a regular equivariant quaternionic vector space. Therefore the map $d\operatorname{Lin}_M$ is bijective at m by Lemma 1.1.7. Since Lin_M is bijective on V, this implies that Lin_M is an open embedding on an open neighborhood $U \subset M$ of the submanifold $V \subset M$.

To finish the proof of proposition, it suffices prove that the linearization map $\operatorname{Lin}_M : M_I \to \overline{T}V$ is injective and étale on the whole M_I. To prove injectivity, consider arbitrary two points $m_1, m_2 \in M_I$ such that $\operatorname{Lin}_M(m_1) = \operatorname{Lin}_M(m_2)$. There exists a point $x \in \Delta \setminus \{0\}$ such that $x \cdot m_1, x \cdot m_2 \in U$. The map Lin_M is $U(1)$-equivariant and holomorphic, therefore it is Δ-equivariant, and we have

$$\operatorname{Lin}_M(x \cdot m_1) = x \cdot \operatorname{Lin}_M(m_1) = x \cdot \operatorname{Lin}_M(m_2) = \operatorname{Lin}_M(x \cdot m_2).$$

Since the map $\operatorname{Lin}_M : U \to \overline{T}V$ is injective, this implies that $x \cdot m_1 = x \cdot m_2$. By Lemma 4.2.1 the action map $x : M_I \to M_I$ is injective. Therefore this is possible only if $m_1 = m_2$, which proves injectivity.

To prove that the linearization map is étale, note that by Lemma 4.2.1 the action map $x : M_I \to M_I$ is not only injective, but also étale. Since Lin_M is ëtale on U, so is the composition $\operatorname{Lin}_M \circ x : M_I \to U \to \overline{T}V$ is étale. Since $\operatorname{Lin}_M \circ x = x \circ \operatorname{Lin}_M$, the map $\operatorname{Lin}_M : M_I \to \overline{T}V$ is étale at the point $m_1 \in M_I$.

Thus the linearization map is also injective and étale on the whole of M_I. Hence it is indeed an open embedding, which proves the propostion. □

4.3. Linear Hodge manifold structures

4.3.1. By Proposition 4.1 every regular Hodge manifold M admits a canonical open embedding $\operatorname{Lin}_M : M \to \overline{T}V$ into the total space $\overline{T}V$ of the (complex-conjugate) tangent bundle to its fixed points submanifold $V \subset M$.

This embedding induces a Hodge manifold structure on a neighborhood of the zero section $V \subset \overline{T}V$.

In order to use the linearization construction, we will need a characterization of all Hodge manifold structures on neighborhoods of $V \subset \overline{T}V$ obtained in this way (see 4.3.5). It is convenient to begin with an invariant characterization of the linearization map $\mathrm{Lin}_M : M \to \overline{T}V$.

4.3.2. Let V be an arbitrary complex manifold, let $\overline{T}V$ be the total space of the complex-conjugate to the tangent bundle $\Theta(V)$ to V, and let $\rho : \overline{T}V \to V$ be the canonical projection. Contraction with the tautological section of the bundle $\rho^*\overline{\Theta(V)}$ defines for every p a bundle map

$$\tau : \rho^*\Lambda^{p+1}(V, \mathbb{C}) \to \rho^*\Lambda^p(V, \mathbb{C}),$$

which we call *the tautological map.* In particular, the induced map

$$\tau : C^\infty(V, \Lambda^{0,1}(V)) \to C^\infty(\overline{T}V, \mathbb{C})$$

identifies the space $C^\infty(V, \Lambda^{0,1}(V))$ of smooth $(0,1)$-forms on V with the subspace in $C^\infty(\overline{T}V, \mathbb{C})$ of function linear along the fibers of the projection $\overline{T}V \to V$.

4.3.3. Let now M be a Hodge manifold. Let $V \subset M_I$ be the complex submanifold of $U(1)$-fixed points, and let $\rho_M : M \to V$ be the canonical projection. Assume that M is equipped with a smooth $U(1)$-equivariant map $f : M \to \overline{T}V$ such that $\rho_M = \rho \circ f$. Let $\bar\partial_I$ be the Dolbeult differential for the preferred complex structure M_I on M, and let $\varphi \in \Theta(M/V)$ be the differential of the $U(1)$-action on M. Let also $j : \Lambda^{0,1}(M_I) \to \Lambda^{1,0}(M_I)$ be the map induced by the quaternionic structure on M.

Lemma. *The map $f : M \to \overline{T}V$ coincides with the linearization map if and only if for every $(0,1)$-form $\alpha \in C^\infty(V, \Lambda^{0,1}(V))$ we have*

$$f^*\tau(\alpha) = \langle \varphi, j(\rho_M^*\alpha)\rangle. \tag{4.1}$$

Moreover, if $f = \mathrm{Lin}_M$, then we have

$$f^*\tau(\beta) = \langle \varphi, j(f^*\beta)\rangle \tag{4.2}$$

for every smooth section $\beta \in C^\infty(\overline{T}V, \rho^*\Lambda^1(V, \mathbb{C}))$.

Proof. Since functions on $\overline{T}V$ linear along the fibers separate points, the correspondence

$$f^* \circ \tau : C^\infty(V, \Lambda^{0,1}(V)) \to C^\infty(M, \mathbb{C})$$

characterizes the map f uniquely, which proves the "only if" part of the first claim. Since by assumption $\rho_M = \rho \circ f$, the equality (4.1) is a particular case of (4.2) with $\beta = \rho^*\alpha$. Therefore the "if" part of the first claim follows from the second claim, which is a rewriting of the definition of the linearization map $\mathrm{Lin}_M : M \to \overline{T}V$ (see 4.2.5). □

4.3.4. Let now $\mathrm{Lin}_M : M \to \overline{T}V$ be the linearization map for the regular Hodge manifold M. Denote by $U \subset \overline{T}V$ the image of Lin_M. The subset $U \subset \overline{T}V$ is open and $U(1)$-invariant. In addition, the isomorphism $\mathrm{Lin}_M : M \to U$ induces a regular Hodge manifold structure on U.

Denote by Lin_U the linearization map for the regular Hodge manifold U. Lemma 4.3.3 implies the following.

Corollary. *We have* $\mathrm{Lin}_M \circ \mathrm{Lin}_U = \mathrm{Lin}_M$*, thus the linearization map*

$$\mathrm{Lin}_U : U \to \overline{T}V$$

coincides with the given embedding $U \hookrightarrow \overline{T}V$*.*

Proof. Let $\alpha \in C^\infty(V, \Lambda^{0,1}(V))$ be a $(0,1)$-form on V. By Lemma 4.3.3 we have $\mathrm{Lin}_U^* \tau(\alpha) = \langle \varphi_U, j_U(\rho_U^*\alpha)\rangle$, and it suffices to prove that

$$\mathrm{Lin}_M^*(\mathrm{Lin}_U^*(\tau(\alpha))) = \langle \varphi_M, j_M(\rho_M^*\alpha)\rangle.$$

By definition we have $\rho_M = \rho_U \circ \mathrm{Lin}_M$. Moreover, the map Lin_M is $U(1)$-equivariant, therefore it sends φ_M to φ_U. Finally, by definition it commutes with the quaternionic structure map j. Therefore

$$\mathrm{Lin}_M^*(\mathrm{Lin}_U^*(\tau(\alpha))) = \mathrm{Lin}_M^*(\langle \varphi_U, j_U(\rho_U^*\alpha)\rangle) = \langle \varphi_M, j_M(\rho_M^*\alpha)\rangle,$$

which proves the corollary. □

4.3.5. Definition. Let $U \subset \overline{T}V$ be an open $U(1)$-invariant neighborhood of the zero section $V \subset \overline{T}V$. A Hodge manifold structure on $\overline{T}V$ is called *linear* if the associated linearization map $\mathrm{Lin}_U : U \to \overline{T}V$ coincides with the given embedding $U \hookrightarrow \overline{T}V$.

By Corollary 4.3.4 every Hodge manifold structure on a subset $U \subset \overline{T}V$ obtained by the linearization construction is linear.

4.3.6. We finish this section with the following simple observation, which we will need in the next section.

Lemma. *Keep the notations of Lemma* 4.3.3*. Moreover, assume given a subspace* $\mathcal{A} \subset C^\infty(\overline{T}V, \rho^*\Lambda^1(V,\mathbb{C}))$ *such that the image of* $\mathcal{A}$ *under the restriction map*

$$\mathrm{Res} : C^\infty(\overline{T}V, \rho^*\Lambda^1(V,\mathbb{C})) \to C^\infty(V, \Lambda^1(V,\mathbb{C}))$$

onto the zero section $V \subset \overline{T}V$ is the whole space $C^\infty(V, \Lambda^1(V, \mathbb{C}))$. If (4.2) *holds for every section $\beta \in \mathcal{A}$, then it holds for every smooh section*

$$\beta \in C^\infty(\overline{T}V, \rho^*\Lambda^1(V, \mathbb{C})).$$

Proof. By assumptions sections $\beta \in \mathcal{A}$ generate the restriction of the bundle $\rho^*\Lambda^1(V, \mathbb{C})$ onto the zero section $V \subset \overline{T}V$. Therefore there exists an open neighborhood $U \subset \overline{T}V$ of the zero section $V \subset \overline{T}V$ such that the $C^\infty(U, \mathbb{C})$-submodule

$$C^\infty(U, \mathbb{C}) \cdot \mathcal{A} \subset C^\infty(U, \rho^*\Lambda^1(V, \mathbb{C}))$$

is dense in the space $C^\infty(U, \rho^*\Lambda^1(V, \mathbb{C}))$ of smooth sections of the pull-back bundle $\rho^*\Lambda^1(V, \mathbb{C})$. Since the equality (4.2) is continuous and linear with respect to multiplication by smooth functions, it holds for all sections $\beta \in C^\infty(U, \rho^*\Lambda^1(V, \mathbb{C}))$. Since it is also compatible with the natural unit disc action on M and $\overline{T}V$, it holds for all sections $\beta \in C^\infty(\overline{T}V, \rho^*\Lambda^1(V, \mathbb{C}))$ as well. □

5. Tangent bundles as Hodge manifolds

5.1. Hodge connections

5.1.1. The linearization construction reduces the study of arbitrary regular Hodge manifolds to the study of linear Hodge manifold structures on a neighborhood $U \subset \overline{T}V$ of the zero section $V \subset \overline{T}V$ of the total space of the complex conjugate to the tangent bundle of a complex manifold V. In this section we use the theory of Hodge bundles developed in Subsection 2.1 in order to describe Hodge manifold structures on U in terms of connections on the locally trivial fibration $U \to V$ of a certain type, which we call *Hodge connections* (see 5.1.7). It is this description, given in Proposition 5.1, which we will use in the latter part of the paper to classify all such Hodge manifold structures.

5.1.2. We begin with some preliminary facts about connections on locally trivial fibrations. Let $f : X \to Y$ be an arbitrary smooth map of smooth manifolds X and Y. Assume that the map f is submersive, that is, the codifferential $\delta_f : f^*\Lambda^1(Y) \to \Lambda^1(X)$ is an injective bundle map. Recall that a *connection* on f is by definition a splitting $\Theta : \Lambda^1(X) \to f^*\Lambda^1(Y)$ of the canonical embedding δ_f.

Let d_X be the de Rham differential on the smooth manifold X. Every connection Θ on $f : X \to Y$ defines an algebra derivation

$$D = \Theta \circ d_X : \Lambda^0(X) \to f^*\Lambda^1(Y),$$

satisfying

$$D\rho^* h = \rho^* d_Y h \tag{5.1}$$

for every smooth function $h \in C^\infty(Y, \mathbb{R})$. Vice versa, by the universal property of the cotangent bundle (Lemma 2.4.1) every algebra derivation $D : \Lambda^0(X) \to \Lambda^1(Y)$ satisfying (5.1) comes from a unique connection Θ on f.

5.1.3. Recall also that a connection Θ is called *flat* if the associated derivation D extends to an algebra derivation

$$D : f^*\Lambda^\bullet(Y) \to f^*\Lambda^{\bullet+1}(Y)$$

so that $D \circ D = 0$. The splitting $\Theta : \Lambda^1(X) \to f^*\Lambda^1(Y)$ extends in this case to an algebra map

$$\Theta : \Lambda^\bullet(X) \to f^*\Lambda^\bullet(Y)$$

compatible with the de Rham differential $d_X : \Lambda^\bullet(X) \to \Lambda^{\bullet+1}(X)$.

5.1.4. We will need a slight generalization of the notion of connection.

Definition. Let $f : X \to Y$ be a smooth submersive morphism of complex manifolds. A *$\mathbb{C}$-valued connection* Θ on f is a splitting $\Theta : \Lambda^1(Y, \mathbb{C}) \to f^*\Lambda^1(X, \mathbb{C})$ of the codifferential map $\delta f : f^*\Lambda^1(Y, \mathbb{C}) \to \Lambda^1(X, \mathbb{C})$ of complex vector bundles. A $\mathbb{C}$-valued connection Θ is called *flat* if the associated algebra derivation

$$D = \Theta \circ d_X : \Lambda^0(X, \mathbb{C}) \to f^*\Lambda^1(Y, \mathbb{C})$$

extends to an algebra derivation

$$D : f^*\Lambda^\bullet(Y, \mathbb{C}) \to f^*\Lambda^{\bullet+1}(Y, \mathbb{C})$$

satisfying $D \circ D = 0$.

As in 5.1.2, every derivation $D : \Lambda^0(X, \mathbb{C}) \to f^*\Lambda^1(Y, \mathbb{C})$ satisfying (5.1) comes from a unique $\mathbb{C}$-valued connection Θ on $f : X \to Y$.

Remark. By definition for every flat connection on $f : X \to Y$ the subbundle of horizontal vectors in the the tangent bundle $\Theta(X)$ is an involutive distribution. By Frobenius Theorem this implies that the connection defines locally a trivialization of the fibration f.

This is no longer true for flat $\mathbb{C}$-valued connections: the subbundle of horizontal vectors in $\Theta(X) \otimes \mathbb{C}$ is only defined over $\mathbb{C}$, and the Frobenius Theorem does not apply. One can try to correct this by replacing the splitting $\Theta : \Lambda^1(X, \mathbb{C}) \to f^*\Lambda^1(Y, \mathbb{C})$ with its real part $\operatorname{Re}\Theta : \Lambda^1(X) \to \Lambda^1(Y)$, but this real part is, in general, no longer flat.

5.1.5. For every $\mathbb{C}$-valued connection $\Theta : \Lambda^1(X,\mathbb{C}) \to f^*\Lambda^1(Y,\mathbb{C})$ on a fibration $f : X \to Y$ the kernel $\operatorname{Ker}\Theta \subset \Lambda^1(X,\mathbb{C})$ is canonically isomorphic to the quotient $\Lambda^1(X,\mathbb{C})/\delta_f(f^*\Lambda^1(Y,\mathbb{C}))$, and the composition

$$R = \Theta \circ d_X : \Lambda^1(X,\mathbb{C})/\delta_f(f^*\Lambda^1(Y,\mathbb{C}) \cong \operatorname{Ker}\Theta \to f^*\Lambda^2(Y,\mathbb{C})$$

is in fact a bundle map. This map is called *the curvature* of the $\mathbb{C}$-valued connection Θ. The connection Θ is flat if and only if its curvature R vanishes.

5.1.6. Let now M be a complex manifold, and let $U \subset \overline{T}M$ be an open neighborhood of the zero section $M \subset \overline{T}M$ in the total space $\overline{T}M$ of the complex-conjugate to the tangent bundle to M. Let $\rho : U \to M$ be the natural projection. Assume that U is invariant with respect to the natural action of the unit disc $\Delta \subset \mathbb{C}$ on $\overline{T}M$.

5.1.7. Since M is complex, by 2.1.7 the bundle $\Lambda^1(M,\mathbb{C})$ is equipped with a Hodge bundle structure of weight 1. The pullback bundle $\rho^*\Lambda^1(M,\mathbb{C})$ is then also equipped with a weight 1 Hodge bundle structure.

Our description of the Hodge manifold structures on the subset $U \in \overline{T}M$ is based on the following notion.

Definition. A *Hodge connection* on the pair $\langle M, U\rangle$ is a $\mathbb{C}$-valued connection on $\rho : U \to M$ such that the associated derivation

$$D : \Lambda^0(U,\mathbb{C}) \to \rho^*\Lambda^1(M,\mathbb{C})$$

is weakly Hodge in the sense of 2.1.3. A Hodge connection is called *flat* if it extends to a weakly Hodge derivation

$$D : \rho^*\Lambda^{\bullet}(M,\mathbb{C}) \to \rho^*\Lambda^{\bullet+1}(M,\mathbb{C})$$

satisfying $D \circ D = 0$.

5.1.8. Assume given a flat Hodge connection $D : \Lambda^0(U,\mathbb{C}) \to \rho^*\Lambda^1(M,\mathbb{C})$ on the pair $\langle U, M\rangle$, and assume in addition that the derivation D is holonomic in the sense of 2.4.2. Then the pair $\langle D, \rho^*\Lambda^1(M,\mathbb{C})\rangle$ defines by Proposition 3.1 a Hodge manifold structure on U.

It turns out that every Hodge manifold structure on U can be obtained in this way. Namely, we have the following.

Proposition 5.1 *There correspondence $D \mapsto \langle \rho^*\Lambda^1(M,\mathbb{C}), D\rangle$ is a bijection between the set of all flat Hodge connections D on the pair $\langle U, M\rangle$ such that $D : \Lambda^0(U,\mathbb{C}) \to \rho^*\Lambda^1(M,\mathbb{C})$ is holonomic in the sense of 2.4.2, and the set of all Hodge manifold structures on the $U(1)$-manifold U such that the projection $\rho : U_I \to M$ is holomorphic for the preferred complex structure U_I on U.*

5.1.9. The crucial part of the proof of Proposition 5.1 is the following observation.

Lemma. *Assume given a Hodge manifold structure on the $U(1)$-manifold $U \subset \overline{T}M$. Let $\delta_\rho : \rho^*\Lambda^1(M, \mathbb{C}) \to \Lambda^1(U, \mathbb{C})$ be the codifferential of the projection $\rho : U \to M$, and let $P : \Lambda^1(U, \mathbb{C}) \to \Lambda^{0,1}(U_J)$ be the canonical projection. The bundle map given by the composition*

$$P \circ \delta_\rho : \rho^*\Lambda^1(M, \mathbb{C}) \to \Lambda^{0,1}(U_J)$$

is an isomorphism of complex vector bundles.

Proof. Since the bundles $\rho^*\Lambda^1(M, \mathbb{C})$ and $\Lambda^{0,1}(U_J)$ are of the same rank, and the maps $\delta_\rho : \rho^*\Lambda^1(M, \mathbb{C}) \to \Lambda^1(U, \mathbb{C})$ and $P \circ \delta_\rho$ are equivariant with respect to the action of the unit disc on U, it suffices to prove the claim on $M \subset U$. Let $m \in M$ be an arbitrary point, and let $V = T^*_m\overline{T}M$ be the cotangent bundle at m to the Hodge manifold $U \subset \overline{T}M$. Let also $V^0 \subset V$ be the subspace of $U(1)$-invariant vectors in V.

The space V is an equivariant quaternionic vector space. Moreover, the fibers of the bundles $\rho^*\Lambda^1(M, \mathbb{C})$ and $\Lambda^{0,1}(U_J)$ at the point m are complex vector spaces, and we have canonical identifications

$$\begin{aligned} \rho^*\Lambda^1(M, \mathbb{C})|_m &\cong V_I^0 \oplus \overline{V_I^0}, \\ \Lambda^{0,1}(U_J)|_m &\cong V_J. \end{aligned}$$

Under these identifications the map $P \circ \delta_\rho$ at the point m coincides with the action map $V_I^0 \oplus \overline{V_I^0} \to V_J$, which is invertible by Lemma 1.1.7. □

5.1.10. By Proposition 3.1 every Hodge manifold structure on U is given by a pair $\langle \mathcal{E}, D \rangle$ of a Hodge bundle $\mathcal{E}$ on U of weight 1 and a holonomic derivation $D : \Lambda^0(U, \mathbb{C}) \to \mathcal{E}$. Lemma 5.1.9 gives an isomorphism $\mathcal{E} \cong \rho^*\Lambda^1(M, \mathbb{C})$, so that D becomes a flat $\mathbb{C}$-valued connection on U over M. To prove Proposition 5.1 it suffices now to prove the following.

Lemma. *The complex vector bundle isomorphism*

$$P \circ \delta_\rho : \rho^*\Lambda^1(M, \mathbb{C}) \to \Lambda^{0,1}(U_J)$$

associated to a Hodge manifold structure on U is compatible with the Hodge bundle structures if and only if the projection $\rho : U_I \to M$ is holomorphic for the preferred complex structure U_I on U.

Proof. The preferred complex structure U_I induces a Hodge bundle structure of weight 1 on $\Lambda^1(U, \mathbb{C})$ by 2.1.7, and the canonical projection $P : \Lambda^1(U, \mathbb{C}) \to \Lambda^{0,1}(U_J)$ is compatible with the Hodge bundle structures by 2.3.4. If the projection $\rho : U_I \to M$ is holomorphic, then the codifferential $\delta_\rho : \rho^*\Lambda^1(M, \mathbb{C}) \to \Lambda^1(U, \mathbb{C})$ sends the subbundles $\rho^*\Lambda^{1,0}(M), \rho^*\Lambda^{0,1}(M) \subset \rho^*\Lambda^1(M, \mathbb{C})$ into, respectively, the subbundles $\Lambda^{1,0}(U_I), \Lambda^{0,1}(U_I) \subset \Lambda^1(U, \mathbb{C})$. Therefore the map $\delta_\rho : \rho^*\Lambda^1(M, \mathbb{C}) \to \Lambda^1(U, \mathbb{C})$ is compatible with the Hodge bundle structures, which implies the "if" part of the lemma.

To prove the "only if" part, assume that $P \circ \delta_\rho$ is a Hodge bundle isomorphism. Since the complex conjugation $\nu : \overline{\Lambda^{0,1}(U_I)} \to \Lambda^{1,0}(U_J)$ is compatible with the Hodge bundle structures, the projection $\overline{P} : \Lambda^1(U, \mathbb{C}) \to \Lambda^{1,0}(U_J)$ and the composition $\overline{P} \circ \delta_\rho : \rho^*\Lambda^1(M, \mathbb{C}) \to \Lambda^{1,0}(U_J)$ are also compatible with the Hodge bundle structures. Therefore the map

$$P \oplus \overline{P} : \Lambda^1(U, \mathbb{C}) \to \Lambda^{1,0}(U_J) \oplus \Lambda^{0,1}(U_J)$$

is a Hodge bundle isomorphism, and the composition

$$\delta_\rho \circ (P \oplus \overline{P}) : \rho^*\Lambda^1(M, \mathbb{C}) \to \Lambda^1(U, \mathbb{C})$$

is a Hodge bundle map. Therefore the codifferential $\delta_\rho : \rho^*\Lambda^1(M, \mathbb{C}) \to \Lambda^1(U, \mathbb{C})$ is compatible with the Hodge bundle structures. This means precisely that the projection $\rho : U_I \to M$ is holomorphic, which finishes the proof of the lemma and of Proposition 5.1. □

5.2. The relative de Rham complex of U over M

5.2.1. Keep the notation of the last subsection. To use Proposition 5.1 in the study of Hodge manifold structures on the open subset $U \subset \overline{T}M$, we will need a way to check whether a given Hodge connection on the pair $\langle U, M \rangle$ is holonomic in the sense of 2.4.2. We will also need to rewrite the linearity condition 4.3.5 for a Hodge manifold structure on U in terms of the associated Hodge connection D. To do this, we will use the so-called *relative de Rham complex* of U over M. For the convenience of the reader, and to fix notation, we recall here its definition and main properties.

5.2.2. Since the projection $\rho : U \to M$ is submersive, the codifferential

$$\delta_\rho : \rho^*\Lambda^1(M, \mathbb{C}) \to \Lambda^1(U, \mathbb{C})$$

is injective. The *relative cotangent bundle* $\Lambda^1(U/M, C)$ is by definition the quotient bundle

$$\Lambda^1(U/M, \mathbb{C}) = \Lambda^1(U, \mathbb{C}) / \delta_\rho(\rho^*\Lambda^1(M, \mathbb{C})).$$

Let $\pi : \Lambda^1(U, \mathbb{C}) \to \Lambda^1(U/M, \mathbb{C})$ be the natural projection. We have by definition the short exact sequence

$$0 \longrightarrow \rho^*\Lambda^1(M, \mathbb{C}) \xrightarrow{\delta_\rho} \Lambda^1(U, \mathbb{C}) \xrightarrow{\pi} \Lambda^1(U/M, \mathbb{C}) \longrightarrow 0 \tag{5.2}$$

of complex vector bundles on U.

5.2.3. The composition $d^r = \pi \circ d_U$ of the de Rham differential d_U with the projection π is an algebra derivation

$$d^r : \Lambda^0(U, \mathbb{C}) \to \Lambda^1(U/M, \mathbb{C}),$$

called *the relative de Rham differential.* It is a first order differential operator, and $d^r f = 0$ if and only if the smooth function $f : U \to \mathbb{C}$ factors through the projection $\rho : U \to M$.

Let $\Lambda^{\cdot}(U/M, \mathbb{C})$ be the exterior algebra of the bundle $\Lambda^1(U/M, \mathbb{C})$. The projection π extends to an algebra map

$$\pi : \Lambda^{\cdot}(U, \mathbb{C}) \to \Lambda^{\cdot}(U/M, \mathbb{C}).$$

The differential d^r extends to an algebra derivation

$$d^r : \Lambda^{\cdot}(U/M, \mathbb{C}) \to \Lambda^{\cdot+1}(U/M, \mathbb{C})$$

satisfying $d^r \circ d^r = 0$, and we have $\pi \circ d_U = d^r \circ \pi$. The differential graded algebra $\langle \Lambda^{\cdot}(U/M, \mathbb{C}), d^r \rangle$ is called *the relative de Rham complex* of U over M.

5.2.4. Since the relative de Rham differential d^r is linear with respect to multiplication by functions of the form $\rho^* f$ with $f \in C^\infty(M, \mathbb{C})$, it extends canonically to an operator

$$d^r : \rho^*\Lambda^i(M, \mathbb{C}) \otimes \Lambda^{\cdot}(U/M, \mathbb{C}) \to \rho^*\Lambda^i(M, \mathbb{C}) \otimes \Lambda^{\cdot+1}(U/M, \mathbb{C}).$$

The two-step filtration $\rho^*\Lambda^1(M, \mathbb{C}) \subset \Lambda^1(U, \mathbb{C})$ induces a filtration on the de Rham complex $\Lambda^{\cdot}(U, \mathbb{C})$, and the i-th associated graded quotient of this filtration is isomorphic to the complex $\langle \rho^*\Lambda^i(M, \mathbb{C}) \otimes \Lambda^{\cdot}(U/M, \mathbb{C}), d^r \rangle$.

5.2.5. Since $U \subset \overline{T}M$ lies in the total space of the complex-conjugate to the tangent bundle to M, we have a canonical algebra isomorphism

$$\mathsf{can} : \overline{\rho^*\Lambda^{\cdot}(M, \mathbb{C})} \to \Lambda^{\cdot}(U/M, \mathbb{C}).$$

Let $\tau : C^\infty(M, \Lambda^1(M, \mathbb{C})) \to C^\infty(U, \mathbb{C})$ be the tautological map sending a 1-form to the corresponding linear function on $\overline{T}M$, as in 4.3.2. Then for every smooth 1-form $\alpha \in C^\infty(M, \Lambda^1(M, \mathbb{C}))$ we have

$$\mathsf{can}(\rho^*\alpha) = d^r \tau(\alpha). \tag{5.3}$$

5.2.6. The complex vector bundle $\Lambda^1(U/M, \mathbb{C})$ has a natural real structure, and it is naturally $U(1)$-equivariant. Moreover, the decomposition $\Lambda^1(M, \mathbb{C}) = \Lambda^{1,0}(M) \oplus \Lambda^{0,1}(M)$ induces a decomposition

$$\Lambda^1(U/M, \mathbb{C}) = \mathsf{can}(\Lambda^{1,0}(M)) \oplus \mathsf{can}(\Lambda^{0,1}(M)).$$

This allows to define, as in 2.1.7, a canonical Hodge bundle structure of weight 1 on $\Lambda^1(U/M, \mathbb{C})$. It gives rise to a Hodge bundle structure on $\Lambda^i(U/M, \mathbb{C})$ of weight i, and the relative de Rham differential

$$d^r : \Lambda^{\bullet}(U/M, \mathbb{C}) \to \Lambda^{\bullet+1}(U/M, \mathbb{C})$$

is weakly Hodge.

5.2.7. The canonical isomorphism $\mathsf{can} : \overline{\rho^*\Lambda^1(M, \mathbb{C})} \to \Lambda^1(U/M, \mathbb{C})$ is not compatible with the Hodge bundle structures. The reason for this is that the real structure on the Hodge bundles $\Lambda^{\bullet}(U/M, \mathbb{C})$ is, by definition 2.1.7, twisted by ι^*, where $\iota : \overline{T}M \to \overline{T}M$ is the action of $-1 \in U(1) \subset \mathbb{C}$. Therefore, while can is $U(1)$-equivariant, it is not real. To correct this, introduce an involution $\zeta : \Lambda^1(M, \mathbb{C}) \to \Lambda^1(M, \mathbb{C})$ by

$$\zeta = \begin{cases} \mathsf{id} & \text{on } \Lambda^{1,0}(M) \subset \Lambda^1(M, \mathbb{C}) \\ -\mathsf{id} & \text{on } \Lambda^{0,1}(M) \subset \Lambda^1(M, \mathbb{C}) \end{cases} \tag{5.4}$$

and set

$$\eta = \mathsf{can} \circ \rho^*\overline{\zeta} : \rho^*\overline{\Lambda^1(M, \mathbb{C})} \to \rho^*\overline{\Lambda^1(M, \mathbb{C})} \to \Lambda^1(U/M, \mathbb{C}) \tag{5.5}$$

Unlike can, the map η preserves the Hodge bundle structures.

It will also be convenient to twist the tautological map $\tau : \rho^*\Lambda^1(M, \mathbb{C}) \to \Lambda^0(U, \mathbb{C})$ by the involution ζ. Namely, define a map $\sigma : \rho^*\Lambda^1(M, \mathbb{C}) \to \Lambda^0(U, \mathbb{C})$ by

$$\sigma = \tau \circ \rho^*\overline{\zeta} : \rho^*\overline{\Lambda^1(M, \mathbb{C})} \to \rho^*\overline{\Lambda^1(M, \mathbb{C})} \to \Lambda^0(U/M, \mathbb{C}) \tag{5.6}$$

By (5.3) the twisted tautological map σ and the canonical map η satisfy

$$\eta(\rho^*\alpha) = d^r\sigma(\alpha) \tag{5.7}$$

for every smooth 1-form $\alpha \in C^\infty(M, \Lambda^1(M, \mathbb{C}))$.

5.2.8. Let $\varphi \in \Theta(U)$ be the differential of the canonical $U(1)$-action on $U \subset \overline{T}M$. The vector field φ is real and tangent to the fibers of the projection $\rho : U \to M$. Therefore the contraction with φ defines an algebra derivation

$$\Lambda^{\bullet+1}(U/M, \mathbb{C}) \to \Lambda^{\bullet}(U/M, \mathbb{C})$$
$$\alpha \mapsto \langle \varphi, \alpha \rangle$$

The following lemma gives a relation between this derivation, the canonical weakly Hodge map $\eta : \rho^*\Lambda^1(M, \mathbb{C}) \to \Lambda^1(U/M, \mathbb{C})$ given by (5.5), and the tautological map $\tau : \rho^*\Lambda^1(M, \mathbb{C}) \to \Lambda^0(U, \mathbb{C})$.

Lemma. *For every smooth section* $\alpha \in C^\infty(U, \rho^*\Lambda^1(M, \mathbb{C}))$ *we have*

$$\sqrt{-1}\tau(\alpha) = \langle \varphi, \eta(\alpha) \rangle \in C^\infty(U, \mathbb{C}).$$

Proof. Since the equality that we are to prove is linear with respect to multiplication by smooth functions on U, we may assume that the section α is the pull-back of a smooth 1-form $\alpha \in C^\infty(M, \Lambda^1(M, \mathbb{C}))$. The Lie derivative $\mathcal{L}_\varphi : \Lambda^\bullet(U, \mathbb{C}) \to \Lambda^\bullet(U, \mathbb{C})$ with respect to the vector field φ is compatible with the projection $\pi : \Lambda^\bullet(U, \mathbb{C}) \to \Lambda^\bullet(U/M, \mathbb{C})$ and defines therefore an algebra derivation $\mathcal{L}_\varphi : \Lambda^\bullet(U/M, \mathbb{C}) \to \Lambda^\bullet(U/M, \mathbb{C})$. The Cartan homotopy formula gives

$$\mathcal{L}_\varphi \tau(\alpha) = \langle \varphi, d^r \tau(\alpha) \rangle. \tag{5.8}$$

The function $\tau(\alpha)$ on $\overline{T}M$ is by definition $\mathbb{R}$-linear along the fibers of the projection $\rho : \overline{T}M \to M$. The subspace $\tau(C^\infty(M, \Lambda^1(M, \mathbb{C}))) \subset C^\infty(U, \mathbb{C})$ of such functions decomposes as

$$\tau(C^\infty(M, \Lambda^1(M, \mathbb{C}))) = \tau(C^\infty(M, \Lambda^{1,0}(M))) \oplus \tau(C^\infty(M, \Lambda^{0,1}(M, \mathbb{C}))),$$

and the group $U(1)$ acts on the components with weight 1 and -1. Therefore the derivative $\mathcal{L}_\varphi$ of the $U(1)$-action acts on the components by multiplication with $\sqrt{-1}$ and $-\sqrt{-1}$. By definition of the involution ς (see (5.4)) this can be written as

$$\mathcal{L}_\varphi \tau(\alpha) = \sqrt{-1}\tau(\varsigma(\alpha)). \tag{5.9}$$

On the other hand, by (5.3) and the definition of the map η we have

$$d^r \tau(\alpha) = \mathsf{can}(\alpha) = \eta(\varsigma(\alpha)). \tag{5.10}$$

Substituting (5.9) and (5.10) into (5.8) gives

$$\sqrt{-1}\tau(\varsigma(\alpha)) = \langle \varphi, \eta(\varsigma(\alpha)),$$

which is equivalent to the claim of the lemma. □

5.3. Holonomic Hodge connections

5.3.1. We will now describe a convenient way to check whether a given Hodge connection D on the pair $\langle U, M\rangle$ is holonomic in the sense of 2.4.2. To do this, we proceed as follows.

Consider the restriction $\Lambda^1(U,\mathbb{C})|_M$ of the bundle $\Lambda^1(U,\mathbb{C})$ to the zero section $M \subset U \subset \overline{T}M$, and let

$$\mathrm{Res} : \Lambda^1(U,\mathbb{C})|_M \to \Lambda^1(M,\mathbb{C})$$

be the restriction map. The kernel of the map Res coincides with the conormal bundle to the zero section $M \subset U$, which we denote by $S^1(M,\mathbb{C})$. The map Res splits the restriction of exact sequence (5.2) onto the zero section $M \subset U$, and we have the direct sum decomposition

$$\Lambda^1(U,\mathbb{C})|_M = S^1(M,\mathbb{C}) \oplus \Lambda^1(M,\mathbb{C}). \tag{5.11}$$

5.3.2. The $U(1)$-action on $U \subset \overline{T}M$ leaves the zero section $M \subset U$ invariant and defines therefore a $U(1)$-action on the conormal bundle $S^1(M,\mathbb{C})$. Together with the usual real structure twisted by the action of the map $\iota : \overline{T}M \to \overline{T}M$, this defines a Hodge bundle structure of weight 0 on the bundle $S^1(M,\mathbb{C})$.

Note that the automorphism $\iota : \overline{T}M \to \overline{T}M$ acts as $-\mathsf{id}$ on the Hodge bundle $S^1(M,\mathbb{C})$, so that the real structure on $S^1(M,\mathbb{C})$ is minus the usual one. Moreover, as a complex vector bundle the conormal bundle $S^1(M,\mathbb{C})$ to $M \subset \overline{T}M$ is canonically isomorphic to the cotangent bundle $\Lambda^1(M,\mathbb{C})$. The Hodge type bigrading on $S^1(M,\mathbb{C})$ is given in terms of this isomorphism by

$$S^1(M,\mathbb{C}) = S^{1,-1}(M) \oplus S^{-1,1}(M) \cong \Lambda^{1,0}(M) \oplus \Lambda^{0,1}(M) = \Lambda^1(M,\mathbb{C}).$$

5.3.3. Let

$$C^\infty_{lin}(U,\mathbb{C}) = \tau(C^\infty(M,\Lambda^1(M,\mathbb{C}))) \subset C^\infty(U,\mathbb{C})$$

be the subspace of smooth functions linear along the fibers of the canonical projection $\rho : U \subset \overline{T}M \to M$. The relative de Rham differential defines an isomorphism

$$d^r : C^\infty_{lin}(U,\mathbb{C}) \to C^\infty(M, S^1(M,\mathbb{C})). \tag{5.12}$$

This isomorphism is compatible with the canonical Hodge structures of weight 0 on both spaces, and it is linear with respect to multiplication by smooth functions $f \in C^\infty(M, \mathbb{C})$.

5.3.4. Let now $D : \Lambda^0(U, \mathbb{C}) \to \rho^*\Lambda^1(M, \mathbb{C})$ be a Hodge connection on the pair $\langle U, M\rangle$, and let $\Theta : \Lambda^1(U, \mathbb{C}) \to \rho^*\Lambda^(M, \mathbb{C})$ be the corresponding bundles map. Since D is a $\mathbb{C}$-valued connection, the restriction $\Theta|_M$ onto the zero section $M \subset M$ decomposes as

$$\Theta = D_0 \oplus \mathrm{id} : S^1(M, \mathbb{C}) \oplus \Lambda^1(M, \mathbb{C}) \to \Lambda^1(M, \mathbb{C}) \tag{5.13}$$

with respect to the direct sum decomposition (5.11) for a certain bundle map $D_0 : S^1(M, \mathbb{C}) \to \Lambda^1(M, \mathbb{C})$.

Definition. The bundle map $D_0 : S^1(M, \mathbb{C}) \to \Lambda^1(M, \mathbb{C})$ is called the *principal part* of the Hodge connection D.

5.3.5. Consider the map $D_0 : C^\infty(M, S^1(M, \mathbb{C})) \to C^\infty(M, \Lambda^1(M, \mathbb{C}))$ on the spaces of smooth sections induced by the principal part D_0 of a Hodge connection D. Under the isomorphism (5.12) this map coincides with the restriction of the composition

$$\mathrm{Res} \circ D : C^\infty(U, \mathbb{C}) \to C^\infty(U, \rho^*\Lambda^1(M, \mathbb{C})) \to C^\infty(M, \Lambda^1(M, \mathbb{C}))$$

onto the subspace $C^\infty_{lin}(U, \mathbb{C}) \subset C^\infty(U, \mathbb{C})$. Each of the maps Res, D is weakly Hodge, so that this composition also is weakly Hodge. Since the isomorphism (5.12) is compatible with the Hodge bundle structures, this implies that the principal part D_0 of the Hodge connection D is a weakly Hodge bundle map. In particular, it is purely imaginary with respect to the usual real structure on the conormal bundle $S^1(M, \mathbb{C})$.

5.3.6. We can now formulate the main result of this subsection.

Lemma. *A Hodge connection $D : \Lambda^0(U, \mathbb{C}) \to \rho^*\Lambda^1(M, \mathbb{C})$ on the pair $\langle U, M\rangle$ is holonomic in the sense of* 2.4.2 *on an open neighborhood $U_0 \subset U$ of the zero section $M \subset U$ if and only if its principal part $D_0 : S^1(M, \mathbb{C}) \to \Lambda^1(M, \mathbb{C})$ is a complex vector bundle isomorphism.*

Proof. By definition the derivation $D : \Lambda^0(U, \mathbb{C}) \to \rho^*\Lambda^1(M, \mathbb{C})$ is holonomic in the sense of 2.4.2 if and only if the corresponding map

$$\Theta : \Lambda^1(U, \mathbb{R}) \to \rho^*\Lambda^1(M, \mathbb{C})$$

is an isomorphism of real vector bundles. This is an open condition. Therefore the derivation D is holonomic on an open neighborhood $U_0 \supset M$ of the zero section $M \subset U$ if and only if the map Θ is an isomorphism on the zero section $M \subset U$ itself.

According to (5.13), the restriction $\Theta|_M$ decomposes as $\Theta_M = D_0 + \mathsf{id}$, and the principal part $D_0 : S^1(M, \mathbb{C}) \to \Lambda^1(M, \mathbb{C})$ of the Hodge connection D is purely imaginary with respect to the usual real structure on $\Lambda^1(U, \mathbb{C})|_M$, while the identity map $\mathsf{id} : \rho^*\Lambda^1(M, \mathbb{C}) \to \rho^*\Lambda^1(M, \mathbb{C})$ is, of course, real. Therefore Θ_M is an isomorphism if and only is D_0 is an isomorphism, which proves the lemma. □

5.4. Hodge connections and linearity

5.4.1. Assume now given a Hodge manifold structure on the subset $U \subset \overline{T}M$, and let $D : \Lambda^0(U, \mathbb{C}) \to \rho^*\Lambda^1(M, \mathbb{C})$ be the associated Hodge connection on the pair $\langle U, M\rangle$ given by Proposition 5.1. We now proceed to rewrite the linearity condition 4.3.5 in terms of the Hodge connection D.

Let $j : \Lambda^1(U, \mathbb{C}) \to \overline{\Lambda^1(U, \mathbb{C})}$ be the canonical map defined by the quaternionic structure on U, and let $\iota^* : \Lambda^1(U, \mathbb{C}) \to \iota^*\Lambda^1(U, \mathbb{C})$ be the action of the canonical involution $\iota : U \to U$. Let also $D^\iota : \Lambda^0(U, \mathbb{C}) \to \rho^*\Lambda^1(M, \mathbb{C})$ be the operator ι^*-conjugate to the Hodge connection D.

We begin with the following identity.

Lemma. *For every smooth function $f \in C^\infty(U, \mathbb{C})$ we have*

$$d^r f = \frac{\sqrt{-1}}{2}\pi(j(\delta_\rho(D - D^\iota)(f))),$$

where $\pi : \Lambda^1(U, \mathbb{C}) \to \Lambda^1(U/M, \mathbb{C})$ is the canonical projection, and

$$\delta_\rho : \rho^*\Lambda^1(M, \mathbb{C}) \to \Lambda^1(U, \mathbb{C})$$

is the codifferential of the projection $\rho : U \to M$.

Proof. By definition of the Hodge connection D the Dolbeault derivative $\bar\partial_J f$ coincides with the $(0,1)$-component of the 1-form $\delta_\rho(Df) \in \Lambda^1(U, \mathbb{C})$ with respect to the complementary complex structure U_J on U. Therefore

$$\bar\partial_J f = \frac{1}{2}\delta_\rho(Df) + \frac{\sqrt{-1}}{2}j(\delta_\rho(Df)).$$

Applying the complex conjugation $\nu : \Lambda^\bullet(U, \mathbb{C}) \to \overline{\Lambda^\bullet(U, \mathbb{C})}$ to this equation, we get

$$\begin{aligned}\partial_J f &= \nu\left(\frac{1}{2}\delta_\rho(D\nu(f)) + \frac{\sqrt{-1}}{2}j(\delta_\rho(D\nu(f))))\right) = \\ &= \frac{1}{2}\nu(\delta_\rho(D\nu(f))) - \frac{\sqrt{-1}}{2}j(\nu(\delta_\rho(D\nu(f)))).\end{aligned}$$

Since the map $\delta_\rho \circ D : \Lambda^0(U, \mathbb{C}) \to \rho^*\Lambda^1(M, \mathbb{C})$ is weakly Hodge, we have

$$\delta_\rho(D(\iota^*\nu(f))) = \iota^*\nu(\delta_\rho(Df)).$$

Therefore $\nu(\delta_\rho(D(f))) = \delta_\rho(D^\iota(\nu(f)))$, and we have

$$\partial_J f = \frac{1}{2}\delta_\rho(D^\iota f) - \frac{\sqrt{-1}}{2} j(\delta_\rho(D^\iota f)).$$

Thus the de Rham derivative $d_U f$ equals

$$d_U f = \partial_J f + \bar{\partial}_J f = \frac{1}{2}\delta_\rho((D + D^\iota)f) + \frac{\sqrt{-1}}{2} j(\delta_\rho((D - D^\iota)f)).$$

Now, by definition $\delta_\rho \circ \pi = 0$. Therefore

$$d^r f = \pi(d_U f) = \frac{\sqrt{-1}}{2}\pi(j(\delta_\rho((D - D^\iota)f))),$$

which is the claim of the lemma. □

5.4.2. We will also need the following fact. It can be derived directly from Lemma 5.4.1, but it is more convenient to use Lemma 5.3.6 and the fact that the Hodge connection $D : \Lambda^0(U, \mathbb{C}) \to \rho^*\Lambda^1(M, \mathbb{C})$ is holonomic.

Lemma. *In the notation of Lemma* 5.4.1, *let*

$$\mathcal{A} = \delta_\rho\left((D - D^\iota)\left(C^\infty_{lin}(U, \mathbb{C})\right)\right) \subset C^\infty(U, \rho^*\Lambda^1(M, \mathbb{C}))$$

be the subspace of sections $\alpha \in C^\infty(U, \rho^*\Lambda^1(M, \mathbb{C}))$ of the form $\alpha = \delta_\rho((D - D^\iota)f)$, where $f \in C^\infty(U, \mathbb{C})$ lies in the subspace $C^\infty_{lin}(U, \mathbb{C}) \subset C^\infty(U, \mathbb{C})$ of smooth functions on U linear along the fibers of the projection $\rho : U \to M$. The restriction $\operatorname{Res}(\mathcal{A}) \subset C^\infty(M, \Lambda^1(M, \mathbb{C}))$ *of the subspace $\mathcal{A}$ onto the zero section $M \subset U$ is the whole space $C^\infty(M, \Lambda^1(M, \mathbb{C}))$.*

Proof. Let $D_0 = \operatorname{Res} \circ D : C^\infty_{lin}(U, \mathbb{C}) \to C^\infty(M, \Lambda^1(M, \mathbb{C}))$ be the principal part of the Hodge connection D in the sense of Definition 5.3.4. Since the canonical automorphism $\iota : \overline{T}M \to \overline{T}M$ acts as $-\mathrm{id}$ on $C^\infty_{lin}(U, \mathbb{C})$, we have $D_0^\iota = -D_0$. Therefore

$$\begin{aligned} \operatorname{Res}(\mathcal{A}) &= \operatorname{Res} \circ (D - D^\iota)\left(C^\infty_{lin}(U, \mathbb{C})\right) = \\ &= (D_0 - D_0^\iota)\left(C^\infty_{lin}(U, \mathbb{C})\right) = D_0\left(C^\infty_{lin}(U, \mathbb{C})\right). \end{aligned}$$

Since the Hodge connection D is holonomic, this space coincides with the whole $C^\infty(M, \Lambda^1(M, \mathbb{C}))$ by Lemma 5.3.6. □

5.4.3. We now apply Lemma 5.4.1 to prove the following criterion for the linearity of the Hodge manifold structure on U defined by the Hodge connection $D : \Lambda^0(U, \mathbb{C}) \to \rho^*\Lambda^1(M, \mathbb{C})$.

Lemma. *The Hodge manifold structure on $U \subset \overline{T}M$ corresponding to a Hodge connection $D : \Lambda^0(U, \mathbb{C}) \to \rho^*\Lambda^1(M, \mathbb{C})$ is linear in the sense of 4.3.5 if and only if for every smooth function $f \in C^\infty(U, \mathbb{C})$ linear along the fibers of the projection $\rho : U \subset \overline{T}M \to M$ we have*

$$f = \frac{1}{2}\sigma\left((D - D^\iota)f\right), \tag{5.14}$$

where $\sigma : \rho^*\Lambda^1(M, \mathbb{C}) \to \Lambda^0(U, \mathbb{C})$ is the twisted tautological map introduced in (5.6), and $D^\iota : \Lambda^0(U, \mathbb{C}) \to \rho^*\Lambda^1(M, \mathbb{C})$ is the operator ι^*-conjugate to D, as in 5.4.1.

Proof. By Lemma 4.3.3 the Hodge manifold structure on U is linear if and only if for every $\alpha \in C^\infty(U, \rho^*\Lambda^1(M, \mathbb{C}))$ we have

$$\langle \varphi, j(\alpha)\rangle = \tau(\alpha), \tag{5.15}$$

where φ is the differential of the $U(1)$-action on U, $j : \Lambda^1(U, \mathbb{C}) \to \overline{\Lambda^1(U, \mathbb{C})}$ is the operator given by the quaternionic structure on U, and $\tau : \rho^*\Lambda^1(M, \mathbb{C}) \to \Lambda^0(U, \mathbb{C})$ is the tautological map sending a 1-form on M to the corresponding linear function on $\overline{T}M$, as in 4.3.2. Moreover, by Lemma 5.4.2 and Lemma 4.3.6 the equality (5.15) holds for all smooth sections $\alpha \in C^\infty(U, \rho^*\Lambda^1(M, \mathbb{C}))$ if and only if it holds for sections of the form

$$\alpha = \frac{\sqrt{-1}}{2}\delta_\rho((D - D^\iota)f), \tag{5.16}$$

where $f \in C^\infty_{lin}(U, \mathbb{C}) \subset C^\infty(U, \mathbb{C})$ is linear along the fibers of $\rho : U \to M$.

Let now $f \in C^\infty(U, \mathbb{C})$ be a smooth function on U linear along the fibers of $\rho : U \to M$, and let α be as in (5.16). Since φ is a vertical vector field on U over M, we have $\langle \varphi, j(\alpha)\rangle = \langle \varphi, \pi(j(\alpha))\rangle$, where $\pi : \Lambda^1(U, \mathbb{C}) \to \Lambda^1(U/M, \mathbb{C})$ is the canonical projection. By Lemma 5.4.1

$$\langle \varphi, j(\alpha)\rangle = \langle \varphi, \pi(j(\alpha))\rangle = \langle \varphi, d^r f\rangle. \tag{5.17}$$

Since the function f is linear along the fibers of $\rho : U \to M$, we can assume that $f = \sigma(\beta)$ for a smooth 1-form $\beta \in C^\infty(M, \Lambda^1(M, \mathbb{C}))$. Then by (5.7) and by Lemma 5.2.8 the right hand side of (5.17) is equal to

$$\langle \varphi, d^r f\rangle = \langle \varphi, d^r(\sigma(\beta))\rangle = \langle \varphi, \eta(\beta)\rangle = \sqrt{-1}\tau(\beta).$$

Therefore, (5.15) is equivalent to

$$\sqrt{-1}\tau(\beta) = \tau\left(\frac{\sqrt{-1}}{2}\delta_\rho((D - D^\iota)\sigma(\beta))\right). \tag{5.18}$$

But we have $\tau = \sigma \circ \zeta$, where $\zeta : \rho^*\Lambda^1(M, \mathbb{C}) \to \rho^*\Lambda^1(M, \mathbb{C})$ is the invulution introduced in (5.4). In particular, the map ζ is invertible, so that (5.18) is in turn equivalent to

$$\sigma(\beta) = \frac{1}{2}\sigma(\delta_\rho((D - D^\iota)\sigma(\beta))),$$

or, substituting back $f = \sigma(\beta)$, to

$$f = \frac{1}{2}\sigma(\delta_\rho((D - D^\iota)f)),$$

which is exactly the condition (5.14). □

5.4.4. Definition. A Hodge connection D on the pair $\langle U, M\rangle$ is called *linear* if it satisfies the condition 5.14.

We can now formulate and prove the following more useful version of Proposition 5.1.

Proposition 5.2 *Every linear Hodge connection*

$$D : \Lambda^0(U, \mathbb{C}) \to \rho^*\Lambda^1(M, \mathbb{C})$$

on the pair $\langle U, M\rangle$ defines a linear Hodge manifold structure on an open neighborhood $V \subset U$ of the zero section $M \subset U$, and the canonical projection $\rho : V_I \to M$ is holomorphic for the preferred complex structure V_I on V. Vice versa, every such linear Hodge manifold structure on U comes from a unique linear Hodge connection D on the pair $\langle U, M\rangle$.

Proof. By Proposition 5.1 and Lemma 5.4.3, to prove this proposition suffices to prove that if a Hodge connection $D : \Lambda^0(U, \mathbb{C}) \to \rho^*\Lambda^1(M, \mathbb{C})$ is linear, then it is holonomic in the sense of 2.4.2 on a open neighborhood $V \subset U$ of the zero section $M \subset U$. Lemma 5.3.6 reduces this to proving that the principal part $D_0 : S^1(M, \mathbb{C}) \to \Lambda^1(M, \mathbb{C})$ of a linear Hodge connection $D : \Lambda^0(U, \mathbb{C}) \to \rho^*\Lambda^1(M, \mathbb{C})$ is a bundle isomorphism.

Let $D : \Lambda^0(U, \mathbb{C}) \to \rho^*\Lambda^1(M, \mathbb{C})$ be such connection. By (5.14) we have

$$\frac{1}{2}\sigma \circ (D_0 - D_0^\iota) = \mathrm{id} : S^1(M, \mathbb{C}) \to \Lambda^1(M, \mathbb{C}) \to S^1(M, \mathbb{C}).$$

Since $\sigma : \Lambda^1(M, \mathbb{C}) \to S^1(M, \mathbb{C})$ is a bundle isomorphism, so is the bundle map $D_0 - D_0^\iota : S^1(M, \mathbb{C}) \to \Lambda^1(M, \mathbb{C})$. As in the proof of Lemma 5.4.2, we have $D_0 = -D_0^\iota$. Thus $D_0 = \frac{1}{2}(D_0 - D_0^\iota) : S^1(M, \mathbb{C}) \to \Lambda^1(M, \mathbb{C})$ also is a bundle isomorphism, which proves the proposition. □

6. Formal completions

6.1. Formal Hodge manifolds

6.1.1. Proposition 5.2 reduces the study of arbitrary regular Hodge manifolds to the study of connections of a certain type on a neighborhood $U \subset \overline{T}M$ of the zero section $M \subset \overline{T}M$ in the total space $\overline{T}M$ of the tangent bundle to a complex manifold M. To obtain further information we will now restrict our attention to the *formal* neighborhood of this zero section. This section contains the appropriate definitions. We study the convergence of our formal series in Section 10.

6.1.2. Let X be a smooth manifold and let $\mathrm{Bun}(X)$ be the category of smooth real vector bundles over X. Let also $\mathrm{Diff}(X)$ be the category with the same objects as $\mathrm{Bun}(X)$ but with differential operators as morphisms.

Consider a closed submanifold $Z \subset X$. For every two real vector bundles $\mathcal{E}$ and $\mathcal{F}$ on X the vector space $\mathrm{Hom}(\mathcal{E}, \mathcal{F})$ of bundle maps from $\mathcal{E}$ to $\mathcal{F}$ is naturally a module over the ring $C^\infty(X)$ of smooth functions on X. Let $\mathfrak{J}_Z \subset C^\infty(X)$ be the ideal of functions that vanish on Z and let $\mathrm{Hom}_Z(\mathcal{E}, \mathcal{F})$ be the $\mathfrak{J}_Z$-adic completion of the $C^\infty(X)$-module $\mathrm{Hom}(\mathcal{E}, \mathcal{F})$.

For any three bundles $\mathcal{E}, \mathcal{F}, \mathcal{G}$ the composition map

$$\mathrm{Mult} : \mathrm{Hom}(\mathcal{E}, \mathcal{F}) \otimes \mathrm{Hom}(\mathcal{F}, \mathcal{G}) \to \mathrm{Hom}(\mathcal{E}, \mathcal{G})$$

is $C^\infty(X)$-linear, hence extends to a map

$$\mathrm{Mult} : \mathrm{Hom}_Z(\mathcal{E}, \mathcal{F}) \otimes \mathrm{Hom}_Z(\mathcal{F}, \mathcal{G}) \to \mathrm{Hom}_Z(\mathcal{E}, \mathcal{G}).$$

Let $\mathrm{Bun}_Z(X)$ be the category with the same objects as $\mathrm{Bun}(X)$ and for every two objects $\mathcal{E}, \mathcal{F} \in \mathrm{Ob}\,\mathrm{Bun}(X)$ with $\mathrm{Hom}_Z(\mathcal{E}, \mathcal{F})$ as the space of maps between $\mathcal{F}$ anf $\mathcal{F}$. The category $\mathrm{Bun}_Z(X)$, as well as $\mathrm{Bun}(X)$, is an additive tensor category.

6.1.3. The space of differential operators $\mathrm{Diff}(\mathcal{E}, \mathcal{F})$ is also a $C^\infty(X)$ module, say, by left multiplication. Let $\mathrm{Diff}_Z(\mathcal{E}, \mathcal{F})$ be its $\mathfrak{J}_Z$-completion. The composition maps in $\mathrm{Diff}(X)$ are no longer $C^\infty(X)$-linear. However, they still are compatible with the $\mathfrak{J}_Z$-adic topology, hence extend to completions. Let $\mathrm{Diff}_Z(X)$ be the category with the same objects as $\mathrm{Bun}(X)$ and with $\mathrm{Diff}_Z(\mathcal{E}, \mathcal{F})$ as the space of maps between two objects $\mathcal{E}, \mathcal{F} \in \mathrm{Ob}\,\mathrm{Bun}(X)$.

By construction we have canonical *Z-adic completion functors*

$$\mathrm{Bun}(X) \to \mathrm{Bun}_Z(X) \text{ and } \mathrm{Diff}(X) \to \mathrm{Diff}_Z(X).$$

Call the categories $\mathrm{Bun}_Z(X)$ and $\mathrm{Diff}_Z(X)$ *the Z-adic completions* of the categories $\mathrm{Bun}(X)$ and $\mathrm{Diff}(X)$.

6.1.4. When the manifold X is equipped with a smooth action of compact Lie group G fixing the submanifold Z, the completion construction extends to the categories of G-equivariant bundles on M. When $G = U(1)$, the categories $\mathcal{WH}odge(X)$ and $\mathcal{WH}odge^{\mathcal{D}}(X)$ defined in 2.1.3 also admit canonical completions, denoted by $\mathcal{WH}odge_Z(X)$ and $\mathcal{WH}odge_Z^{\mathcal{D}}(X)$.

6.1.5. Assume now that the manifold X is equipped with a smooth $U(1)$-action fixing the smooth submanifold $Z \subset X$.

Definition. A *formal quaternionic structure* on X along the submanifold $Z \subset X$ is given by an algebra map

$$\mathrm{Mult} : \mathbb{H} \to \mathcal{E}nd_{\mathrm{Bun}_Z(X)}\left(\Lambda^1(X)\right)$$

from the algebra $\mathbb{H}$ to the algebra $\mathcal{E}nd_{\mathrm{Bun}_Z(X)}\left(\Lambda^1(X)\right)$ of endomorphisms of the cotangent bundle $\Lambda^1(X)$ in the category $\mathrm{Bun}_Z(X)$. A formal quaternionic structure is called *equivariant* if the map Mult is equivariant with respect to the natural $U(1)$-action on both sides.

6.1.6. Note that Lemma 2.4.1 still holds in the situation of formal completions. Consequently, everything in Section 2 carries over word-by-word to the case of formal quaternionic structures. In particular, by Lemma 2.4.3 giving a formal equivariant quaternionic structure on X along Z is equivalent to giving a pair $\langle \mathcal{E}, D\rangle$ of a Hodge bundle $\mathcal{E}$ on X and a holonomic algebra derivation $D : \Lambda^0(X) \to \mathcal{E}$ in $\mathcal{WH}odge_Z^{\mathcal{D}}(X)$.

6.1.7. The most convenient way to define Hodge manifold structures on X in a formal neighborhood of Z is by means of the Dolbeault complex, as in Proposition 3.1.

Definition. A *formal Hodge manifold structure* on X along Z is a pair of a Hodge bundle $\mathcal{E} \in \mathrm{Ob}\,\mathcal{WH}odge_Z(X)$ of weight 1 and an algebra derivation $D^{\bullet} : \Lambda^{\bullet}\mathcal{E} \to \Lambda^{\bullet+1}\mathcal{E}$ in $\mathcal{WH}odge_Z^{\mathcal{D}}(X)$ such that $D^0 : \Lambda^0(\mathcal{E}) \to \mathcal{E}$ is holonomic and $D^0 \circ D^1 = 0$.

6.1.8. Let $U \subset X$ be an open subset containing $Z \subset X$. For every Hodge manifold structure on U the Z-adic completion functor defines a formal Hodge manifold structure on X along Z. Call it *the Z-adic completion* of the given structure on U.

Remark. Note that a Hodge manifold structure on U is completely defined by the preferred and the complementary complex structures U_I, U_J, hence always real-analytic by the Newlander-Nirenberg Theorem. Therefore, if two Hodge manifold structures on U have the same completion, they coincide on every connected component of U intersecting Z.

6.2. Formal Hodge manifold structures on tangent bundles

6.2.1. Let now M be a complex manifold, and let $\overline{T}M$ be the total space of the complex-conjugate to the tangent bundle to M equipped with an action of $U(1)$by dilatation along the fibers of the projection $\rho : \overline{T}M \to M$. All the discussion above applies to the case $X = \overline{T}M$, $Z = M \subset \overline{T}M$. Moreover, the linearity condition in the form given in Lemma 4.3.3 generalizes immediately to the formal case.

Definition. A formal Hodge manifold structure on $\overline{T}M$ along M is called *linear* if for every smooth $(0,1)$-form $\alpha \in C^\infty(M, \Lambda^{0,1}(M))$ we have

$$\tau(\alpha) = \langle \varphi, j(\rho^*) \rangle \in C^\infty_M(\overline{T}M, \mathbb{C}),$$

where j is the map induced by the formal quaternionic structure on $\overline{T}M$ and φ and τ are as in Lemma 4.3.3.

6.2.2. As in the non-formal case, linear Hodge manifold structures on $\overline{T}M$ along $M \subset \overline{T}M$ can be described in terms of differential operators of certain type.

Definition. A *formal Hodge connection* on $\overline{T}M$ along $M \subset \overline{T}M$ is an algebra derivation

$$D : \Lambda^0(\overline{T}M, \mathbb{C}) \to \rho^*\Lambda^1(M, \mathbb{C})$$

in $\mathcal{WH}odge^{\mathcal{D}}_M(\overline{T}M)$ such that for every smooth function $f \in C^\infty(M, \mathbb{C})$ we have $D\rho^* = \rho^* d_M f$, as in (5.1). A formal Hodge connection is called *flat* if it extends to an algebra derivation

$$D : \rho^*\Lambda^{\cdot}(M, \mathbb{C}) \to \Lambda^{\cdot+1}(M, \mathbb{C})$$

in $\mathcal{WH}odge^{\mathcal{D}}_M(\overline{T}M)$ such that $D \circ D = 0$. A formal Hodge connection is called *linear* if it satisfies the condition (5.14) of Lemma 5.4.3, that is, for every function $f \in C^\infty_{lin}(\overline{T}M, \mathbb{C})$ linear along the fibers of the projection $\rho : \overline{T}M \to M$ we have

$$f = \frac{1}{2}\sigma\left((D - D^\iota)f\right),$$

where $\sigma : \rho^*\Lambda^1(M, \mathbb{C}) \to \Lambda^0(\overline{T}M, \mathbb{C})$ is the twisted tautological map introduced in (5.6), the automorphism $\iota : \overline{T}M \to \overline{T}M$ is the multiplication by $-1 \in \mathbb{C}$ on every fiber of the projection $\rho : \overline{T}M \to M$, and $D^\iota : \Lambda^0(\overline{T}M, \mathbb{C}) \to \rho^*\Lambda^1(M, \mathbb{C})$ is the operator ι^*-conjugate to D, as in 5.4.1.

The discussion in Section 5 generalizes immediately to the formal case and gives the following.

Lemma. *Linear formal Hodge manifold structures on $\overline{T}M$ along the zero section $M \subset \overline{T}M$ are in a natural one-to-one correspondence with linear flat formal Hodge connections on $\overline{T}M$ along M.*

6.3. The Weil algebra

6.3.1. Let, as before, M be a complex manifold and let $\overline{T}M$ be the total space of the complex conjugate to its tangent bundle, as in 4.2.5. In the remaining part of this section we give a description of the set of all formal Hodge connections on $\overline{T}M$ along M in terms of certain differential operators on M rather than on $\overline{T}M$. We call such operators *extended connections* on M (see 6.4.1 for the definition). Together with a complete classification of extended connections given in the next Section, this description provides a full classification of regular Hodge manifolds "in the formal neighborhood of the subset of $U(1)$-fixed points".

6.3.2. Before we define extended connections in Subsection 6.4), we need to introduce a certain algebra bundle in $\mathcal{WH}odge(M)$ which we call *the Weil algebra.* We begin with some preliminary facts.

Recall (see, e.g., [D1]) that every additive category $\mathcal{A}$ admits a canonical completion $\underleftarrow{\operatorname{Lim}} \mathcal{A}$ with respect to filtered projective limits. The category $\underleftarrow{\operatorname{Lim}} \mathcal{A}$ is also additive, and it is tensor if $\mathcal{A}$ was tensor. Objects of the canonical completion $\underleftarrow{\operatorname{Lim}} \mathcal{A}$ are called *pro-objects in $\mathcal{A}$.*

6.3.3. Let $\rho : \overline{T}M \to M$ be the canonical projection. Extend the pullback functor $\rho^* : \operatorname{Bun}(M) \to \operatorname{Bun}(\overline{T}M)$ to a functor

$$\rho^* : \operatorname{Bun}(M) \to \operatorname{Bun}_M(\overline{T}M)$$

to the M-adic completion $\operatorname{Bun}_M(\overline{T}M)$. The functor ρ^* admits a right adjoint direct image functor

$$\rho_* : \operatorname{Bun}_M(\overline{T}M) \to \underleftarrow{\operatorname{Lim}} \operatorname{Bun}(M).$$

Moreover, the functor ρ_* extends to a functor

$$\rho_* : \operatorname{Diff}_M(\overline{T}M) \to \underleftarrow{\operatorname{Lim}} \operatorname{Diff}(M).$$

Denote by $\mathcal{B}^0(M, \mathbb{C}) = \rho_*\Lambda^0(\overline{T}M)$ the direct image under the projection $\rho : \overline{T}M \to M$ of the trivial bundle $\Lambda^0(\overline{T}M)$ on $\overline{T}M$.

The compact Lie group $U(1)$ acts on $\overline{T}M$ by dilatation along the fibers, and the functor $\rho_* : \operatorname{Diff}_M(\overline{T}M) \to \underleftarrow{\operatorname{Lim}} \operatorname{Diff}(M)$ obviously extends to a functor $\rho_* : \mathcal{WH}odge^{\mathcal{D}}_M(\overline{T}M) \to \varprojlim \mathcal{WH}odge(M)$. The restriction of ρ_* to the subcategory $\mathcal{WH}odge_M(\overline{T}M) \subset \mathcal{WH}odge^{\mathcal{D}}_M(\overline{T}M)$ is adjoint on the right to the pullback functor $\rho^* : \mathcal{WH}odge(M) \to \mathcal{WH}odge_M(\overline{T}M)$.

6.3.4. The constant bundle $\Lambda^0(\overline{T}M)$ is canonically a Hodge bundle of weight 0. Therefore $\mathcal{B}^0(M,\mathbb{C}) = \rho_*\Lambda^0(M,\mathbb{C})$ is also a Hodge bundle of weight 0. Moreover, it is a commutative algebra bundle in $\varprojlim \mathcal{WH}odge_0(M)$. Let $S^1(M,\mathbb{C})$ be the conormal bundle to the zero section $M \subset \overline{T}M$ equipped with a Hodge bundle structure of weight 0 as in 5.3.2, and denote by $S^i(M,\mathbb{C})$ the i-th symmetric power of the Hodge bundle $S^1(M,\mathbb{C})$. Then the algebra bundle $\mathcal{B}^0(M,\mathbb{C})$ in $\varprojlim \mathcal{WH}odge_0(M)$ is canonically isomorphic

$$\mathcal{B}^0(M,\mathbb{C}) \cong \widehat{S}^{\bullet}(M,\mathbb{C})$$

to the completion $\widehat{S}^{\bullet}(M,\mathbb{C})$ of the symmetric algebra $S^{\bullet}(M,\mathbb{C})$ of the Hodge bundle $S^1(M,\mathbb{C})$ with respect to the augmentation ideal $S^{>0}(M,\mathbb{C})$.

Since the $U(1)$-action on M is trivial, the category $\mathcal{WH}odge(M)$ of Hodge bundles on M is equivalent to the category of pairs $\langle \mathcal{E}, \bar{-} \rangle$ of a complex bundle $\mathcal{E}$ equipped with a Hodge type bigrading

$$\mathcal{E} = \bigoplus_{p,q} \mathcal{E}^{p,q}$$

and a real structure $\bar{-} : \mathcal{E}^{p,q} \to \overline{\mathcal{E}^{q,p}}$. The Hodge type bigrading on $\mathcal{B}^0(M,\mathbb{C})$ is induced by the Hodge type bograding $S^1(M,\mathbb{C}) = S^{1,-1}(M) \oplus S^{-1,1}(M)$ on the generators subbundle $S^1(M,\mathbb{C}) \subset \mathcal{B}^0(M,\mathbb{C})$, which was described in 5.3.2.

Remark. The complex vector bundle $S^1(M,\mathbb{C})$ is canonically isomorphic to the cotangent bundle $\Lambda^1(M,\mathbb{C})$. However, the Hodge bundle structures on these two bundles are different (in fact, they have different weights).

6.3.5. Consider the pro-bundles

$$\mathcal{B}^{\bullet}(M,\mathbb{C}) = \rho_*\rho^*\Lambda^{\bullet}(M,\mathbb{C})$$

on M. The direct sum $\oplus\mathcal{B}^{\bullet}(M,\mathbb{C})$ is a graded algebra in $\varprojlim \operatorname{Bun}(M,\mathbb{C})$. Moreover, since for every $i \geq 0$ the bundle $\Lambda^i(M,\mathbb{C})$ is a Hodge bundle of weight i (see 2.1.7), $\mathcal{B}^i(M,\mathbb{C})$ is also a Hodge bundle of weight i. Denote by

$$\mathcal{B}^i(M,\mathbb{C}) = \bigoplus_{p+q=i} \mathcal{B}^{p,q}(M,\mathbb{C})$$

the Hodge type bigrading on $\mathcal{B}^i(M,\mathbb{C})$.

The Hodge bundle structures on $\mathcal{B}^\bullet(M,\mathbb{C})$ are compatible with the multiplication. By the projection formula

$$\mathcal{B}^\bullet(M,\mathbb{C}) \cong \mathcal{B}^0(M,\mathbb{C}) \otimes \Lambda^\bullet(M,\mathbb{C}),$$

and this isomorphism is compatible with the Hodge bundle structures on both sides.

Definition. Call the algebra $\mathcal{B}^\bullet(M,\mathbb{C})$ in $\underleftarrow{\mathrm{Lim}}\, \mathcal{WH}odge(M)$ *the Weil algebra* of the complex manifold M.

6.3.6. The canonical involution $\iota : \overline{T}M \to \overline{T}M$ induces an algebra involution $\iota^* : \mathcal{B}^\bullet(M,\mathbb{C}) \to \mathcal{B}^\bullet(M,\mathbb{C})$. It acts on generators as follows

$$\iota^* = -\mathsf{id} : S^1(M,\mathbb{C}) \to S^1(M,\mathbb{C}) \qquad \iota^* = \mathsf{id} : \Lambda^1(M,\mathbb{C}) \to \Lambda^1(M,\mathbb{C}).$$

For every operator $N : \mathcal{B}^p(M,\mathbb{C}) \to \mathcal{B}^q(M,\mathbb{C})$, p and q arbitrary, we will denote by

$$N^\iota = \iota^* \circ N \circ \iota^* : \mathcal{B}^p(M,\mathbb{C}) \to \mathcal{B}^q(M,\mathbb{C})$$

the operator ι^*-conjugate to N.

6.3.7. The twisted tautological map $\sigma : \rho^*\Lambda^1(M,\mathbb{C}) \to \Lambda^0(\overline{T}M,\mathbb{C})$ introduced in 5.6 induces via the functor ρ_* a map $\sigma : \mathcal{B}^1(M,\mathbb{C}) \to \mathcal{B}^0(M,\mathbb{C})$. Extend this map to a derivation

$$\sigma : \mathcal{B}^{\bullet+1}(M,\mathbb{C}) \to \mathcal{B}^\bullet(M,\mathbb{C})$$

by setting $\sigma = 0$ on $S^1(M,\mathbb{C}) \subset \mathcal{B}^0(M,\mathbb{C})$. The derivation $\mathcal{B}^{\bullet+1}(M,\mathbb{C}) \to \mathcal{B}^\bullet(M,\mathbb{C})$ is not weakly Hodge. However, it is real with respect to the real structure on the Weil algebra $\mathcal{B}^\bullet(M,\mathbb{C})$.

6.3.8. By definition of the twisted tautological map (5.6, 4.3.2), the derivation $\sigma : \mathcal{B}^{\bullet+1}(M,\mathbb{C}) \to \mathcal{B}^\bullet(M,\mathbb{C})$ maps the subbundle $\Lambda^1(M,\mathbb{C}) \subset \mathcal{B}^1(M,\mathbb{C})$ to the subbundle $S^1(M,\mathbb{C}) \subset \mathcal{B}^0(M,\mathbb{C})$ and defines a complex vector bundle isomorphism $\sigma : \Lambda^1(M,\mathbb{C}) \to S^1(M,\mathbb{C})$. To describe this isomorphism explicitly, recall that sections of the bundle $\mathcal{B}^0(M,\mathbb{C})$ are the same as formal germs along $M \subset \overline{T}M$ of smooth functions on the manifold $\overline{T}M$. The sections of the subbundle $S^1(M,\mathbb{C}) \subset \mathcal{B}^0(M,\mathbb{C})$ form the subspace of functions linear along the fibers of the canonical projection $\rho : \overline{T}M \to M$. The isomorphism $\sigma : \Lambda^1(M,\mathbb{C}) \to S^1(M,\mathbb{C})$ induces an isomorphism between the space of smooth 1-forms on the manifold M and the space of smooth functions on $\overline{T}M$ linear long the fibers of $\rho : \overline{T}M \to M$. This isomorphism coincides with the tautological one on the subbundle $\Lambda^{1,0} \subset \Lambda^1(M,\mathbb{C})$, and it is minus the tautological isomorphism on the subbundle $\Lambda^{0,1} \subset \Lambda^1(M,\mathbb{C})$.

Denote by

$$C = \sigma^{-1} : S^1(M, \mathbb{C}) \to \Lambda^1(M, \mathbb{C})$$

the bundle isomorphism inverse to σ. Note that the complex vector bundle isomorphism $\sigma : \Lambda^1(M, \mathbb{C}) \to S^1(M, \mathbb{C})$ is real. Moreover, it sends the subbundle $\Lambda^{1,0}(M) \subset \Lambda^1(M, \mathbb{C})$ to $S^{1,-1}(M) \subset S^1(M, \mathbb{C})$, and it sends $\Lambda^{0,1}(M)$ to $S^{-1,1}(M)$. Therefore the inverse isomorphism $C : S^1(M, \mathbb{C}) \to \Lambda^1(M, \mathbb{C})$ is weakly Hodge. It coincides with the tautological isomorphism on the subbundle $S^{1,-1} \subset S^1(M, \mathbb{C})$, and it equals minus the tautological isomorphism on the subbundle $S^{-1,1} \subset S^1(M, \mathbb{C})$.

6.4. Extended connections

6.4.1. We are now ready to introduce the extended connections. Keep the notation of the last subsection.

Definition. An *extended connection* on a complex manifold M is a differential operator $D : S^1(M, \mathbb{C}) \to \mathcal{B}^1(M, \mathbb{C})$ which is weakly Hodge in the sense of 2.1.3 and satisfies

$$D(fa) = fDa + a \otimes df \tag{6.1}$$

for any smooth function f and a local section a of the pro-bundle $\mathcal{B}^0(M, \mathbb{C})$.

6.4.2. Let D be an extended connection on the manifold M. By 6.3.5 we have canonical bundle isomorphisms

$$\mathcal{B}^1(M, \mathbb{C}) \cong \mathcal{B}^0(M, \mathbb{C}) \otimes \Lambda^1(M, \mathbb{C}) \cong \bigoplus_{i \geq 0} S^i(M, \mathbb{C}) \otimes \Lambda^1(M, \mathbb{C}).$$

Therefore the operator $D : S^1 \to \mathcal{B}^1$ decomposes

$$D = \sum_{p \geq 0} D_p, \quad D_p : S^1(M, \mathbb{C}) \to S^i(M, \mathbb{C}) \otimes \Lambda^1(M, \mathbb{C}). \tag{6.2}$$

By (6.1) all the components D_p except for the D_1 are weakly Hodge bundle maps on M, while

$$D_1 : S^1(M, \mathbb{C}) \to S^1(M, \mathbb{C}) \otimes \Lambda^1(M, \mathbb{C})$$

is a connection in the usual sense on the Hodge bundle $S^1(M, \mathbb{C})$.

Definition. The weakly Hodge bundle map $D_0 : S^1(M, \mathbb{C}) \to \Lambda^1(M, \mathbb{C})$ is called *the principal part* of the extended connection D on M. The connection D_1 is called *the reduction* of the extended connection D.

6.4.3. Extended connection on M are related to formal Hodge connections on the total space $\overline{T}M$ of the complex-conjugate to the tangent bundle to M by means of the direct image functor

$$\rho_* : \mathcal{WH}odge^{\mathcal{D}}_M(\overline{T}M) \to \underset{\longleftarrow}{\mathrm{Lim}}\, \mathcal{WH}odge^{\mathcal{D}}(M).$$

Namely, let $D : \Lambda^0(M, \mathbb{C}) \to \rho^*\Lambda^1(M, \mathbb{C})$ be a formal Hodge connection on $\overline{T}M$ along M in the sense of 6.2.2. The restriction of the operator

$$\rho_* D : \mathcal{B}^0(M, \mathbb{C}) \to \mathcal{B}^1(M, \mathbb{C})$$

to the generators subbundle $S^1(M, \mathbb{C}) \subset \mathcal{B}^0(M, \mathbb{C})$ is then an extended connection on M in the sense of 6.4.1. The principal part $D_0 : S^1(M, \mathbb{C}) \to \Lambda^1(M, \mathbb{C})$ of the Hodge connection D in the sense of 5.3.4 coincides with the principal part of the extended connection $\rho_* D$.

6.4.4. We now write down the counterparts of the flatness and linearity conditions on a Hodge connection on $\overline{T}M$ for the associated extended connection on M. We begin with the linearity condition 5.4.4. Let $D : S^1(M, \mathbb{C}) \to \mathcal{B}^1(M, \mathbb{C})$ be an extended connection on M, let $\sigma : \mathcal{B}^{\bullet+1}(M, \mathbb{C}) \to \mathcal{B}^{\bullet}(M, \mathbb{C})$ be the algebra derivation introduced in 6.3.7, and let

$$D^{\iota} : \mathcal{B}^{\bullet}(M, \mathbb{C}) \to \mathcal{B}^{\bullet+1}(M, \mathbb{C})$$

be the operator ι^*-conjugate to D as in 6.3.6.

Definition. An extended connection D is called *linear* if for every local section f of the bundle $S^1(M, \mathbb{C})$ we have

$$f = \frac{1}{2}\sigma((D - D^{\iota})f).$$

This is, of course, the literal rewriting of Definition 5.4.4. In particular, a formal Hodge connection D on $\overline{T}M$ is linear if and only if so is the extended connection $\rho_* D$ on M.

6.4.5. Next we rewrite the flatness condition 5.1.7. Again, let

$$D : S^1(M, \mathbb{C}) \to \mathcal{B}^1(M, \mathbb{C})$$

be an extended connection on M. Since the algebra pro-bundle $\mathcal{B}^0(M, \mathbb{C})$ is freely generated by the subbundle $S^1(M, \mathbb{C}) \subset \mathcal{B}^1(M, \mathbb{C})$, by (6.1) the operator $D : S^1(M, \mathbb{C}) \to \mathcal{B}^1(M, \mathbb{C})$ extends uniquely to an algebra derivation

$$D : \mathcal{B}^0(M, \mathbb{C}) \to \mathcal{B}^1(M, \mathbb{C}).$$

Moreover, we can extend this derivation even further to a derivation of the Weil algebra

$$D : \mathcal{B}^{\bullet}(M, \mathbb{C}) \to \mathcal{B}^{\bullet+1}(M, \mathbb{C})$$

by setting

$$D(f \otimes \alpha) = Df \otimes \alpha + f \otimes d\alpha \tag{6.3}$$

for any smooth section $f \in C^{\infty}(M, \mathcal{B}^0(M, \mathbb{C}))$ and any smooth form $\alpha \in C^{\infty}(M, \Lambda^{\bullet}(M, \mathbb{C}))$. We will call this extension *the derivation, associated to the extended connection D*.

Vice versa, the Weil algebra $\mathcal{B}^{\bullet}(M, \mathbb{C})$ is generated by the subbundles

$$S^1(M, \mathbb{C}), \Lambda^1(M, \mathbb{C}) \subset \mathcal{B}^{\bullet}(M, \mathbb{C}).$$

Moreover, for every algebra derivation $D : \mathcal{B}^{\bullet}(M, \mathbb{C}) \to \mathcal{B}^{\bullet+1}(M, \mathbb{C})$ the condition (6.3) completely defines the restriction of D to the generator subbundle $\Lambda^1(M, \mathbb{C}) \subset \mathcal{B}^1(M, \mathbb{C})$. Therefore an algebra derivation $D : \mathcal{B}^{\bullet}(M, \mathbb{C}) \to \mathcal{B}^{\bullet}(M, \mathbb{C})$ satisfying (6.3) is completely determined by its restriction to the generators subbundle $S^1(M, \mathbb{C}) \to \mathcal{B}^1(M, \mathbb{C})$. If the derivation D is weakly Hodge, then this restriction is an extended connection on M.

6.4.6. Definition. The extended connection D is called *flat* if the associated derivation satisfies $D \circ D = 0$.

If a formal Hodge connection D on $\overline{T}M$ is flat in the sense of 5.1.7, then we have a derivation $D : \rho^*\Lambda^{\bullet}(M, \mathbb{C}) \to \rho^*\Lambda^{\bullet+1}(M, \mathbb{C})$ such that $D \circ D = 0$. The associated derivation $\rho_* D : \mathcal{B}^{\bullet}(M, \mathbb{C}) \to \mathcal{B}^{\bullet+1}(M, \mathbb{C})$ satisfies (6.3). Therefore the extended connection $\rho_* D : S^1(M, \mathbb{C}) \to \mathcal{B}^1(M, \mathbb{C})$ is also flat.

6.4.7. It turns out that one can completely recover a Hodge connection D on $\overline{T}M$ from the corresponding extended connection $\rho_* D$ on M. More precisely, we have the following.

Lemma. *The correspondence $D \mapsto \rho_* D$ is a bijection between the set of formal Hodge connections on $\overline{T}M$ along $M \subset \overline{T}M$ and the set of extended connections on M. A connection D is flat, resp. linear if and only if $\rho_* D$ is flat, resp. linear.*

Proof. To prove the first claim of the lemma, it suffices to prove that every extended connection on M comes from a unique formal Hodge connection on the pair $\langle \overline{T}M, M \rangle$. In general, the functor ρ_* is not fully faithful on the category $\operatorname{Diff}(M)$, in other words, it does not induce an isomorphism on the spaces of differential operators between vector bundles on $\overline{T}M$. However, for every complex vector bundle $\mathcal{E}$ on $\overline{T}M$ the functor ρ_* does induce an isomorphism

$$\rho_* : \mathrm{Der}_M(\Lambda^0(M,\mathbb{C}),\mathcal{E}) \cong \mathrm{Der}_{\mathcal{B}^0(M,\mathbb{C})}(\mathcal{B}^0(M,\mathbb{C}),\rho_*\mathcal{E})$$

between the space of *derivations* from $\Lambda^0(M,\mathbb{C})$ to $\mathcal{F}$ completed along $M \subset \overline{T}M$ and the space of derivations from the algebra $\mathcal{B}^0(M,\mathbb{C}) = \rho_*\Lambda^0(M,\mathbb{C})$ to the $\mathcal{B}^0(M,\mathbb{C})$-module $\rho_*\mathcal{E}$. Therefore every derivation

$$D' : \mathcal{B}^0(M,\mathbb{C}) \to \mathcal{B}^1(M,\mathbb{C}) = \rho_*\rho^*\Lambda^1(M,\mathbb{C})$$

must be of the form $D' = \rho_* D$ for some derivation

$$D : \Lambda^0(\overline{T}M,\mathbb{C}) \to \rho^*\Lambda^1(M,\mathbb{C})$$

It is easy to check that D is a Hodge connection if and only if $D' = \rho_* D$ is weakly Hodge and satisfies (6.1). By 6.4.5 the space of all such derivations $D' : \mathcal{B}^0(M,\mathbb{C}) \to \mathcal{B}^1(M,\mathbb{C})$ coincides with the space of all extended connections on M, which proves the first claim of the lemma.

Analogously, for every extended connection $D' = \rho_* D$ on M, the canonical extension of the operator D' to an algebra derivation $D' : \mathcal{B}^\bullet(M,\mathbb{C}) \to \mathcal{B}^{\bullet+1}(M,\mathbb{C})$ constructed in 6.4.5 must be of the form $\rho_* D$ for a certain weakly Hodge differential operator $D : \rho^*\Lambda^\bullet(M,\mathbb{C}) \to \Lambda^{\bullet+1}(M,\mathbb{C})$. If the extended connection D' is flat, then $D' \circ D' = 0$. Therefore $D \circ D = 0$, which means that the Hodge connection D is flat. Vice versa, if the Hodge connection D is flat, then it extends to a weakly Hodge derivation $D : \rho^*\Lambda^\bullet(M,\mathbb{C}) \to \Lambda^{\bullet+1}(M,\mathbb{C})$ so that $D \circ D = 0$. The equality $D \circ D = 0$ implies, in particular, that the operator $\rho_* D$ vanishes on the sections of the form

$$Df = df \in C^\infty(M,\Lambda^1(M,\mathbb{C})) \subset C^\infty(M,\mathcal{B}^1(M,\mathbb{C})),$$

where $f \in C^\infty(M,\mathbb{C})$ is a smooth function on M. Therefore $\rho_* D$ coincides with the de Rham differential on the subbundle $\Lambda^1(M,\mathbb{C}) \subset \mathcal{B}^1(M,\mathbb{C})$. Hence by 6.4.5 it is equal to the canonical derivation $D' : \mathcal{B}^\bullet(M,\mathbb{C}) \to \mathcal{B}^{\bullet+1}(M,\mathbb{C})$. Since $D \circ D = 0$, we have $D' \circ D' = 0$, which means that the extended connection D' is flat.

Finally, the equivalence of the linearity conditions on the Hodge connection D and on the extended connection $D' = \rho_* D$ is trivial and has already been noted in 6.4.4. □

This lemma together with Lemma 6.2.2 reduces the classification of linear formal Hodge manifold structures on $\overline{T}M$ along the zero section $M \subset \overline{T}M$ to the classification of extended connections on the manifold M itself.

7. Preliminaries on the Weil algebra

7.1. The total de Rham complex

7.1.1. Before we proceed further in the study of extended connections on a complex manifold M, we need to establish some linear-algebraic facts on the structure of the Weil algebra $\mathcal{B}^\bullet(M, \mathbb{C})$ defined in 6.3.5. We also need to introduce an auxiliary Hodge bundle algebra on M which we call *the total Weil algebra*. This is the subject of this section. Most of the facts here are of a technical nature, and the reader is advised to skip this section until needed.

7.1.2. We begin with introducing and studying a version of the de Rham complex of a complex manifold M which we call *the total de Rham complex*. Let M be a smooth complex $U(1)$-manifold. Recall that by 2.1.7 the de Rham complex $\Lambda^\bullet(M, \mathbb{C})$ of the complex manifold M is canonically a Hodge bundle algebra on M. Let $\Lambda^\bullet_{tot}(M) = \Gamma(\Lambda^\bullet(M, \mathbb{C}))$ be the weight 0 Hodge bundle obtained by applying the functor Γ defined in 2.1.4 to the de Rham algebra $\Lambda^\bullet(M, \mathbb{C})$. By 1.4.9 the bundle $\Lambda^\bullet_{tot}(M)$ carries a canonical algebra structure. By 2.1.7 the de Rham differential d_M is weakly Hodge. Therefore it induces an algebra derivation $d_M : \Lambda^\bullet_{tot}(M) \to \Lambda^{\bullet+1}_{tot}(M)$ which is compatible with the Hodge bundle structure and satisfies $d_M \circ d_M = 0$.

Definition. The weight 0 Hodge bundle algebra $\Lambda^\bullet_{tot}(M)$ is called *the total de Rham complex* of the complex manifold M.

7.1.3. By definition

$$\Lambda^i_{tot}(M) = \Gamma(\Lambda^i(M, \mathbb{C})) = \Lambda^i(M, \mathbb{C}) \otimes \mathcal{W}^*_i,$$

where $\mathcal{W}^*_i = S^i\mathcal{W}^*_1$ is the symmetric power of the $\mathbb{R}$-Hodge structure $\mathcal{W}^*_1$, as in 1.4.4. To describe the structure of the algebra $\Lambda^\bullet_{tot}(M)$, we will use the following well-known general fact. (For the sake of completeness, we have included a sketch of its proof, see 7.1.12.)

Lemma. *Let A, B be two objects in an arbitrary $\mathbb{Q}$-linear symmetric tensor category $\mathcal{A}$, and let $C^\bullet = S^\bullet(A \otimes B)$ be the sum of symmetric powers of the object $A \otimes B$. Note that the object $C^\bullet$ is naturally a commutative algebra in $\mathcal{A}$ in the obvious sense. Let also $\widetilde{C}^\bullet = \bigoplus_k S^k A \otimes S^k B$ with the obvious commutative algebra structure. The isomorphism $C^1 \cong \widetilde{C}^1 \cong A \otimes B$ extends to a surjective algebra map $C^\bullet \to \widetilde{C}^\bullet$, and its kernel $\mathfrak{J}^\bullet \subset C^\bullet$ is the ideal generated by the subobject $\mathfrak{J}^2 = \Lambda^2(A) \otimes \Lambda^2(B) \subset S^2(A \otimes B)$.*

7.1.4. The category of complexes of Hodge bundles on M is obviously $\mathbb{Q}$-linear and tensor. Applying Lemma 7.1.3 to $A = \mathcal{W}^*_1$, $B = \Lambda^1(M, \mathbb{C})[1]$

immediately gives the following.

Lemma. *The total de Rham complex $\Lambda^{\bullet}_{tot}(M)$ of the complex manifold M is generated by its first component $\Lambda^1_{tot}(M)$, and the kernel of the canonical surjective algebra map*

$$\Lambda^{\bullet}(\Lambda^1_{tot}(M)) \to \Lambda^{\bullet}_{tot}(M)$$

from the exterior algebra of the bundle $\Lambda^1_{tot}(M)$ to $\Lambda^{\bullet}_{tot}(M)$ is the ideal generated by the subbundle

$$\Lambda^2\mathcal{W}_1 \otimes S^2(\Lambda^1(M,\mathbb{C})) \subset S^2(\Lambda^1_{tot}(M)).$$

7.1.5. We can describe the Hodge bundle $\Lambda^1_{tot}(M)$ more explicitly in the following way. By definition, as a $U(1)$-equivariant complex vector bundle it equals

$$\Lambda^1_{tot}(M) = \Lambda^1(M,\mathbb{C}) \otimes \mathcal{W}_1^* = (\Lambda^{1,0}(M)(1) \oplus \Lambda^{0,1}(M)(0)) \otimes (\mathbb{C}(0) \oplus \mathbb{C}(-1)),$$

where $\Lambda^{p,q}(M)(i)$ is the $U(1)$-equivariant bundle $\Lambda^{p,q}(M)$ tensored with the 1-dimensional representation of weight i, and $\mathbb{C}(i)$ is the constant $U(1)$-bundle corresponding to the representation of weight i. If we denote

$$\begin{aligned} S^1(M,\mathbb{C}) &= \Lambda^{1,0}(M)(1) \oplus \Lambda^{0,1}(M)(-1) \subset \Lambda^1_{tot}(M), \\ \Lambda^1_{ll}(M) &= \Lambda^{1,0}(M) \subset \Lambda^1_{tot}(M), \\ \Lambda^1_{rr}(M) &= \Lambda^{0,1}(M) \subset \Lambda^1_{tot}(M), \end{aligned}$$

then we have

$$\Lambda^1_{tot}(M) = S^1(M,\mathbb{C}) \oplus \Lambda^1_{ll}(M) \oplus \Lambda^1_{rr}(M).$$

The complex conjugation $\bar{-} : \Lambda^1_{tot}(M) \to \iota^*\overline{\Lambda^1_{tot}(M)}$ preserves the subbundle

$$S^1(M,\mathbb{C}) \subset \Lambda^1_{tot}(M,\mathbb{C})$$

and interchanges $\Lambda^1_{ll}(M)$ and $\Lambda^1_{rr}(M)$.

7.1.6. If the $U(1)$-action on the manifold M is trivial, then Hodge bundles are the same as bigraded complex vector bundles with a real structure. In this case the Hodge bigrading on the Hodge bundle $\Lambda^1_{tot}(M,\mathbb{C})$ is given by

$$\begin{aligned} (\Lambda^1_{tot}(M))^{1,-1} &= S^{1,-1}(M,\mathbb{C}) = \Lambda^{1,0}(M)(1), \\ (\Lambda^1_{tot}(M))^{-1,1} &= S^{-1,1}(M,\mathbb{C}) = \Lambda^{0,1}(M)(-1), \\ (\Lambda^1_{tot}(M))^{0,0} &= \Lambda^1_{ll}(M) \oplus \Lambda^1_{rr}(M) = \Lambda^1(M,\mathbb{C}). \end{aligned}$$

Under these identifications, the real structure on $\Lambda^1_{tot}(M, \mathbb{C})$ is minus the one induced by the usual real structure on the complex vector bundle $\Lambda^1(M, \mathbb{C})$.
Remark. The Hodge bundle $S^1(M, \mathbb{C})$ is canonically isomorphic to the conormal bundle to the zero section $M \subset \overline{T}M$, which we have described in 5.3.2.
7.1.7. Recall now that we have defined in 1.4.7 canonical embeddings $\gamma_l, \gamma_r : \mathcal{W}_p^* \to \mathcal{W}_k^*$ for every $0 \le p \le k$. Since $\mathcal{W}_0^* = \mathbb{C}$, for every $p, q \ge 0$ we have by (1.3) a short exact sequence

$$0 \longrightarrow \mathbb{C} \longrightarrow \mathcal{W}_p^* \oplus \mathcal{W}_q^* \xrightarrow{\gamma_l \oplus \gamma_r} \mathcal{W}_{p+q}^* \longrightarrow 0 \tag{7.1}$$

of complex vector spaces. Recall also that the embeddings γ_l, γ_r are compatible with the natural maps $\mathsf{can} : \mathcal{W}_p^* \otimes \mathcal{W}_q^* \to \mathcal{W}_{p+q}^*$. Therefore the subbundles defined by

$$\Lambda_l^k(M) = \bigoplus_{0 \le p \le k} \gamma_l(\mathcal{W}_p^*) \otimes \Lambda^{p,k-p}(M) \subset \Lambda_{tot}^k(M) = \bigoplus_{0 \le p \le k} \mathcal{W}_k^* \otimes \Lambda^{p,k-p}(M)$$
$$\Lambda_r^k(M) = \bigoplus_{0 \le p \le k} \gamma_r(\mathcal{W}_p^*) \otimes \Lambda^{k-p,p}(M) \subset \Lambda_{tot}^k(M) = \bigoplus_{0 \le p \le k} \mathcal{W}_k^* \otimes \Lambda^{k-p,p}(M)$$

are actually subalgebras in the total de Rham complex $\Lambda_{tot}^{\bullet}(M)$.
7.1.8. To describe the algebras $\Lambda_l^{\bullet}(M)$ and $\Lambda_r^{\bullet}(M)$ explicitly, note that we obviously have $\Lambda_{tot}^{\bullet}(M) = \Lambda_l^{\bullet}(M) + \Lambda_r^{\bullet}(M)$. Moreover, in the notation of 7.1.5 we have

$$\Lambda_l^1(M) = S^1(M, \mathbb{C}) \oplus \Lambda_{ll}^1(M) \subset \Lambda_{tot}^1(M),$$
$$\Lambda_r^1(M) = S^1(M, \mathbb{C}) \oplus \Lambda_{rr}^1(M) \subset \Lambda_{tot}^1(M).$$

By Lemma 7.1.3, the algebra

$$\Lambda_l^{\bullet}(M) = \left(\bigoplus_p \mathcal{W}_p^* \otimes \Lambda^{p,0}(M) \right) \otimes \left(\bigoplus_q \Lambda^{0,q}(M) \right)$$

is the subalgebra in the total de Rham complex $\Lambda_{tot}^{\bullet}(M)$ generated by $\Lambda_l^1(M)$, and the ideal of relations is generated by the subbundle

$$S^2(\Lambda^{1,0}(M)) \otimes \Lambda^2(\mathcal{W}_1^*) \subset \Lambda^2(\Lambda_l^1(M)).$$

Analogously, the subalgebra $\Lambda_r^{\bullet}(M) \subset \Lambda_{tot}^1(M)$ is generated by $\Lambda_r^1(M)$, and the relations are generated by

$$S^2(\Lambda^{0,1}(M)) \otimes \Lambda^2(\mathcal{W}_1^*) \subset \Lambda^2(\Lambda_r^1(M)).$$

7.1.9. We will also need to consider the ideals in these algebras defined by

$$\Lambda^k_{ll}(M) = \bigoplus_{1 \leq p \leq k} \gamma_l(\mathcal{W}^*_p) \otimes \Lambda^{p,k-p}(M) \subset \Lambda^k_l(M)$$

$$\Lambda^k_{rr}(M) = \bigoplus_{1 \leq p \leq k} \gamma_r(\mathcal{W}^*_p) \otimes \Lambda^{k-p,p}(M) \subset \Lambda^k_r(M)$$

The ideal $\Lambda^{\bullet}_{ll}(M) \subset \Lambda^{\bullet}_l(M)$ is generated by the subbundle $\Lambda^1_{ll}(M) \subset \Lambda^1_l(M)$, and the ideal $\Lambda^{\bullet}_{rr}(M) \subset \Lambda^{\bullet}_r(M)$ is generated by the subbundle $\Lambda^1_{rr}(M) \subset \Lambda^1_r(M)$.

7.1.10. Denote by $\Lambda^{\bullet}_o(M) = \Lambda^{\bullet}_l(M) \cap \Lambda^{\bullet}_r(M) \subset \Lambda^{\bullet}_{tot}(M)$ the intersection of the subalgebras $\Lambda^{\bullet}_l(M)$ and $\Lambda^{\bullet}_r(M)$. Unlike either of these subalgebras, the subalgebra $\Lambda^{\bullet}_o(M) \subset \Lambda^{\bullet}_{tot}(M)$ is compatible with the weight 0 Hodge bundle structure on the total de Rham complex. By (7.1) we have a short exact sequence

$$0 \longrightarrow \Lambda^{\bullet}(M, \mathbb{C}) \longrightarrow \Lambda^{\bullet}_l(M) \oplus \Lambda^{\bullet}_r(M) \longrightarrow \Lambda^{\bullet}_{tot}(M) \longrightarrow 0 \tag{7.2}$$

of complex vector bundles on M. Therefore the algebra $\Lambda^{\bullet}_o(M)$ is isomorphic, as a complex bundle algebra, to the usual de Rham complex $\Lambda^{\bullet}(M, \mathbb{C})$. As a Hodge bundle algebra it is canonically isomorphic to the exterior algebra of the Hodge bundle $S^1(M, \mathbb{C})$ of weight 0 on the manifold M.

Finally, note that the short exact sequence (7.2) induces a direct sum decomposition

$$\Lambda^{\bullet}_{tot}(M) \cong \Lambda^{\bullet}_{ll}(M) \oplus \Lambda^{\bullet}_o(M) \oplus \Lambda^{\bullet}_{rr}(M).$$

7.1.11. Remark. The total de Rham complex $\Lambda^{\bullet}_{tot}(M)$ is related to Simpson's theory of Higgs bundles (see [S3]) in the following way. Recall that Simpson has proved that every (sufficiently stable) complex bundle $\mathcal{E}$ on a compact complex manifold M equipped with a flat connection ∇ admits a unique Hermitian metric h such that ∇ and the 1-form $\theta = \nabla - \nabla^h \in C^{\infty}(M, \Lambda^1(\mathcal{E}nd\,\mathcal{E}))$ satisfy the so-called *harmonicity condition.* He also has shown that this condition is equivalent to the vanishing of a certain curvature-like tensor $R \in \Lambda^2(M, \mathcal{E}nd\,\mathcal{E})$ which he associated canonically to every pair $\langle \nabla, \theta \rangle$.

Recall that flat bundles $\langle \mathcal{E}, \nabla \rangle$ on the manifold M are in one-to-one correspondence with free differential graded modules $\mathcal{E} \otimes \Lambda^{\bullet}(M, \mathbb{C})$ over the de Rham complex $\Lambda^{\bullet}(M, \mathbb{C})$. It turns out that complex bundles $\mathcal{E}$ equipped with a flat connection ∇ and a 1-form $\theta \in C^{\infty}(M, \Lambda^1(\mathcal{E}nd\,\mathcal{E}))$ such that

Simpson's tensor R vanishes are in natural one-to-one correspondence with free differential graded modules $\mathcal{E}\otimes\Lambda^{\bullet}_{tot}(M)$ over the total de Rham complex $\Lambda^{\bullet}_{tot}(M)$. Moreover, a pair $\langle\theta,\nabla\rangle$ comes from a variation of pure $\mathbb{R}$-Hodge structure on $\mathcal{E}$ if and only if there exists a Hodge bundle structure on $\mathcal{E}$ such that the product Hodge bundle structure on the free module $\mathcal{E}\otimes\Lambda^{\bullet}_{tot}(M)$ is compatible with the differential.

7.1.12. *Proof of Lemma* 7.1.3. For every $k\geq 0$ let $G=\Sigma_k\times\Sigma_k$ be the product of two copies of the symmetric group Σ_k on k letters. Let $\mathcal{V}_k$ be the $\mathbb{Q}$-representation of G_k induced from the trivial representation of the diagonal subgroup $\Sigma_k\subset G_k$. The representation $\mathcal{V}_k$ decomposes as

$$\mathcal{V}_k=\bigoplus_V V\boxtimes V,$$

where the sum is over the set of irreducible representations V of Σ_k. We obviously have

$$\mathcal{C}^k=\operatorname{Hom}_{G_k}\left(\mathcal{V}_k,A^{\otimes k}\otimes B^{\otimes k}\right)=\bigoplus_V\operatorname{Hom}_{\Sigma_k}\left(V,A^{\otimes k}\right)\otimes\operatorname{Hom}_{\Sigma_k}\left(V,A^{\otimes k}\right).$$

Let $\mathfrak{J}^{\bullet}\subset\mathcal{C}^{\bullet}$ be the ideal generated by $\Lambda^2A\otimes\Lambda^2B\subset S^2(A\otimes B)$. It is easy to see that

$$\mathfrak{J}^k=\sum_{1\leq l\leq k-1}\bigoplus_V\operatorname{Hom}_{\Sigma_k}\left(V,A^{\otimes k}\right)\otimes\operatorname{Hom}_{\Sigma_k}\left(V,A^{\otimes k}\right)\subset\mathcal{C}^k,$$

where the first sum is taken over the set of $k-1$ subgroups $\Sigma_2\subset\Sigma_k$, the l-th one transposing the l-th and the $l+1$-th letter, while the second sum is taken over all irreducible constituents V of the representation of Σ_k induced from the sign representation of the corresponding $\Sigma_2\subset\Sigma_k$. Now, there is obviously only one irreducible representation of Σ_k which is not encountered as an index in this double sum, namely, the trivial one. Hence $\mathcal{C}^k/\mathfrak{J}^k=S^kA\otimes S^kB$, which proves the lemma. □

7.2. The total Weil algebra

7.2.1. Assume from now on that the $U(1)$-action on the complex manifold M is trivial. We now turn to studying the Weil algebra of the manifold M. Let $S^1(M,\mathbb{C})=S^{1,-1}(M,\mathbb{C})\oplus S^{-1,1}(M,\mathbb{C})$ be the weight 0 Hodge bundle on M introduced in 5.3.2. To simplify notation, denote

$$S^{\bullet}=\widehat{S}^{\bullet}(S^1(M,\mathbb{C}))$$
$$\Lambda^{\bullet}=\Lambda^{\bullet}(M,\mathbb{C}),$$

where $\widehat{S}^{\bullet}$ is the completed symmetric power, and let

$$\mathcal{B}^{\bullet} = \mathcal{B}^{\bullet}(M, \mathbb{C}) = S^{\bullet} \otimes \Lambda^{\bullet}$$

be the Weil algebra of the complex manifold M introduced in 6.3.5. Recall that the algebra $\mathcal{B}^{\bullet}$ carries a natural Hodge bundle structure. In particular, it is equipped with a Hodge type bigrading $\mathcal{B}^i = \sum_{p+q=i} \mathcal{B}^{p,q}$.

7.2.2. We now introduce a different bigrading on the Weil algebra $\mathcal{B}^{\bullet}$. The commutative algebra $\mathcal{B}^{\bullet}$ is freely generated by the subbundles

$$S^1 = S^{1,-1} \oplus S^{-1,1} \subset \mathcal{B}^0 \quad \text{and} \quad \Lambda^1 = \Lambda^{1,0} \oplus \Lambda^{0,1} \subset \mathcal{B}^1,$$

therefore to define a multiplicative bigrading on the algebra $\mathcal{B}^{\bullet}$ it suffices to assign degrees to these generator subbundles $S^{1,-1}, S^{-1,1}, \Lambda^{1,0}, \Lambda^{0,1} \subset \mathcal{B}^{\bullet}$.

Definition. The *augmentation bigrading* on $\mathcal{B}^{\bullet}$ is the multiplicative bigrading defined by setting

$$\deg S^{1,-1} = \deg \Lambda^{1,0} = (1,0)$$
$$\deg S^{-1,1} = \deg \Lambda^{0,1} = (0,1)$$

on generators $S^{1,-1}, S^{-1,1}, \Lambda^{1,0}, \Lambda^{0,1} \subset \mathcal{B}^{\bullet,\bullet}$.

We will denote by $\mathcal{B}^{\bullet,\bullet}_{p,q}$ the component of the Weil algebra of augmentation bidegree (p,q). For any linear map $a : \mathcal{B}^{\bullet} \to \mathcal{B}^{\bullet}$ we will denote by $a = \sum_{p,q} a_{p,q}$ its decomposition with respect to the augmentation bidegree.

It will also be useful to consider a coarser *augmentation grading* on $\mathcal{B}^{\bullet}$, defined by $\deg \mathcal{B}^{\bullet}_{p,q} = p+q$. We will denote by $\mathcal{B}^{\bullet}_k = \bigoplus_{p+q=k} \mathcal{B}^{\bullet}_{p,q}$ the component of $\mathcal{B}^{\bullet}$ of augmentation degree k.

7.2.3. Note that the Hodge bidegree and the augmentation bidegree are, in general, independent. Moreover, the complex conjugation $\bar{\ } : \mathcal{B}^{\bullet} \to \overline{\mathcal{B}^{\bullet}}$ sends $\mathcal{B}^{\bullet}_{p,q}$ to $\overline{\mathcal{B}^{\bullet}_{q,p}}$. Therefore the augmentation bidegree components $\mathcal{B}^{\bullet}_{p,q} \subset \mathcal{B}^{\bullet}$ are not Hodge subbundles. However, the coarser augmentation grading is compatible with the Hodge structures, and the augmentation degree k-component $\mathcal{B}^i_k \subset \mathcal{B}^i$ carries a natural Hodge bundle structure of weight i. Moreover, the sum $\mathcal{B}^{\bullet}_{p,q} + \mathcal{B}^{\bullet}_{q,p} \subset \mathcal{B}^{\bullet}$ is also a Hodge subbundle.

7.2.4. We now introduce an auxiliary weight 0 Hodge algebra bundle on M, called the total Weil algebra. Recall that we have defined in 2.1.4 a functor $\Gamma : \mathcal{WH}odge_{\geq 0}(M) \to \mathcal{WH}odge_0(M)$ adjoint on the right to the canonical embedding. Consider the Hodge bundle $\mathcal{B}^{\bullet}_{tot} = \Gamma(\mathcal{B}^{\bullet})$ of weight 0 on M. By 1.4.9 the multiplication on $\mathcal{B}^{\bullet}$ induces an algebra structure on $\Gamma(\mathcal{B}^{\bullet})$.

Definition. The Hodge algebra bundle $\mathcal{B}^{\bullet}_{tot}$ of weight 0 is called *the total Weil algebra* of the complex manifold M.

Remark. For a more conceptual description of the functor Γ and the total Weil algebra, see Appendix.

7.2.5. By definition of the functor Γ we have $\mathcal{B}^k_{tot} = \mathcal{B}^k \otimes \mathcal{W}^*_k = S^\bullet \otimes \Lambda^k \otimes \mathcal{W}^*_k = S^\bullet \otimes \Lambda^k_{tot}$, where $\Lambda^k_{tot} = \Lambda^k \otimes \mathcal{W}^*_\bullet = \Gamma(\Lambda^k)$ is the total de Rham complex introduced in Subsection 7.1. We have also introduced in Subsection 7.1 Hodge bundle subalgebras $\Lambda^\bullet_o, \Lambda^\bullet_l, \Lambda^\bullet_r \subset \Lambda^\bullet_{tot}$ in the total de Rham complex $\Lambda^\bullet_{tot}$ and ideals $\Lambda^\bullet_{ll} \subset \Lambda^\bullet_l$, $\Lambda^\bullet_{rr} \subset \Lambda^\bullet_r$ in the algebras $\Lambda^\bullet_l$, $\Lambda^\bullet_r$. Let

$$\begin{aligned}\mathcal{B}^k_o &= S^\bullet \otimes \Lambda^k_o \subset \mathcal{B}^k_{tot}\\ \mathcal{B}^k_l &= S^\bullet \otimes \Lambda^k_l \subset \mathcal{B}^k_{tot}\\ \mathcal{B}^k_r &= S^\bullet \otimes \Lambda^k_r \subset \mathcal{B}^k_{tot}\end{aligned}$$

be the associated subalgebras in the total Weil algebra $\mathcal{B}^\bullet_{tot}$ and let

$$\begin{aligned}\mathcal{B}^k_{ll} &= S^\bullet \otimes \Lambda^k_{ll} \subset \mathcal{B}^k_l\\ \mathcal{B}^k_{rr} &= S^\bullet \otimes \Lambda^k_{rr} \subset \mathcal{B}^k_r\end{aligned}$$

be the corresponding ideals in the Hodge bundle algebras $\mathcal{B}^\bullet_l$, $\mathcal{B}^\bullet_r$.

By 7.1.10 we have bundle isomorphisms $\Lambda^\bullet_{tot} = \Lambda^\bullet_l + \Lambda^\bullet_r$ and $\Lambda^\bullet_o = \Lambda^\bullet_l \cap \Lambda^\bullet_r$, and the direct sum decomposition $\Lambda^\bullet_{tot} \cong \Lambda^\bullet_{ll} \oplus \Lambda^\bullet_o \oplus \Lambda^\bullet_{rr}$. Therefore we also have

$$\begin{aligned}\mathcal{B}^\bullet_{tot} &= \mathcal{B}^\bullet_l + \mathcal{B}^\bullet_r = \mathcal{B}^\bullet_{ll} \oplus \mathcal{B}^\bullet_o \oplus \mathcal{B}^\bullet_{rr}\\ \mathcal{B}^\bullet_o &= \mathcal{B}^\bullet_l \bigcap \mathcal{B}^\bullet_r \subset \mathcal{B}^\bullet_{tot}\end{aligned} \tag{7.3}$$

Moreover, the algebra $\Lambda^\bullet_o$ is isomorphic to the usual de Rham complex $\Lambda^\bullet$, therefore the subalgebra $\mathcal{B}^\bullet_o \subset \mathcal{B}^\bullet_{tot}$ is isomorphic to the usual Weil algebra $\mathcal{B}^\bullet$. These isomorphisms are *not* weakly Hodge.

7.2.6. The total Weil algebra carries a canonical weight 0 Hodge bundle structure, and we will denote the corresponding Hodge type grading by upper indices: $\mathcal{B}^\bullet_{tot} = \oplus_p (\mathcal{B}^\bullet_{tot})^{p,-p}$. The augmentation bigrading on the Weil algebra introduced in 7.2.2 extends to a bigrading of the total Weil algebra, which we will denote by lower indices. In general, both these grading and the direct sum decomposition (7.3) are independent, so that, in general, for every $i \geq 0$ we have a decomposition

$$\mathcal{B}^i_{tot} = \bigoplus_{n,p,q} \left(\mathcal{B}^i_{ll}\right)^{n,-n}_{p,q} \oplus \left(\mathcal{B}^i_o\right)^{n,-n}_{p,q} \oplus \left(\mathcal{B}^i_{rr}\right)^{n,-n}_{p,q}.$$

We would like to note, however, that some terms in this decomposition vanish when $i = 0, 1$. Namely, we have the following fact.

Lemma. *Let n, k be arbitrary integers such that $k \geq 0$.*

(i) *If $n+k$ is odd, then $\left(\mathcal{B}^0_{tot}\right)^{n,-n}_k = 0$.*

(ii) *If $n+k$ is even, then $\left(\mathcal{B}^1_{ll}\right)^{n,-n}_k = \left(\mathcal{B}^1_{rr}\right)^{n,-n}_k = 0$, while if $n+k$ is odd, then $\left(\mathcal{B}^1_o\right)^{n,-n}_k = 0$.*

Proof.

(i) The bundle $\mathcal{B}^0_{tot}$ by definition coincides with $\mathcal{B}^0$, and it is generated by the subbundles $\mathcal{S}^{1,-1}, \mathcal{S}^{-1,1} \subset \mathcal{B}^0$. Both these subbundles have augmentation degree 1 and Hodge degree ± 1, so that the sum $n+k$ of the Hodge degree with the augmentation degree is even. Since both gradings are multiplicative, for all non-zero components $\mathcal{B}^{n,-n}_k \subset \mathcal{B}^0$ the sum $n+k$ must also be even.

(ii) By definition we have $\mathcal{B}^1_{tot} = \mathcal{B}^0 \otimes \Lambda^1_{tot}$. The subbundle $\Lambda^1_{tot} \subset \mathcal{B}^1_{tot}$ has augmentation degree 1, and it decomposes

$$\Lambda^1_{tot} = \Lambda^1_o \oplus \Lambda^1_{ll} \oplus \Lambda^1_{rr}.$$

By 7.1.5 we have $\Lambda^1_o \cong \mathcal{S}^1 = \mathcal{S}^{1,-1} \oplus \mathcal{S}^{-1,1}$ as Hodge bundles, so that the Hodge degrees on $\Lambda^1_o \subset \Lambda^1_{tot}$ are odd. On the other hand, the subbundles $\Lambda^1_{ll}, \Lambda^1_{rr} \subset \Lambda^1_{tot}$ are by 7.1.6 of Hodge bidegree $(0,0)$. Therefore the sum $n+k$ of the Hodge and the augmentation degrees is even for Λ^1_o and odd for Λ^1_{ll} and Λ^1_{rr}. Together with (i) this proves the claim. □

7.3. Derivations of the Weil algebra

7.3.1. We will now introduce certain canonical derivations of the Weil algebra $\mathcal{B}^{\bullet}(M, \mathbb{C})$ which will play an important part in the rest of the paper. First of all, to simplify notation, for any two linear maps a, b let

$$\{a, b\} = a \circ b + b \circ a$$

be their anticommutator, and for any linear map $a : \mathcal{B}^{\bullet} \to \mathcal{B}^{\bullet+i}$ let $a = \sum_{p+q=i} a^{p,q}$ be the Hodge type decomposition. The following fact is well-known, but we have included a proof for the sake of completeness.

Lemma. *For every two odd derivations P, Q of a graded-commutative algebra $\mathcal{A}$, their anticommutator $\{P, Q\}$ is an even derivation of the algebra $\mathcal{A}$.*

Proof. Indeed, for every $a, b \in \mathcal{A}$ we have

$$\begin{aligned}
\{P, Q\}(ab) &= P(Q(ab)) + Q(P(ab)) \\
&= P(Q(a)b + (-1)^{\deg a} aQ(b)) + Q(P(a)b + (-1)^{\deg a} aP(b)) \\
&= P(Q(a))b + (-1)^{\deg Q(a)} Q(a)P(b) + (-1)^{\deg a} P(a)Q(b) \\
&\quad + aP(Q(b)) + Q(P(a))b + (-1)^{\deg P(a)} P(a)Q(b) \\
&\quad + (-1)^{\deg a} Q(a)P(b) + aQ(P(b)) \\
&= P(Q(a))b - (-1)^{\deg a} Q(a)P(b) + (-1)^{\deg a} P(a)Q(b) \\
&\quad + aP(Q(b)) + Q(P(a))b - (-1)^{\deg a} P(a)Q(b) \\
&\quad + (-1)^{\deg a} Q(a)P(b) + aQ(P(b)) \\
&= P(Q(a))b + aP(Q(b)) + Q(P(a))b + aQ(P(b)) \\
&= \{P, Q\}(a)b + a\{P, Q\}(b).
\end{aligned}$$

□

7.3.2. Let $C : S^1 \to \Lambda^1$ be the canonical weakly Hodge map introduced in 6.3.8. Extend C to an algebra derivation $C : \mathcal{B}^{\bullet} \to \mathcal{B}^{\bullet+1}$ by setting $C = 0$ on $\Lambda^1 \subset \mathcal{B}^1$. By 6.3.8 the derivation C is weakly Hodge. The composition

$$C \circ C = \frac{1}{2}\{C, C\} : \mathcal{B}^{\bullet} \to \mathcal{B}^{\bullet+2}$$

is also an algebra derivation, and it obviously vanishes on the generator subbundles $S^1, \Lambda^1 \subset \mathcal{B}^{\bullet}$. Therefore $C \circ C = 0$ everywhere.

Let also $\sigma : \mathcal{B}^{\bullet+1} \to \mathcal{B}^{\bullet}$ be the derivation introduced in 6.3.7. The derivation σ is not weakly Hodge; however, it is real and admits a decomposition $\sigma = \sigma^{-1,0} + \sigma^{0,-1}$ into components of Hodge types $(-1, 0)$ and $(0, -1)$. Both these components are algebra derivations of the Weil algebra $\mathcal{B}^{\bullet}$. We obviously have $\sigma \circ \sigma = \sigma^{-1,0} \circ \sigma^{-1,0} = \sigma^{0,-1} \circ \sigma^{0,-1} = 0$ on generators $S^1, \Lambda^1 \subset \mathcal{B}^{\bullet}$, and, therefore, on the whole Weil algebra.

Remark. Up to a sign the derivations C, σ and their Hodge bidegree components coincide with the so-called *Koszul differentials* on the Weil algebra $\mathcal{B}^{\bullet} = S^{\bullet} \otimes \Lambda^{\bullet}$.

7.3.3. The derivation $C : \mathcal{B}^{\bullet} \to \mathcal{B}^{\bullet+1}$ is by definition weakly Hodge. Applying the functor Γ to it, we obtain a derivation $C : \mathcal{B}^{\bullet}_{tot} \to \mathcal{B}^{\bullet+1}_{tot}$ of the total Weil algebra $\mathcal{B}^{\bullet}_{tot}$ preserving the weight 0 Hodge bundle structure on $\mathcal{B}^{\bullet}_{tot}$. The canonical identification $\mathcal{B}^{\bullet} \cong \mathcal{B}^{\bullet}_{o} \subset \mathcal{B}^{\bullet}_{tot}$ is compatible with the derivation $C : \mathcal{B}^{\bullet}_{tot} \to \mathcal{B}^{\bullet+1}_{tot}$. Moreover, by 7.3.2 this derivation satisfies $C \circ C = 0 : \mathcal{B}^{\bullet}_{tot} \to \mathcal{B}^{\bullet+2}_{tot}$. Therefore the total Weil algebra $\mathcal{B}^{\bullet}_{tot}$ equipped with the derivation C is a complex of Hodge bundles of weight 0.

The crucial linear algebraic property of the total Weil algebra $\mathcal{B}^{\bullet}_{tot}$ of the manifold M which will allow us to classify flat extended connections on M is the following.

Proposition 7.1 *Consider the subbundle*

$$\bigoplus_{p,q\geq 1} (\mathcal{B}^{\bullet}_{tot})_{p,q} \subset \mathcal{B}^{\bullet}_{tot} \tag{7.4}$$

of the total Weil algebra $\mathcal{B}^{\bullet}_{tot}$ consisting of the components of augmentation bidegrees (p,q) with $p,q \geq 1$. This subbundle equipped with the differential $C : (\mathcal{B}^{\bullet}_{tot})_{\bullet,\bullet} \to (\mathcal{B}^{\bullet+1}_{tot})_{\bullet,\bullet}$ is an acyclic complex of Hodge bundles of weight 0 *on M.*

7.3.4. We sketch a more or less simple and conceptual proof of Proposition 7.1 in the Appendix. However, in order to be able to study in Section 10 the analytic properties of our formal constructions, we will need an explicit contracting homotopy for the complex (7.4), which we now introduce.

The restriction of the derivation $C : \mathcal{B}^0_{tot} \to \mathcal{B}^1_{tot}$ to the subbundle $S^1 \subset \mathcal{B}_0 \cong \mathcal{B}^0_{tot}$ induces a Hodge bundle isomorphism

$$C : S^1 \to \Lambda^1_o \subset \Lambda^1_{tot} \subset \mathcal{B}^1_{tot}.$$

Define a map $\sigma_{tot} : \Lambda^1_{tot} \to S^1$ by

$$\sigma_{tot} = \begin{cases} 0 & \text{on} \quad \Lambda^1_{ll}, \Lambda^1_{rr} \subset \Lambda^1_{tot}, \\ C^{-1} & \text{on} \quad \Lambda^1_o \subset \Lambda^1_{tot}. \end{cases} \tag{7.5}$$

The map $\sigma_{tot} : \Lambda^1_{tot} \to S^1$ preserves the Hodge bundle structures of weight 0 on both sides. Moreover, its restriction to the subbundle $\Lambda^1 \cong \Lambda^1_o \subset \Lambda^1_{tot}$ coincides with the canonical map $\sigma : \Lambda^1 \to S^1$ introduced in 7.3.2.

7.3.5. Unfortunately, unlike $\sigma : \Lambda^1 \to S^1$, the map $\sigma_{tot} : \Lambda^1_{tot} \to S^1$ does *not* admit an extension to a derivation $\mathcal{B}^{\bullet+1}_{tot} \to \mathcal{B}^{\bullet}_{tot}$ of the total Weil algebra $\mathcal{B}^{\bullet}_{tot}$. We will extend it to a *bundle map* $\sigma_{tot} : \mathcal{B}^{\bullet+1}_{tot} \to \mathcal{B}^{\bullet}_{tot}$ in a somewhat roundabout way. To do this, define a map $\sigma_l : \Lambda^1_l \to S^0 \subset \mathcal{B}^0$ by

$$\sigma_l = \begin{cases} 0 & \text{on} \quad \Lambda^1_{ll} \subset \Lambda^1_l \quad \text{and on} \quad (\Lambda^1_o)^{-1,1} \subset \Lambda^1_{tot}, \\ C^{-1} & \text{on} \quad (\Lambda^1_o)^{1,-1} \subset \Lambda^1_{tot}. \end{cases}$$

and set $\sigma_l = 0$ on S^1. By 7.1.7 we have

$$\Lambda^1_l = \Lambda^{1,0} \oplus \left(\Lambda^{0,1} \otimes \mathcal{W}^*_1\right).$$

The map $\sigma_l : \Lambda^1_l \to S^1$ vanishes on the second summand in this direct sum, and it equals $C^{-1} : \Lambda^{1,0} \to S^{1,-1} \subset S^1$ on the first summand. The restriction of the map σ_l to the subbundle

$$\Lambda^{1,0} \oplus \Lambda^{0,1} = \Lambda^1 \cong \Lambda^1_o \subset \Lambda^1_l$$

vanishes on $\Lambda^{0,1}$ and equals C^{-1} on $\Lambda^{1,0}$. Thus it is equal to the Hodge type-$(0,-1)$ component $\sigma^{0,-1} : \Lambda^1 \to S^1$ of the canonical map $\sigma : \Lambda^1 \to S^1$.

7.3.6. By 7.1.8 the algebra $\mathcal{B}^{\bullet}_l$ is generated by the bundles S^1 and Λ^1_l, and the ideal of relations is generated by the subbundle

$$S^2\left(\Lambda^{0,1}\right) \otimes \Lambda^2\left(\mathcal{W}^*_1\right) \subset \Lambda^2\left(\Lambda^1_l\right). \tag{7.6}$$

Since the map $\sigma_l : \Lambda^1_l \to S^1$ vanishes on $\Lambda^{0,1} \otimes \mathcal{W}^*_1 \subset \Lambda^1_l$, it extends to an algebra derivation $\sigma_l : \mathcal{B}^{\bullet+1}_l \to \mathcal{B}^{\bullet}_l$. The restriction of the derivation σ_l to the subalgebra $\mathcal{B}^{\bullet} \cong \mathcal{B}^{\bullet}_0 \subset \mathcal{B}^{\bullet}_{tot}$ coincides with the $(0,-1)$-component $\sigma^{0,-1}$ of the derivation $\sigma : \mathcal{B}^{\bullet+1} \to \mathcal{B}^{\bullet}$.

Analogously, the $(-1,0)$-component $\sigma^{-1,0}$ of the derivation $\sigma : \mathcal{B}^{\bullet+1} \to \mathcal{B}^{\bullet}$ extends to an algebra derivation $\sigma_r : \mathcal{B}^{\bullet+1}_r \to \mathcal{B}^{\bullet}_r$ of the subalgebra $\mathcal{B}^{\bullet}_r \subset \mathcal{B}^{\bullet}_{tot}$. By definition, the derivation σ_l preserves the decomposition $\mathcal{B}^{\bullet}_l = \mathcal{B}_{ll} \oplus \mathcal{B}^{\bullet}_o$, while the derivation σ_r preserves the decomposition $\mathcal{B}^{\bullet}_r = \mathcal{B}^{\bullet}_{rr} \oplus \mathcal{B}^{\bullet}_o$. Both these derivations vanish on $\mathcal{B}^0_{tot}$, therefore both are maps of $\mathcal{B}^0_{tot}$-modules. In addition, the compositions $\sigma_l \circ \sigma_l$ and $\sigma_r \circ \sigma_r$ vanish on generator and, therefore, vanish identically.

7.3.7. Extend both σ_l and σ_r to the whole $\mathcal{B}^{\bullet}_{tot}$ by setting

$$\sigma_l = 0 \text{ on } \mathcal{B}^{\bullet}_r \qquad \sigma_r = 0 \text{ on } \mathcal{B}^{\bullet}_l, \tag{7.7}$$

and let

$$\sigma_{tot} = \sigma_l + \sigma_r : \mathcal{B}^{\bullet+1}_{tot} \to \mathcal{B}^{\bullet}_{tot}.$$

On $\Lambda^1_{tot} \subset \mathcal{B}^1_{tot}$ this is the same map as in (7.5). The bundle map $\sigma_{tot} : \mathcal{B}^{\bullet+1}_{tot} \to \mathcal{B}^{\bullet}_{tot}$ preserves the direct sum decomposition (7.3), and its restriction to $\mathcal{B}^{\bullet}_o \subset \mathcal{B}^{\bullet}_{tot}$ coincides with the derivation σ. Note that neither of the maps $\sigma_l, \sigma_r, \sigma_{tot}$ is a derivation of the total Weil algebra $\mathcal{B}^{\bullet}_{tot}$. However, all these maps are linear with respect to the $\mathcal{B}^0_{tot}$-module structure on $\mathcal{B}^{\bullet}_{tot}$ and preserve the decomposition (7.3). The map $\sigma_{tot} : \mathcal{B}^{\bullet+1}_{tot} \to \mathcal{B}^{\bullet}_{tot}$ is equal to σ_l on $\mathcal{B}^{\bullet}_{ll} \subset \mathcal{B}^{\bullet}_{tot}$, to σ_r on $\mathcal{B}^{\bullet}_{rr}$ and to $\sigma : \mathcal{B}^{\bullet+1} \to \mathcal{B}^{\bullet}$ on $\mathcal{B}^{\bullet} \cong \mathcal{B}^{\bullet}_o \subset \mathcal{B}^{\bullet}_{tot}$. Since $\sigma_l \circ \sigma_l = \sigma_r \circ \sigma_r = \sigma \circ \sigma = 0$, we have $\sigma_{tot} \circ \sigma_{tot} = 0$.

7.3.8. The commutator

$$h = \{C, \sigma_{tot}\} : \mathcal{B}^{\bullet}_{tot} \to \mathcal{B}^{\bullet}_{tot}$$

of the maps C and σ_{tot} also preserves the decomposition (7.3), and we have the following.

Lemma. *The map h acts as multiplication by p on $(\mathcal{B}^{\bullet}_{ll})_{p,q}$, as multiplication by q on $(\mathcal{B}^{\bullet}_{rr})_{p,q}$ and as multiplication by $(p+q)$ on $(\mathcal{B}^{\bullet}_{o})_{p,q}$.*

Proof. It suffices to prove the claim separately on each term in the decomposition (7.3). By definition $\sigma_{tot} = \sigma_l + \sigma_r$, and $h = h_l + h_r$, where $h_l = \{\sigma_l, C\}$ and $h_r = \{\sigma_r, C\}$. Moreover, h_l vanishes on $\mathcal{B}^{\bullet}_{rr}$ and h_r vanishes on $\mathcal{B}^{\bullet}_{ll}$. Therefore it suffices to prove that $h_l = p\,\mathsf{id}$ on $(\mathcal{B}^{\bullet}_{l})_{p,q}$ and that $h_r = q\,\mathsf{id}$ on $(\mathcal{B}^{\bullet}_{r})_{p,q}$. The proofs of these two identities are completely symmetrical, and we will only give a proof for h_l.

The algebra $\mathcal{B}^{\bullet}_{l}$ is generated by the subbundles $S^1 \subset \mathcal{B}^0_l$ and $\Lambda^1_l \subset \mathcal{B}^1_l$. The augmentation bidegree decomposition of S^1 is by definition given by

$$S^1_{1,0} = S^{1,-1} \qquad S^1_{0,1} = S^{-1,1},$$

while the augmentation bidegree decomposition of Λ^1_l is given by

$$(\Lambda^1_l)_{1,0} = \Lambda^{1,0} \qquad (\Lambda^1_l)_{0,1} = \Lambda^{0,1} \otimes \mathcal{W}^*_1.$$

By the definition of the map $\sigma_l : \Lambda^1_l \to S^1$ (see 7.3.5) we have $h_l = \{C, \sigma_l\} = \mathsf{id}$ on $\Lambda^{1,0}$ and $S^{1,-1}$, and $h_l = 0$ on $\Lambda^{0,1} \otimes \mathcal{W}^*_1$ and on $S^{-1,1}$. Therefore for every $p, q \geq 0$ we have $h_l = p\,\mathsf{id}$ on the generator subbundles $S^1_{p,q}$ and on $(\Lambda^1_l)_{p,q}$. Since the map h_l is a derivation and the augmentation bidegree is multiplicative, the same holds on the whole algebra $\mathcal{B}^{\bullet}_l = \oplus (\mathcal{B}^{\bullet}_l)_{p,q}$. □

Lemma 7.3.8 shows that the map σ_{tot} is a homotopy, contracting the subcomplex (7.4) in the total Weil algebra $\mathcal{B}^{\bullet}_{tot}$, which immediately implies Proposition 7.1.

Remark. In fact, in our classification of flat extended connections given in Section 8 it will be more convenient for us to use Lemma 7.3.8 directly rather than refer to Proposition 7.1.

7.3.9. We finish this section with the following corollary of Lemma 7.2.6 and Lemma 7.3.8, which we will need in Section 10.

Lemma. *Let $n = \pm 1$. If the integer $k \geq 1$ is odd, then the map $h : \mathcal{B}^{\bullet}_{tot} \to \mathcal{B}^{\bullet}_{tot}$ acts on $(\mathcal{B}^1_{tot})^{n,-n}_k$ by multiplication by k. If $k = 2m \geq 1$ is even, then the endomorphism $h : (\mathcal{B}^1_{tot})^{n,-n}_k \to (\mathcal{B}^1_{tot})^{n,-n}_k$ is diagonalizable, and its only eigenvalues are m and $m-1$.*

Proof. If k is odd, then by Lemma 7.2.6 we have $(\mathcal{B}^1_{tot})^{n,-n}_k = (\mathcal{B}^1_o)^{n,-n}_k$, and Lemma 7.3.8 immediately implies the claim. Assume that the integer $k = 2m$ is even. By Lemma 7.2.6 we have

$$(\mathcal{B}^{\bullet}_{tot})^{n,-n}_k = (\mathcal{B}^{\bullet}_{ll})^{n,-n}_k \oplus (\mathcal{B}^{\bullet}_{rr})^{n,-n}_k = \mathcal{B}^{n,-n}_{k-1} \otimes \left(\Lambda^1_{ll} \oplus \Lambda^1_{rr}\right).$$

The bundle $\mathcal{B}^0$ is generated by subbundles $S^{1,-1}$ and $S^{-1,1}$. The first of these subbundles has augmentation bidegree $(1,0)$, while the second one has augmentation bidegree $(0,1)$. Therefore for every augmentation bidegree component $\mathcal{B}^{n,-n}_{p,q} \subset \mathcal{B}^{n,-n}_{k-1}$ we have $p-q = n$ and $p+q = k-1$. This implies that $\mathcal{B}^{n,-n}_{k-1} = \mathcal{B}^{n,-n}_{p,q}$ with $p = m - (1-n)/2$ and $q = m - (1+n)/2$.

By definition the augmentation bidegrees of the bundles Λ^1_{ll} and Λ^1_{rr} are, respectively, $(0,1)$ and $(1,0)$. Lemma 7.3.8 shows that the only eigenvalue of the map h on $\left(\mathcal{B}^1_{ll}\right)_{k+1}$ is $p = (m - (1-n)/2))$, while its only eigenvalue on $\left(\mathcal{B}^1_{rr}\right)_{k+1}$ is $q = (m - (1+n)/2)$. Since $n = \pm 1$, one of these numbers equals m and the other one equals $m-1$. □

8. Classification of flat extended connections

8.1. Kählerian connections

8.1.1. Let M be a complex manifold. In Section 6 we have shown that formal Hodge manifold structures on the tangent bundle $\overline{T}M$ are in one-to-one correspondence with linear flat extended connections on the manifold M (see 6.4.1–6.4.6 for the definitions). It turns out that flat linear extended connections on M are, in turn, in natural one-to-one correspondence with differential operators of a much simpler type, namely, connections on the cotangent bundle $\Lambda^{1,0}(M)$ satisfying certain vanishing conditions (Theorem 8.1). We call such connections *Kählerian*. In this section we use the results of Section 7 establish the correspondence between extended connections on M and Kählerian connections on $\Lambda^{1,0}(M)$.

8.1.2. We first give the definition of Kählerian connections. Assume that the manifold M is equipped with a connection

$$\nabla : \Lambda^1(M) \to \Lambda^1(M) \otimes \Lambda^1(M)$$

on its cotangent bundle $\Lambda^1(M)$. Let

$$T = \mathrm{Alt} \circ \nabla - d_M : \Lambda^1(M) \to \Lambda^2(M)$$
$$R = \mathrm{Alt}\, \nabla \circ \nabla : \Lambda^1(M) \to \Lambda^1(M) \otimes \Lambda^2(M)$$

be its torsion and curvature, and let $R = R^{2,0} + R^{1,1} + R^{0,2}$ be the decomposition of the curvature according to the Hodge type.

Definition. The connection ∇ is called *Kählerian* if

$$T = 0 \tag{i}$$

$$R^{2,0} = R^{0,2} = 0 \tag{ii}$$

Example. The Levi-Civita connection on a Kähler manifold is Kählerian.

Remark. The condition $T = 0$ implies, in particular, that the component

$$\nabla^{0,1} : \Lambda^{1,0}(M) \to \Lambda^{1,1}(M)$$

of the connection ∇ coincides with the Dolbeault differential. Therefore a Kählerian connection is always holomorphic .

8.1.3. Recall that in 6.4.2 we have associated to any extended connection D on M a connection ∇ on the cotangent bundle $\Lambda^1(M, \mathbb{C})$ called the reduction of D. We can now formulate the main result of this section.

Theorem 8.1 (i) *If an extended connection D on M is flat and linear, then its reduction ∇ is Kählerian.*

(ii) *Every Kählerian connection ∇ on the cotangent bundle $\Lambda^1(M, \mathbb{C})$ is the reduction of a unique linear flat extended connection D on M.*

The rest of this section is taken up with the proof of Theorem 8.1. To make it more accessible, we first give an informal outline. The actual proof starts with Subsection 8.2, and it is independent from the rest of this subsection.

8.1.4. Assume given a Kählerian connection ∇ on the manifold M. To prove Theorem 8.1, we have to construct a flat linear extended connection D on M with reduction ∇. Every extended connection decomposes into a series $D = \sum_{k \geq 0} D_k$ as in (6.2), and, since ∇ is the reduction of D, we must have $D_1 = \nabla$. We begin by checking in Lemma 8.2.1 that if D is linear, then $D_0 = C$, where $C : S^1(M, \mathbb{C}) \to \Lambda^1(M, \mathbb{C})$ is as in 6.3.8. The sum $C + \nabla$ is already a linear extended connection on M. By 6.4.5 it extends to a derivation $D_{\leq 1}$ of the Weil algebra $\mathcal{B}^{\bullet}(M, \mathbb{C})$ of the manifold M, but this derivation does not necessarily satisfy $D_{\leq 1} \circ D_{\leq 1} = 0$, thus the extended connection $D_{\leq 1}$ is not necessarily flat.

We have to show that one can add the "correction terms" $D_k, k \geq 2$ to $D_{\leq 1}$ so that $D = \sum_k D_k$ satisfies all the conditions of Theorem 8.1. To do this, we introduce in 8.3.3 a certain quotient $\widetilde{\mathcal{B}}^{\bullet}(M, \mathbb{C})$ of the Weil algebra

$\mathcal{B}^\bullet(M,\mathbb{C})$, called the reduced Weil algebra. The reduced Weil algebra is defined in such a way that for every extended connection D the associated derivation $D : \mathcal{B}^\bullet(M,\mathbb{C}) \to \mathcal{B}^{\bullet+1}(M,\mathbb{C})$ preserves the kernel of the surjection $\mathcal{B}^\bullet \to \widetilde{\mathcal{B}}^\bullet$, thus inducing a derivation $\widetilde{D} : \widetilde{\mathcal{B}}^\bullet(M,\mathbb{C}) \to \widetilde{\mathcal{B}}^{\bullet+1}(M,\mathbb{C})$. Moreover, the algebra $\widetilde{\mathcal{B}}^\bullet(M,\mathbb{C})$ has the following two properties:

(i) The derivation $\widetilde{D} : \widetilde{\mathcal{B}}^\bullet(M,\mathbb{C}) \to \widetilde{\mathcal{B}}^{\bullet+1}(M,\mathbb{C})$ satisfies $\widetilde{D} \circ \widetilde{D} = 0$ if and only if the connection D_1 is Kählerian.

(ii) Let $\widetilde{D}$ be the weakly Hodge derivation of the quotient algebra $\widetilde{\mathcal{B}}^\bullet(M,\mathbb{C})$ induced by an arbitrary linear extended connection $D_{\leq 1}$ and such that $\widetilde{D} \circ \widetilde{D} = 0$. Then the derivation $\widetilde{D}$ lifts uniquely to a weakly Hodge derivation D of the Weil algebra $\mathcal{B}^\bullet(M,\mathbb{C})$ such that $D \circ D = 0$, and the derivation also D comes from a linear extended connection on M (see Proposition 8.1 for a precise formulation of this statement).

8.1.5. The property (i) is relatively easy to check, and we do it in the end of the proof, in Subsection 8.4. The rest is taken up with establishing the property (ii). The actual proof of this statement is contained in Proposition 8.1, and Subsection 8.2 contains the necessary preliminaries.

Recall that we have introduced in 7.2.2 a new grading on the Weil algebra $\mathcal{B}^\bullet(M,\mathbb{C})$, called the augmentation grading, so that the component D_k in the decomposition $D = \sum_k D_k$ is of augmentation degree k. In order to lift $\widetilde{D}$ to a derivation D so that $D \circ D = 0$, we begin with the given lifting $D_{\leq 1}$ and then add components $D_k, k \geq 2$, one by one, so that on each step for $D_{\leq k} = D_{\leq 1} + \sum_{2 \leq p \leq k}$ the composition $D_{\leq k} \circ D_{\leq k}$ is zero in augmentation degrees from 0 to k. In order to do it, we must find for each k a solution to the equation

$$D_0 \circ D_k = -R_k, \tag{8.1}$$

where R_k is the component of augmentation degree k in the composition $D_{\leq k-1} \circ D_{\leq k-1}$. This solution must be weakly Hodge, and the extended connection $D_{\leq k} = D_{\leq k-1} + D_k$ must be linear.

We prove in Lemma 8.2.1 that since $D_{\leq 0}$ is linear, we may assume that $D_0 = C$. In addition, since $\widetilde{D} \circ \widetilde{D} = 0$, we may assume by induction that the image of R_k lies in the kernel $\mathcal{I}^\bullet$ of the quotient map $\mathcal{B}^\bullet(M,\mathbb{C}) \to \widetilde{\mathcal{B}}^\bullet(M,\mathbb{C})$.

8.1.6. In order to analyze weakly Hodge maps from $S^1(M,\mathbb{C})$ to the Weil algebra $\mathcal{B}^\bullet(M,\mathbb{C})$, we apply the functor $\Gamma : \mathcal{WH}odge_{\geq 0}(M) \to \mathcal{WH}odge_0(M)$ constructed in 2.1.4 to the bundle $\mathcal{B}^\bullet(M,\mathbb{C})$ to obtain the total Weil algebra $\mathcal{B}^\bullet_{tot}(M,\mathbb{C}) = \Gamma(\mathcal{B}^\bullet(M,\mathbb{C}))$ of weight 0, which we studied in Subsection 7.2. The Hodge bundle $S^1(M,\mathbb{C})$ on the manifold M is of weight 0, and, by the universal property of the functor Γ, weakly Hodge maps from $S^1(M,\mathbb{C})$ to $\mathcal{B}^{\bullet}(M,\mathbb{C})$ are in one-to-one correspondence with Hodge bundle maps from $S^1(M,\mathbb{C})$ to the total Weil algebra $\mathcal{B}^{\bullet}_{tot}(M,\mathbb{C})$. The canonical map $C : S^1(M,\mathbb{C}) \to \mathcal{B}^1(M,\mathbb{C})$ extends to a derivation $C : \mathcal{B}^{\bullet}_{tot}(M,\mathbb{C}) \to \mathcal{B}^{\bullet+1}_{tot}(M,\mathbb{C})$. Moreover, the weakly Hodge map $R_k : S^1(M,\mathbb{C}) \to \mathcal{B}^2(M,\mathbb{C})$ defines a Hodge bundle map $R_k^{tot} : S^1(M,\mathbb{C}) \to \mathcal{B}^2_{tot}(M,\mathbb{C})$, and solving (8.1) is equivalent to finding a Hodge bundle map $D_k : S^1(M,\mathbb{C}) \to \mathcal{B}^1_{tot}(M,\mathbb{C})$ such that

$$C \circ D_k = -R_k. \tag{8.2}$$

8.1.7. Recall that by 7.3.2 the derivation

$$C : \mathcal{B}^{\bullet}_{tot}(M,\mathbb{C}) \to \mathcal{B}^{\bullet+1}_{tot}(M,\mathbb{C})$$

satisfies $C \circ C = 0$, so that the total Weil algebra $\mathcal{B}^{\bullet}_{tot}(M,\mathbb{C})$ becomes a complex with differential C. The crucial part of the proof of Theorem 8.1 consists in noticing that the subcomplex $\mathcal{I}^{\bullet}_{tot}(M,\mathbb{C}) = \Gamma(\mathcal{I}^{\bullet}(M,\mathbb{C})) \subset \mathcal{B}^{\bullet}_{tot}(M,\mathbb{C})$ of the total Weil algebra $\mathcal{B}^{\bullet}_{tot}(M,\mathbb{C})$ corresponding to the kernel $\mathcal{I}^{\bullet}(M,\mathbb{C}) \subset \mathcal{B}^{\bullet}(M,\mathbb{C})$ of the quotient map $\mathcal{B}^{\bullet}(M,\mathbb{C}) \to \widetilde{\mathcal{B}}^{\bullet}(M,\mathbb{C})$ is canonically contractible. This statement is analogous to Proposition 7.1, and we prove it in the same way. Namely, we check that the subcomplex $\mathcal{I}^{\bullet}_{tot} \subset \mathcal{B}^{\bullet}_{tot}$ is preserved by the bundle map $\sigma_{tot} : \mathcal{B}^{\bullet+1}_{tot} \to \mathcal{B}^{\bullet}_{tot}$ constructed in 7.3.7, and that the anticommutator $h = \{\sigma_{tot}, C\} : \mathcal{B}^{\bullet}_{tot}(M,\mathbb{C}) \to \mathcal{B}^{\bullet}_{tot}(M,\mathbb{C})$ is invertible on the subcomplex $\mathcal{I}^{\bullet}_{tot}(M,\mathbb{C}) \subset \mathcal{B}^{\bullet}_{tot}(M,\mathbb{C})$ (Corollary 8.3.4 of Lemma 7.3.8). We also check that $C \circ R_k^{tot} = 0$, which implies that the Hodge bundle map

$$D_k = -h^{-1} \circ \sigma_{tot} \circ R_k^{tot} : S^1(M,\mathbb{C}) \to \mathcal{B}^1_{tot}(M,\mathbb{C}) \tag{8.3}$$

provides a solution to the equation (8.2).

8.1.8. To establish the property (ii), we have to insure additionally that the extended connection $D = D_{\leq k}$ is linear, and and we have to show that the solution D_k of (8.2) with this property is unique. This turns out to be pretty straightforward. We show in Lemma 8.2.1 that $D_{\leq k}$ is linear if and only if

$$\sigma_{tot} \circ D_k = 0. \tag{8.4}$$

Moreover, we show that the homotopy $\sigma_{tot} : \mathcal{B}^{\bullet+1}_{tot}(M,\mathbb{C}) \to \mathcal{B}^{\bullet}_{tot}(M,\mathbb{C})$ satisfies $\sigma_{tot} \circ \sigma_{tot} = 0$. Therefore the solution D_k to (8.2) given by (8.3) satisfies (8.4) automatically.

The uniqueness of such a solution D_k follows from the invertibility of $h = C \circ \sigma_{tot} + \sigma_{tot} \circ C$. Indeed, for every two solutions D_k, D'_k to (8.1), both satisfying (8.4), their difference $P = D_k - D'_k$ satisfies $C \circ P = 0$. If, in addition, both D_k and D'_k were to satisfy (8.4), we would have had $\sigma_{tot} \circ P = 0$. Therefore $h \circ P = 0$, and P has to vanish.

8.1.9. These are the main ideas of the proof of Theorem 8.1. The proof itself begins in the next subsection, and it is organized as follows. In Subsection 8.2 we express the linearity condition on an extended connection D in terms of the associated derivation $D_{tot} : \mathcal{B}^{\bullet}_C \to \mathcal{B}^{\bullet+1}_{tot}$ of the total Weil algebra $\mathcal{B}^{\bullet}_{tot}$ of the manifold M. After that, we introduce in Subsection 8.3 the reduced Weil algebra $\widetilde{\mathcal{B}}^{\bullet}(M, \mathbb{C})$ and prove Proposition 8.1, thus reducing Theorem 8.1 to a statement about derivations of the reduced Weil algebra. Finally, in Subsection 8.4 we prove this statement.

Remark. In the Appendix we give, following Deligne and Simpson, a more geometric description of the functor $\Gamma : \mathcal{WH}odge_{\geq 0} \to \mathcal{WH}odge$ and of the total Weil algebra $\mathcal{B}^{\bullet}_{tot}(M, \mathbb{C})$, which allows to give a simpler and more conceptual proof for the key parts of Theorem 8.1.

8.2. Linearity and the total Weil algebra

8.2.1. Assume given an extended connection $D : S^1 \to \mathcal{B}^1$ on the manifold M, and extend it to a derivation $D : \mathcal{B}^{\bullet} \to \mathcal{B}^{\bullet+1}$ of the Weil algebra as in 6.4.5. Let $D = \sum_{k\geq 0} D_k$ be the augmentation degree decomposition. The derivation D is weakly Hodge and defines therefore a derivation $D = \sum_{k\geq 0} D_k : \mathcal{B}^{\bullet}_{tot} \to \mathcal{B}^{\bullet+1}_{tot}$ of the total Weil algebra $\mathcal{B}^{\bullet}_{tot}$.

Before we begin the proof of Theorem 8.1, we give the following rewriting of the linearity condition 6.4.4 on the extended connection D in terms of the total Weil algebra.

Lemma. *The extended connection D is linear if and only if $D_0 = C$ and $\sigma_{tot} \circ D_k = 0$ on $S^1 \subset \mathcal{B}^0_{tot}$ for every $k \geq 0$.*

Proof. Indeed, by Lemma 7.2.6 for odd integers k and $n = \pm 1$ the subbundle $\left(\mathcal{B}^1_o\right)^{n,-n}_{k+1} \subset \mathcal{B}^1_{tot}$ vanishes. Therefore the map $D_k : S^1 \to \mathcal{B}^1_{tot}$ factors through $\mathcal{B}^1_{ll} \oplus \mathcal{B}^1_{rr}$. Since by definition (7.3.7) we have $\sigma_{tot} = 0$ on both $\mathcal{B}^1_{ll}$ and $\mathcal{B}^1_{rr}$, for odd k we have $\sigma_{tot} \circ D_k = 0$ on S^1 regardless of the extended connection D. On the other hand, for even k we have $\left(\mathcal{B}^1_{tot}\right)^{n,-n}_{k+1} = \left(\mathcal{B}^1_o\right)^{n,-n}_{k+1}$. Therefore on S^1 we have $\sigma_{tot} \circ D_k = \sigma \circ D_k$ (where $\sigma : \mathcal{B}^{\bullet+1} \to \mathcal{B}^{\bullet}$ is as in 7.3.2). Moreover, since $\sigma : \Lambda^1 \to S^1$ is an isomorphism, $D_0 = C$ is equivalent to $\sigma \circ D_0 = \mathsf{id} : S^1 \to \Lambda^1 \to S^1$. Therefore the condition of the lemma is equivalent to the following

$$\sigma \circ D_k = \begin{cases} \mathsf{id}, & \text{for } k = 0 \\ 0, & \text{for even integers } k > 0. \end{cases} \tag{8.5}$$

Let now $\iota^* : \mathcal{B}^\bullet \to \mathcal{B}^\bullet$ be the operator given by the action of the canonical involution $\iota : \overline{T}M \to \overline{T}M$, as in 6.3.6, and let $D^\iota = \sum_{k \geq 0} D^\iota_k = \iota^* \circ D \circ (\iota^*)^{-1}$ be the operator ι^*-conjugate to the derivation D. The operator ι^* acts as $-\mathsf{id}$ on $S^1 \subset \mathcal{B}^0$ and as id on $\Lambda^1 \subset \mathcal{B}^1$. Since it is an algebra automorphism, it acts as $(-1)^{i+k}$ on $\mathcal{B}^i_k \subset \mathcal{B}^\bullet$. Therefore $D^\iota_k = (-1)^{k+1} D_k$, and (8.5) is equivalent to

$$\sigma \circ \frac{1}{2}(D - D^\iota) = \mathsf{id} : S^1 \to \mathcal{B}^1 \to S^1,$$

which is precisely the definition of a linear extended connection. □

8.3. The reduced Weil algebra

8.3.1. We now begin the proof of Theorem 8.1. Our first step is to reduce the classification of linear flat extended connections $D : S^1 \to \mathcal{B}^1$ on the manifold M to the study of derivations of a certain quotient $\widetilde{\mathcal{B}}^\bullet$ of the Weil algebra $\mathcal{B}^\bullet$. We introduce this quotient in this subsection under the name of reduced Weil algebra. We then show that every extended connection D on M induces a derivation $\widetilde{D} : \widetilde{\mathcal{B}}^\bullet \to \widetilde{\mathcal{B}}^{\bullet+1}$ of the reduced Weil algebra, and that a linear flat extended connection D on M is completely defined by the derivation $\widetilde{D}$.

8.3.2. By Lemma 7.3.8 the anticommutator $h = \{C, \sigma_{tot}\} : \mathcal{B}^\bullet_{tot} \to \mathcal{B}^\bullet_{tot}$ of the canonical bundle endomorphisms C, σ_{tot} of the total Weil algebra $\mathcal{B}^\bullet_{tot}$ is invertible on every component $(\mathcal{B}^\bullet_{tot})_{p,q}$ of augmentation bidegree (p, q) with $p, q \geq 1$.

The direct sum $\oplus_{p,q \geq 1} (\mathcal{B}^\bullet_{tot})_{p,q}$ is an ideal in the total Weil algebra $\mathcal{B}^\bullet_{tot}$, and it is obtained by applying the functor Γ to the ideal $\oplus_{p,q\geq 1} \mathcal{B}^\bullet_{p,q}$ in the Weil algebra $\mathcal{B}^\bullet$. For technical reasons, it will be more convenient for us to consider the smaller subbundle

$$\mathcal{I}^\bullet = \bigoplus_{p \geq 2, q \geq 1} \mathcal{B}^\bullet_{p,q} + \bigoplus_{p \geq 1, q \geq 2} \mathcal{B}^\bullet_{p,q} \subset \mathcal{B}^\bullet$$

of the Weil algebra $\mathcal{B}^\bullet$. The subbundle $\mathcal{I}^\bullet$ is a Hodge subbundle in $\mathcal{B}^\bullet$, and it is an ideal with respect to the multiplication in $\mathcal{B}^\bullet$.

8.3.3. Definition. The reduced Weil algebra $\widetilde{\mathcal{B}}^\bullet = \widetilde{\mathcal{B}}^\bullet(M, \mathbb{C})$ of the manifold M is the quotient

$$\widetilde{\mathcal{B}}^\bullet = \mathcal{B}^\bullet / \mathcal{I}^\bullet$$

of the full Weil algebra $\mathcal{B}^{\bullet}$ by the ideal $\mathcal{I}^{\bullet}$.

The reduced Weil algebra decomposes as

$$\widetilde{\mathcal{B}}^{\bullet} = \mathcal{B}^{\bullet}_{1,1} \oplus \bigoplus_{p \geq 0} \mathcal{B}^{\bullet}_{p,0} \oplus \bigoplus_{q \geq 0} \mathcal{B}^{\bullet}_{0,q}$$

with respect to the augmentation bigrading on the Weil algebra $\mathcal{B}^{\bullet}$. The two summands on the right are equal to

$$\bigoplus_{p \geq 0} \mathcal{B}^{\bullet}_{p,0} = \bigoplus_{pgeq0} S^p \left(S^{1,-1}\right) \otimes \Lambda^{\bullet,0},$$

$$\bigoplus_{q \geq 0} \mathcal{B}^{\bullet}_{0,q} = \bigoplus_{qgeq0} S^q \left(S^{-1,1}\right) \otimes \Lambda^{0,\bullet},$$

8.3.4. Since $\mathcal{I}^{\bullet}$ is a Hodge subbundle, the reduced Weil algebra carries a canonical Hodge bundle structure compatible with the multiplication. It also obviously inherits the augmentation bigrading, and defines an ideal $\mathcal{I}^{\bullet}_{tot} = \Gamma(\mathcal{I}^{\bullet}) \subset \mathcal{B}^{\bullet}_{tot}$ in the total Weil algebra $\mathcal{B}^{\bullet}_{tot}$. Lemma 7.3.8 immediately implies the following fact.

Corollary. *The map $h = \{C, \sigma_{tot}\} : \mathcal{B}^{\bullet}_{tot} \to \mathcal{B}^{\bullet}_{tot}$ is invertible on $\mathcal{I}^{\bullet}_{tot} \subset \mathcal{B}^{\bullet}_{tot}$.*

8.3.5. Let now $D : \mathcal{B}^{\bullet} \to \mathcal{B}^{\bullet}$ be the derivation associated to the extended connection D as in 6.4.5. The derivation D does not increase the augmentation bidegree, it preserves the ideal $\mathcal{I}^{\bullet} \subset \mathcal{B}^{\bullet}$ and defines therefore a weakly Hodge derivation of the reduced Weil algebra $\widetilde{\mathcal{B}}^{\bullet}$, which we denote by $\widetilde{D}$. If the extended connection D is flat, then the derivation $\widetilde{D}$ satisfies $\widetilde{D} \circ \widetilde{D} = 0$.

We now prove that every derivation $\widetilde{D} : \widetilde{\mathcal{B}}^{\bullet} \to \widetilde{\mathcal{B}}^{\bullet+1}$ of this type comes from a linear flat extended connection D, and that the connection D is completely defined by $\widetilde{D}$. More precisely, we have the following.

Proposition 8.1 *Let $D : S^1 \to \mathcal{B}^1$ be a linear but not necessarily flat extended connection on M, and let $\widetilde{D} : \widetilde{\mathcal{B}}^{\bullet} \to \widetilde{\mathcal{B}}^{\bullet+1}$ be the associated weakly Hodge derivation of the reduced Weil algebra $\widetilde{\mathcal{B}}^{\bullet}$. Assume that $\widetilde{D} \circ \widetilde{D} = 0$.*

There exists a unique weakly Hodge bundle map $P : S^1 \to \mathcal{I}^1$ such that the extended connection $D' = D + P : S^1 \to \mathcal{B}^1$ is linear and flat.

8.3.6. *Proof.* Assume given a linear extended connection D satisfying the condition of Proposition 8.1. To prove the proposition, we have to construct a weakly Hodge map $P : S^1 \to \mathcal{I}^1$ such that the extended connection $D + P$ is linear and flat. We do it by induction on the augmentation degree, that is, we construct one-by-one the terms P_k in the augmentation degree decomposition $P = \sum_k P_k$. The identity $\widetilde{D} \circ \widetilde{D} = 0$ is the base of the induction, and the induction step is given by applying the following lemma to $D + \sum_{i=2}^{k} P_k$, for each $k \geq 1$ in turn.

Lemma. *Assume given a linear extended connection $D : S^1 \to \mathcal{B}^{\bullet+1}$ on M and let $D : \mathcal{B}^{\bullet} \to \mathcal{B}^{\bullet+1}$ also denote the associated derivation. Assume also that the composition $D \circ D : \mathcal{B}^{\bullet} \to \mathcal{B}^{\bullet+2}$ maps S^1 into $\mathcal{I}^2_{>k} = \oplus_{p>k}\mathcal{I}^2_p$.*

There exists a unique weakly Hodge bundle map $P_k : S^1 \to \mathcal{I}^1_{k+1}$ such that the extended connection $D' = D + P_k : S^1 \to \mathcal{B}^1$ is linear, and for the associated derivation $D' : \mathcal{B}^{\bullet} \to \mathcal{B}^{\bullet+1}$ the composition $D' \circ D'$ maps S^1 into $\mathcal{I}^2_{>k+1} = \oplus_{p>k+1}\mathcal{I}^2_p$.

Proof. Let $D : \mathcal{B}^{\bullet}_{tot} \to \mathcal{B}^{\bullet}_{tot}$ be the derivation of the total Weil algebra associated to the extended connection D, and let

$$R : (\mathcal{B}^{\bullet}_{tot})_{\bullet} \to \left(\mathcal{B}^{\bullet+2}_{tot}\right)_{\bullet+k}$$

be the component of augmentation degree k of the composition $D \circ D : S^1 \to \mathcal{B}^2_{tot}$. Note that by 6.4.5 the map R vanishes on the subbundle $\Lambda^1_{tot} \subset \left(\mathcal{B}^1_{tot}\right)_1$. Moreover, the composition $C \circ R : \mathcal{B}^{\bullet}_{tot} \to \mathcal{B}^{\bullet+3}$ of the map R with the canonical derivation $C : \mathcal{B}^{\bullet}_{tot} \to \mathcal{B}^{\bullet+1}_{tot}$ vanishes on the subbundle $S^1 \subset \mathcal{B}^0_{tot}$. Indeed, since C maps S^1 into Λ^1_{tot}, the composition $C \circ R$ is equal to the commutator $[C, R] : S^1 \to \mathcal{B}^3_{tot}$. Since the extended connection D is by assumption linear, we have $D_0 = C$, and

$$\begin{aligned} C \circ R = [C, R] &= \sum_{0 \leq p \leq k} [C, D_p \circ D_{k-p}] = \\ &= [C, \{C, D_k\}] + \sum_{1 \leq p \leq k-1} [C, D_p \circ D_{k-p}]. \end{aligned}$$

Since $C \circ C = 0$, the first term in the right hand side vanishes. Let $\Theta = \sum_{1 \leq p \leq k-1} D_p : \mathcal{B}^{\bullet}_{tot} \to \mathcal{B}^{\bullet+1}_{tot}$. Then the second term is the component of augmentation degree k in the commutator $[C, \Theta \circ \Theta] : S^1 \to \mathcal{B}^3_{tot}$. By assumption $\{D, D\} = 0$ in augmentation degrees $< k$. Therefore we have $\{C, \Theta\} = -\{\Theta, \Theta\}$ in augmentation degrees $< k$. Since Θ increases the augmentation degree, this implies that in augmentation degree k

$$[C, \Theta \circ \Theta] = \{C, \Theta\} \circ \Theta - \Theta \circ \{C, \Theta\} = [\Theta, \{\Theta, \Theta\}],$$

which vanishes tautologically.

The set of all weakly Hodge maps $P : S^1 \to \mathcal{I}_k^1$ coincides with the set of all maps $P : S^1 \to (\mathcal{I}_{tot}^1)_k$ preserving the Hodge bundle structures. Let P be such a map, and let $D' : \mathcal{B}_{tot}^{\bullet} \to \mathcal{B}_{tot}^{\bullet+1}$ be the derivation associated to the extended connection $D' = D + P$.

Since the extended connection D is by assumption linear, by Lemma 8.2.1 the extended connection D' is linear if and only if $\sigma_{tot} \circ P = 0$. Moreover, since the augmentation degree-0 component of the derivation D equals C, the augmentation degree-k component $Q : S^1 \to \mathcal{B}_{tot}^2$ in the composition $D' \circ D'$ is equal to

$$Q = R + \{C, (D' - D)\}.$$

By definition $D' - D : \mathcal{B}_{tot}^{\bullet} \to \mathcal{B}_{tot}^{\bullet+1}$ equals P on $S^1 \subset \mathcal{B}_{tot}^0$ and vanishes on $\Lambda_{tot}^1 \subset \mathcal{B}_{tot}^1$. Since C maps S^1 into $\Lambda_{tot}^1 \subset \mathcal{B}_{tot}^1$, we have $Q = R + C \circ P$. Thus, a map P satisfies the condition of the lemma if and only if

$$\begin{cases} C \circ P = -R \\ \sigma_{tot} \circ P = 0 \end{cases}$$

To prove that such a map P is unique, note that these equations imply

$$h \circ P = (\sigma_{tot} \circ C + C \circ \sigma_{tot}) \circ P = \sigma_{tot} \circ R,$$

and h is invertible by Corollary 8.3.4. To prove that such a map P exists, define P by

$$P = -h^{-1} \circ \sigma_{tot} \circ R : S^1 \to \mathcal{I}_{k+1}^1.$$

The map $h = \{C, \sigma_{tot}\}$ and its inverse h^{-1} commute with C and with σ_{tot}. Since $\sigma_{tot} \circ \sigma_{tot} = C \circ C0$, we have $\sigma_{tot} \circ P = h^{-1} \circ \sigma_{tot} \circ \sigma_{tot} \circ R = 0$. On the other hand, $C \circ R = 0$. Therefore

$$\begin{aligned} C \circ P = -C \circ h^{-1} \circ \sigma_C \circ R = -h^{-1} \circ C \circ \sigma_{tot} \circ R = \\ = h^{-1} \circ \sigma_{tot} \circ C \circ R - h^{-1} \circ h \circ R = -R. \end{aligned}$$

This finishes the proof of the lemma and of Proposition 8.1. □

8.4. Reduction of extended connections

8.4.1. We now complete the proof of Theorem 8.1. First we will need to identify explicitly the low Hodge bidegree components of the reduced Weil algebra $\widetilde{\mathcal{B}}^{\bullet}$. The following is easily checked by direct inspection.

Lemma. *We have*

$$\widetilde{\mathcal{B}}^{2,-1} \oplus \widetilde{\mathcal{B}}^{1,0} \oplus \widetilde{\mathcal{B}}^{0,1} \oplus \widetilde{\mathcal{B}}^{-1,2} = \Lambda^1 \oplus \left(S^1 \otimes \Lambda^1\right) \subset \widetilde{\mathcal{B}}^1$$

$$\widetilde{\mathcal{B}}^{3,-2} \oplus \widetilde{\mathcal{B}}^{2,-1} \oplus \widetilde{\mathcal{B}}^{1,1} \oplus \widetilde{\mathcal{B}}^{-1,2} \oplus \widetilde{\mathcal{B}}^{-2,3} =$$
$$= \left(S^{1,-1} \otimes \Lambda^{2,0}\right) \oplus \left(S^{-1,1} \otimes \Lambda^{0,2}\right) \oplus \Lambda^2 \subset \widetilde{\mathcal{B}}^2$$

8.4.2. Let now $\nabla : S^1 \to S^1 \otimes \Lambda^1$ be an arbitrary real connection on the bundle S^1. The operator

$$D = C + \nabla : S^1 \to \Lambda^1 \oplus \left(\Lambda^1 \otimes S^1\right) \subset \mathcal{B}^1$$

is then automatically weakly Hodge and defines therefore an extended connection on M. This connection is linear by Lemma 8.2.1. Extend D to a derivation $D : \mathcal{B}^{\bullet} \to \mathcal{B}^{\bullet+1}$ as in 6.4.5, and let $\widetilde{D} : \widetilde{\mathcal{B}}^{\bullet} \to \widetilde{\mathcal{B}}^{\bullet+1}$ be the associated derivation of the reduced Weil algebra.

Lemma. *The derivation $\widetilde{D}$ satisfies $\widetilde{D} \circ \widetilde{D} = 0$ if and only if the connection ∇ is Kählerian.*

Proof. Indeed, the operator $\widetilde{D} \circ \widetilde{D}$ is weakly Hodge, hence factors through a bundle map

$$\widetilde{D} \circ \widetilde{D} : S^1 \to \widetilde{\mathcal{B}}^{3,-1} \oplus \widetilde{\mathcal{B}}^{2,0} \oplus \widetilde{\mathcal{B}}^{1,1} \oplus \widetilde{\mathcal{B}}^{0,2} \oplus \widetilde{\mathcal{B}}^{-1,3} =$$
$$= \left(S^{1,-1} \otimes \Lambda^{2,0}\right) \oplus \left(S^{-1,1} \otimes \Lambda^{0,2}\right) \oplus \Lambda^2 \subset \mathcal{B}^2.$$

By definition we have

$$\widetilde{D} \circ \widetilde{D} = (C + \nabla) \circ (C + \nabla) = \{C, \nabla\} + \{\nabla, \nabla\}.$$

An easy inspection shows that the sum is direct, and the first summand equals

$$\{C, \nabla\} = T \circ C : S^1 \to \Lambda^2,$$

where T is the torsion of the connection ∇, while the second summand equals

$$\{\nabla, \nabla\} = R^{2,0} \oplus R^{0,2} : S^{1,-1} \oplus S^{-1,1} \to \left(S^{1,-1} \otimes \Lambda^{2,0}\right) \oplus \left(S^{-1,1} \otimes \Lambda^{0,2}\right),$$

where $R^{2,0}$, $R^{0,2}$ are the Hodge type components of the curvature of the connection ∇. Hence $\widetilde{D} \circ \widetilde{D} = 0$ if and only if $R^{2,0} = R^{0,2} = T = 0$, which proves the lemma and finishes the proof of Theorem 8.1. □

8.4.3. We finish this section with the following corollary of Theorem 8.1 which gives an explicit expression for the augmentation degree-2 component D_2 of a flat linear extended connection D on the manifold M. We will need this expression in Section 9.

Corollary. *Let $D = \sum_{k \geq 0} D_k : S^1 \to \mathcal{B}^1$ be a flat linear extended connection on M, so that $D_0 = C$ and D_1 is a Kählerian connection on M. We have*

$$D_2 = \frac{1}{3}\sigma \circ R,$$

where $\sigma : \mathcal{B}^{\bullet+1} \to \mathcal{B}^{\bullet}$ is the canonical derivation introduced in 6.3.7, and $R = D_1 \circ D_1 : S^1 \to S^1 \otimes \Lambda^{1,1} \subset \mathcal{B}^2$ is the curvature of the Kählerian connection D_1.

Proof. Extend the connection D to a derivation $D_C = \sum_{k \geq 0} D_k^{tot} : \mathcal{B}_{tot}^{\bullet} \to \mathcal{B}_{tot}^{\bullet+1}$ of the total Weil algebra. By the construction used in the proof of Lemma 8.3.6 we have $D_2^{tot} = h^{-1} \circ \sigma_{tot} \circ R_{tot} : S^1 \to \left(\mathcal{B}_{tot}^1\right)_3$, where $h : \mathcal{B}_{tot}^{\bullet} \to \mathcal{B}_{tot}^{\bullet}$ is as in Lemma 7.3.8, the map $\sigma_{tot} : \mathcal{B}_{tot}^{\bullet+1} \to \mathcal{B}_{tot}^{\bullet}$ is the canonical map constructed in 7.3.7, and $R^{tot} : S^1 \to \left(\mathcal{B}_{tot}^2\right)_3$ is the square $R_{tot} = D_1^{tot} \circ D_1^{tot}$ of the derivation $D_1^{tot} : \mathcal{B}_{tot}^{\bullet} \to \mathcal{B}_{tot}^{\bullet+1}$. By Lemma 7.3.9 the map h acts on $\left(\mathcal{B}_{tot}^1\right)_3$ by multiplication by 3. Moreover, it is easy to check that

$$\left(\mathcal{B}_{tot}^2\right)_3 = \left(S^1 \otimes \Lambda^2\right) \oplus \left(S^{-1,1} \otimes \Lambda^{2,0}\right) \oplus \left(S^{1,-1} \otimes \Lambda^{0,2}\right),$$

and the map $R^{tot} : S^1 \to \left(\mathcal{B}_{tot}^2\right)_3$ sends S^1 into the first summand in this decomposition and coincides with the curvature $R : S^1 \to S^1 \otimes \Lambda^2 \subset \left(\mathcal{B}_{tot}^2\right)_3$ of the Kählerian connection D_1. Therefore $\sigma_{tot} \circ R^{tot} = \sigma \circ R$, which proves the claim. □

9. Metrics

9.1. Hyperkähler metrics on Hodge manifolds

9.1.1. Let M be a complex manifold equipped with a Kählerian connection ∇, and consider the associated linear formal Hodge manifold structure on the tangent bundle $\overline{T}M$. In this section we construct a natural bijection between the set of all polarizations on the Hodge manifold $\overline{T}M$ in the sense of Subsection 3.3 and the set of all Kähler metrics on M compatible with the given connection ∇.

9.1.2. Let h be a hyperkähler metric on $\overline{T}M$, or, more generally, a formal germ of such a metric in the neighborhood of the zero section $M \subset \overline{T}M$.

Assume that the metric h is compatible with the given hypercomplex structure and Hermitian-Hodge in the sense of 1.5.2, and let ω_I be the Kähler form associated to h in the preferred complex structure $\overline{T}M_I$ on $\overline{T}M$.

Let h_M be the restriction of the metric h to the zero section $M \subset \overline{T}M$, and let $\omega \in C^\infty(M, \Lambda^{1,1}(M))$ be the associated real $(1,1)$-form on the complex manifold M. Since the embedding $M \subset \overline{T}M_I$ is holomorphic, the form ω is the restriction onto M of the form ω_I. In particular, it is closed, and the metric h_M is therefore Kähler.

9.1.3. The main result of this section is the following.

Theorem 9.1 *Restriction onto the zero section $M \subset \overline{T}M$ defines a one-to-one correspondence between*

(i) *Kähler metrics on M compatible with the Kählerian connection ∇, and*

(ii) *formal germs in the neighborhood on $M \subset \overline{T}M$ of Hermitian-Hodge hyperkähler metrics on $\overline{T}M$ compatible with the given formal Hodge manifold structure.*

The rest of this section is devoted to the proof of this theorem.

9.1.4. In order to prove Theorem 9.1, we reformulate it in terms of polarizations rather than metrics. Recall (see 3.3.3) that a *polarization* of the formal Hodge manifold $\overline{T}M$ is by definition a $(2,0)$-form $\Omega \in C^\infty_M(\overline{T}M, \Lambda^{2,0}(\overline{T}M_J))$ for the complementary complex structure $\overline{T}M_J$ which is holomorphic, real and of H-type $(1,1)$ with respect to the canonical Hodge bundle structure on $\Lambda^{2,0}(\overline{T}M_J)$, and satisfies a certain positivity condition (3.3).

9.1.5. By Lemma 3.3.4 Hermitian-Hodge hyperkähler metrics on $\overline{T}M$ are in one-to-one correspondence with polarizations. Let h be a Kähler metric on M, and let ω_I and ω be the Kähler forms for h on $\overline{T}M_I$ and on $M \subset \overline{T}M_I$. The corresponding polarization $\Omega \in C^\infty(\overline{T}M, \Lambda^{2,0}(\overline{T}M_J))$ satisfies by (1.4)

$$\omega_I = \frac{1}{2}(\Omega + \nu(\Omega)) \in \Lambda^2(\overline{T}M, \mathbb{C}), \tag{9.1}$$

where $\nu : \Lambda^2(\overline{T}M, \mathbb{C}) \to \overline{\Lambda^2(\overline{T}M, \mathbb{C})}$ is the usual complex conjugation.

9.1.6. Let $\rho : \overline{T}M \to M$ be the natural projection, and let

$$\mathrm{Res} : \rho_*\Lambda^{\bullet,0}(\overline{T}M_J) \to \Lambda^\bullet(M, \mathbb{C})$$

be the map given by the restriction onto the zero section $M \subset \overline{T}M$. Both bundles are naturally Hodge bundles of the same weight on M in the sense of 2.1.2, and the bundle map Res preserves the Hodge bundle structures.

Since Ω is of H-type $(1,1)$, the form $\operatorname{Res}\Omega \in C^\infty(M, \Lambda^2(M,\mathbb{C}))$ is real and of Hodge type $(1,1)$. By (9.1)

$$\operatorname{Res}\Omega = \frac{1}{2}\left(\operatorname{Res}\Omega + \overline{\operatorname{Res}\Omega}\right) = \frac{1}{2}(\Omega + \nu(\Omega))|_{M\subset\overline{T}M} =$$
$$= \omega_I|_{M\subset\overline{T}M} = \omega \in \Lambda^{1,1}(M).$$

Therefore to prove Theorem 9.1, it suffices to prove the following.

- For every polarization Ω of the formal Hodge manifold $\overline{T}M$ the restriction $\omega = \operatorname{Res}\Omega \in C^\infty(\Lambda^{1,1}(M))$ is compatible with the connection ∇, that is, $\nabla\omega = 0$. Vice versa, every real positive $(1,1)$-form $\omega \in C^\infty(\Lambda^{1,1}(M))$ satisfying $\nabla\omega = 0$ extends to a polarization Ω of $\overline{T}M$, and such an extension is unique.

This is what we will actually prove.

9.2. Preliminaries

9.2.1. We begin with introducing a convenient model for the holomorphic de Rham algebra $\Lambda^{\bullet,0}(\overline{T}M_J)$ of the complex manifold $\overline{T}M_J$, which would be independent of the Hodge manifold structure on $\overline{T}M$. To construct such a model, consider the relative de Rham complex $\Lambda^\bullet(\overline{T}M/M,\mathbb{C})$ of $\overline{T}M$ over M (see 5.2 for a reminder of its definition and main properties). Let $\pi : \Lambda^\bullet(\overline{T}M,\mathbb{C}) \to \Lambda^\bullet(\overline{T}M/M,\mathbb{C})$ be the canonical projection. Recall that the bundle $\Lambda^i(\overline{T}M/M,\mathbb{C})$ of relative i-forms on $\overline{T}M$ over M carries a natural structure of a Hodge bundle of weight i. Moreover, we have introduced in (5.5) a Hodge bundle isomorphism

$$\eta : \rho^*\Lambda^\bullet(M,\mathbb{C}) \to \Lambda^\bullet(\overline{T}M/M,\mathbb{C})$$

between $\Lambda^i(\overline{T}M/M,\mathbb{C})$ and the pullback $\rho^*\Lambda^\bullet(M,\mathbb{C})$ of the bundle of $\mathbb{C}$-valued i-forms on M.

Lemma. (i) *The projection π induces an algebra isomorphism*

$$\pi : \Lambda^{\bullet,0}(\overline{T}M_J) \to \Lambda^\bullet(\overline{T}M/M,\mathbb{C})$$

compatible with the natural Hodge bundle structures.

(ii) *Let $\alpha \in C^\infty_M(\overline{T}M, \Lambda^{i,0}(\overline{T}M_J))$ be a smooth $(i,0)$-form on $\overline{T}M_J$, and consider the smooth i-form*

$$\beta = \eta^{-1}\pi(\alpha) \in C^\infty_M(\overline{T}M, \Lambda^i(\overline{T}M,\mathbb{C}))$$

on $\overline{T}M$. The forms α and β have the same restriction to the zero section $M \subset \overline{T}M$.

Proof. Since η, π and the restriction map are compatible with the algebra structure on $\Lambda^{\bullet}(M, \mathbb{C})$, it suffices to prove both claims for $\Lambda^1(M, \mathbb{C})$. For every bundle $\mathcal{E}$ on $\overline{T}M$ denote by $\mathcal{E}|_{M \subset \overline{T}M}$ the restriction of $\mathcal{E}$ to the zero section $M \subset \overline{T}M$. Consider the bundle map

$$\chi = \eta \circ \mathrm{Res} : \Lambda^{1,0}(\overline{T}M_J)|_{M \subset \overline{T}M} \to \Lambda^1(M, \mathbb{C}) \to \Lambda^1(\overline{T}M/M, \mathbb{C})|_{M \subset \overline{T}M}.$$

The second claim of the lemma is then equivalent to the identity $\chi = \pi$. Moreover, note that the contraction with the canonical vector field φ on $\overline{T}M$ defines an injective map $i_\varphi : C^\infty(M, \Lambda^1(\overline{T}M/M, \mathbb{C})|_M) \to C^\infty(\overline{T}M, \mathbb{C})$. Therefore it suffices to prove that $i_\varphi \circ \chi = i_\varphi \circ \pi$.

Every smooth section s of the bundle $\Lambda^{1,0}(\overline{T}M_J)|_{M \subset \overline{T}M}$ is of the form

$$s = (\rho^*\alpha + \sqrt{-1} j \rho^*\alpha)|_{M \subset \overline{T}M},$$

where $\alpha \in C^\infty(M, \Lambda^1(M, \mathbb{C}))$ is a smooth 1-form on M, and

$$j : \Lambda^1(\overline{T}M, \mathbb{C}) \to \overline{\Lambda^1(\overline{T}M, \mathbb{C})}$$

is the map induced by the quaternionic structure on $\overline{T}M$. For such a section s we have $\mathrm{Res}(s) = \alpha$, and by (5.2.8) $i_\varphi(\chi(s)) = \sqrt{-1}\tau(\alpha)$, where $\tau : C^\infty(M, \Lambda^1(M, \mathbb{C})) \to C^\infty(\overline{T}M, \mathbb{C})$ is the tautological map introduced in 4.3.2. On the other hand, since $\pi \circ \rho^* = 0$, we have

$$i_\varphi(\pi(s)) = i_\varphi(\pi(\sqrt{-1} j \rho^*\alpha)) = i_\varphi(\sqrt{-1} j \rho^*\alpha).$$

Since the Hodge manifold structure on $\overline{T}M$ is linear, this equals

$$i_\varphi(\pi(s)) = \sqrt{-1} i_\varphi(j(\rho^*\alpha)) = \sqrt{-1}\tau(\alpha) = \chi(s),$$

which proves the second claim of the lemma. Moreover, it shows that the restriction of the map π to the zero section $M \subset \overline{T}M$ is an isomorphism. As in the proof of Lemma 5.1.9, this implies that the map π is an isomorphism on the whole $\overline{T}M$, which proves the first claim and finishes the proof of the lemma. □

9.2.2. Lemma 9.2.1 (i) allows to define the bundle isomorphism

$$\pi^{-1} \circ \eta : \rho^*\Lambda^{\bullet}(M, \mathbb{C}) \to \Lambda^{\bullet}(\overline{T}M/M, \mathbb{C}) \to \Lambda^{\bullet,0}(\overline{T}M_J),$$

between $\rho^*\Lambda^{\bullet}(M, \mathbb{C})$ and $\Lambda^{\bullet,0}(\overline{T}M_J)$, and it induces an isomorphism

$$\rho_*(\eta \circ \pi^{-1}) : \rho_*\rho^*\Lambda^{\bullet}(M, \mathbb{C}) \cong \rho_*\Lambda^{\bullet,0}(\overline{T}M_J)$$

between the direct images of these bundles under the canonical projection $\rho : \overline{T}M \to M$.

On the other hand, by adjunction we have the canonical embedding

$$\Lambda^{\bullet}(M,\mathbb{C}) \hookrightarrow \rho_*\rho^*\Lambda^{\bullet}(M,\mathbb{C}),$$

and by the projection formula it extends to an isomorphism

$$\rho_*\rho^*\Lambda^{\bullet}(M,\mathbb{C}) \cong \Lambda^{\bullet}(M,\mathbb{C}) \otimes \mathcal{B}^0,$$

where $\mathcal{B}^0 = \rho_*\Lambda^0(\overline{T}M,\mathbb{C})$ is the 0-th component of the Weil algebra $\mathcal{B}^{\bullet}$ of M. All these isomorphisms are compatible with the Hodge bundle structures and with the multiplication.

9.2.3. It will be convenient to denote the image $\rho_*(\eta \circ \pi^{-1})(\Lambda^{\bullet}(M,\mathbb{C})) \subset \rho_*\Lambda^{\bullet,0}(\overline{T}M_J)$ by $L^{\bullet}(M,\mathbb{C})$ or, to simplify the notation, by $L^{\bullet}$. (The algebra $L^{\bullet}(M,\mathbb{C})$ is, of course, canonically isomorphic to $\Lambda^{\bullet}(M,\mathbb{C})$.) We then have the identification

$$L^{\bullet} \otimes \mathcal{B}^0 \cong \rho_*\rho^*\Lambda^{\bullet}(M,\mathbb{C}) \cong \rho_*\Lambda^{\bullet,0}(\overline{T}M_J). \tag{9.2}$$

This identification is independent of the Hodge manifold structure on $\overline{T}M$. Moreover, by Lemma 9.2.1 (ii) the restriction map $\mathrm{Res} : \rho_*\Lambda^{\bullet,0}(\overline{T}M_J) \to \Lambda^{\bullet}(M,\mathbb{C})$ is identified under (9.2) with the canonical projection $L^{\bullet} \otimes \mathcal{B}^0 \to L^{\bullet} \otimes \mathcal{B}^0_0 \cong L^{\bullet}$.

By Lemma 5.1.9 we also have the identification $\rho_*\Lambda^{0,\bullet}(\overline{T}M_J) \cong \mathcal{B}^{\bullet}$. Therefore (9.2) extends to an algebra isomorphism

$$\rho_*\Lambda^{\bullet,\bullet}(\overline{T}M_J) \cong \rho_*\Lambda^{\bullet}(\overline{T}M/M,\mathbb{C}) \otimes \Lambda^{\bullet}(M,\mathbb{C}) \cong L^{\bullet} \otimes \mathcal{B}^{\bullet}. \tag{9.3}$$

This isomorphism is also compatible with the Hodge bundle structures on both sides.

9.3. The Dolbeult differential on $\overline{T}M_J$

9.3.1. Our next goal is to express the Dolbeult differential $\bar\partial_J$ of the complex manifold $\overline{T}M_J$ in terms of the model for the de Rham complex $\Lambda^{\bullet,\bullet}(\overline{T}M_J)$ given by (9.3). For every $k \geq 0$ denote by

$$D : L^k \otimes \mathcal{B}^{\bullet} \to L^k \otimes \mathcal{B}^{\bullet+1}.$$

the differential operator induced by $\bar\partial_J : \Lambda^{\bullet,\bullet}(\overline{T}M_J) \to \Lambda^{\bullet,\bullet+1}(\overline{T}M_J)$ under (9.3). The operator D is weakly Hodge. It satisfies the Leibnitz rule with respect to the algebra structure on $L^{\bullet} \otimes \mathcal{B}^{\bullet}$, and we have $D \circ D = 0$. By definition for $k = 0$ it coincides with the derivation $D : \mathcal{B}^{\bullet} \to \mathcal{B}^{\bullet+1}$ defined by the Hodge manifold structure on $\overline{T}M$. For $k > 0$ the complex $\langle L^k \otimes \mathcal{B}^{\bullet}, D\rangle$ is a free differential graded module over the Weil algebra $\langle \mathcal{B}^{\bullet}, D\rangle$.

9.3.2. The relative de Rham differential d^r (see Subsection 5.2) induces under the isomorphism (9.3) an algebra derivation

$$d^r : L^{\bullet} \otimes \mathcal{B}^{\bullet} \to L^{\bullet+1} \otimes \mathcal{B}^{\bullet}.$$

The derivation d^r also is weakly Hodge, and we have the following.

Lemma. *The derivations D and d^r commute, that is,*

$$\{D, d^r\} = 0 : L^{\bullet} \otimes \mathcal{B}^{\bullet} \to L^{\bullet+1} \otimes \mathcal{B}^{\bullet+1}.$$

Proof. The operator $\{D, d^r\}$ satisfies the Leibnitz rule, so it suffices to prove that it vanishes on $\mathcal{B}^0$, $\mathcal{B}^1$ and $L^1 \otimes \mathcal{B}^0$. Moreover, the $\mathcal{B}^0$-modules $\mathcal{B}^1$ and $L^1 \otimes \mathcal{B}^0$ are generated, respectively, by local sections of the form Df and $d^r f$, $f \in \mathcal{B}^0$. Since $\{D, d^r\}$ commutes with D and d^r, it suffices to prove that it vanishes on $\mathcal{B}^0$. Finally, $\{D, d^r\}$ is continuous in the adic topology on $\mathcal{B}^0$. Since the subspace

$$\{fg | f, g \in \mathcal{B}^0, Df = d^r g = 0\} \subset \mathcal{B}^0$$

is dense in this topology, it suffices to prove that for a local section $f \in \mathcal{B}^0$ we have $\{D, d^r\}f = 0$ if either $d^r f = 0$ of $Df = 0$.

It is easy to see that for every local section $\alpha \in \mathcal{B}^{\bullet}$ we have $d^r \alpha = 0$ if and only if $\alpha \in \mathcal{B}^{\bullet}_0$ is of augmentation degree 0. By definition the derivation D preserves the component $\mathcal{B}^{\bullet}_0 \subset \mathcal{B}^{\bullet}$ of augmentation degree 0 in $\mathcal{B}^0$. Therefore $d^r f = 0$ implies $d^r Df = 0$ and consequently $\{D, d^r\}f = 0$. This handles the case $d^r f = 0$. To finish the proof, assume given a local section $f \in \mathcal{B}^0$ such that $Df = 0$. Such a section by definition comes from a germ at $M \subset \overline{T}M$ of a holomorphic function $\widehat{f}$ on $\overline{T}M_J$. Since $\widehat{f}$ is holomorphic, we have $\bar{\partial}_J \widehat{f} = 0$ and $d\widehat{f} = \partial_J \widehat{f}$. Therefore $d^r f = \pi(d\widehat{f}) = \pi(\partial_J f)$, and

$$Dd^r f = \pi(\bar{\partial}_J \partial_J \widehat{f}) = -\pi(\partial_J \bar{\partial}_J \widehat{f}) = 0,$$

which, again, implies $\{D, d^r\}f = 0$. □

9.3.3. Let now $\nabla = D_1 : S^1 \to S^1 \otimes \Lambda^1$ be the reduction of the extended connection D. It induces a connection on the bundle $L^1 \cong S^1$, and this connection extends by the Leibnitz rule to a connection on the exterior algebra $L^{\bullet}$ of the bundle L^1, which we will also denote by ∇.

Denote by $R = \nabla \circ \nabla : L^{\bullet} \to L^{\bullet} \otimes \Lambda^2$ the curvature of the connection ∇. Since $\nabla \circ \nabla = \frac{1}{2}\{\nabla, \nabla\}$, the operator R also satisfies the Leibnitz rule with respect to the multiplication in $L^{\bullet}$.

9.3.4. Introduce the augmentation grading on the bundle $L^{\bullet} \otimes \mathcal{B}^{\bullet}$ by setting $\deg L^{\bullet} = 0$. The derivation $D : L^{\bullet} \otimes \mathcal{B}^{\bullet} \to L^{\bullet} \otimes \mathcal{B}^{\bullet+1}$ obviously does not increase the augmentation degree, and we have the decomposition $D = \sum_{k \geq 0} D_k$. On the other hand, the derivation d^r preserves the augmentation degree. Therefore Lemma 9.3.2 implies that for every $k \geq 0$ we have $\{D_k, d^r\} = 0$. This in turn implies that $D_0 = 0$ on L^p for $p > 0$, and therefore $D_0 = \mathsf{id} \otimes C : L^p \otimes \mathcal{B}^{\bullet} \to L^p \otimes \mathcal{B}^{\bullet+1}$. Moreover, this allows to identify explicitly the components D_1 and D_2 of the derivation $D : L^{\bullet} \to L^{\bullet} \otimes \mathcal{B}^1$. Namely, we have the following.

Lemma. *We have*

$$D_1 = \nabla : L^{\bullet} \to L^{\bullet} \otimes \mathcal{B}^1_1 = L^{\bullet} \otimes \Lambda^1$$
$$D_2 = \frac{1}{3}\sigma \circ R : L^{\bullet} \to L^{\bullet} \otimes \mathcal{B}^1_2,$$

where $\sigma = \mathsf{id} \otimes \sigma : L^{\bullet} \otimes \mathcal{B}^{\bullet+1} \to L^{\bullet} \otimes \mathcal{B}^{\bullet}$ *is as in* 7.3.2.

Proof. Since both sides of these identities satisfy the Leibnitz rule with respect to the multiplication in $L^{\bullet}$, it suffices to prove them for L^1. But $d^r : \mathcal{B}^0 \to L^1 \otimes \mathcal{B}^0$ restricted to $S^1 \subset \mathcal{B}^0$ becomes an isomorphism $d^r : S^1 \to L^1$. Since $\{D_1, d^r\} = \{D_2, d^r\} = 0$, it suffices to prove the identities with L^1 replaced with S^1. The first one then becomes the definition of ∇, and the second one is Corollary 8.4.3. □

9.4. The proof of Theorem 9.1

9.4.1. We can now prove Theorem 9.1 in the form 9.1.6. We begin with the following corollary of Lemma 9.3.4.

Corollary. *Let* $\mathcal{I}^{\bullet} \subset \mathcal{B}^{\bullet}$ *be the ideal introduced in* 8.3.3. *An arbitrary smooth section* $\alpha \in C^{\infty}(M, L^{\bullet})$ *satisfies*

$$D\alpha \in C^{\infty}(M, L^{\bullet} \otimes \mathcal{I}^1) \subset C^{\infty}(M, L^{\bullet} \otimes \mathcal{B}^1) \tag{9.4}$$

if and only if $\nabla\alpha = 0$.

Proof. Again, both the identity (9.4) and the equality $\nabla\alpha$ are compatible with the Leibnitz rule with respect to the multiplication in α. Therefore it suffices to prove that they are equivalent for every $\alpha \in L^1$. By definition of the ideal $\mathcal{I}^{\bullet}$ the equality (9.4) holds if and only if $D_1\alpha = D_2\alpha = 0$. By Lemma 9.3.4 this is equivalent to $\nabla\alpha = \sigma \circ R(\alpha) = 0$. But since $R = \nabla \circ \nabla$, $\nabla\alpha = 0$ implies $\sigma \circ R(\alpha) = 0$, which proves the claim. □

9.4.2. Let now $\Omega \in C^\infty(M, \rho_*\Lambda^{2,0}(\overline{T}M_J) \cong L^2 \otimes \mathcal{B}^0$ be a polarization of the Hodge manifold $\overline{T}M_J$, so that Ω is of Hodge type $(1,1)$ and $D\Omega = 0$. Let $\omega = \operatorname{Res}\Omega \in C^\infty(M, \Lambda^{1,1}(M, \mathbb{C}))$ be its restriction, and let $\Omega = \sum_{k \geq 0} \Omega_k$ be its augmentation degree decomposition.

As noted in 9.2.3, the restriction map $\operatorname{Res} : \rho_*\Lambda^{\bullet,0}(\overline{T}M_J) \to \Lambda^{\bullet}(M, \mathbb{C})$ is identified under the isomorphism (9.3) with the projection $\mathcal{L}^{\bullet} \otimes \mathcal{B}^0 \to L^{\bullet}$ onto the component of augmentation degree 0. Therefore $\omega = \Omega_0$. Since the augmentation degree-1 component $\left(L^2 \otimes B^0_{tot}\right)^{1,1}_1 = 0$ and $D\Omega = 0$, we have $\nabla\omega = D_1\Omega_0 = 0$, which proves the first claim of Theorem 9.1.

9.4.3. To prove the second claim of the theorem, let ω be a Kähler form on M compatible with the connection ∇, so that $\nabla\omega = 0$. We have to show that there exists a unique section $\Omega = \sum_{k \geq 0} \Omega_k \in C^\infty(M, L^2 \otimes \mathcal{B}^0)$ which is of Hodge type $(1,1)$ and satisfies $D\Omega = 0$ and $\Omega_0 = \omega$. As in the proof of Theorem 8.1, we will use induction on k. Since Ω is of Hodge type $(1,1)$, we must have $\Omega_1 = 0$, and by Corollary 9.4.1 we have $D(\Omega_0 + \Omega_1) \in C^\infty(M, L^2 \otimes \mathcal{I}^1)$, which gives the base of our induction. The induction step is given by applying the following proposition to $\sum_{0 \geq p \geq k} \Omega_k$ for each $k \geq 1$ in turn.

Proposition 9.1 *Assume given integers $p, q, k; p, q \geq 0, k \geq 1$ and assume given a section $\alpha \in C^\infty(M, L^{p+q} \otimes \mathcal{B}^0)$ of Hodge type (p,q) such that*

$$D\alpha \in C^\infty(M, L^{p+q} \otimes \mathcal{I}^1_{\geq k}),$$

where $\mathcal{I}^{\bullet}_{\geq k} = \oplus_{m \geq k} \mathcal{I}^{\bullet}_m$. Then there exists a unique section

$$\beta \in C^\infty(M, L^{p+q} \otimes \mathcal{B}_k)$$

of the same Hodge type (p,q) and such that

$$D(\alpha + \beta) \in C^\infty(M, L^{p+q} \otimes \mathcal{I}^1_{\geq k+1}).$$

Proof. Let $\mathcal{B}^{\bullet}_{tot}$ be the total Weil algebra introduced in 7.2.4, and consider the free module $L^{p+q} \otimes \mathcal{B}^{\bullet}_{tot}$ over $\mathcal{B}^{\bullet}_{tot}$ generated by the Hodge bundle L^{p+q}. This module carries a canonical Hodge bundle structure of weight $p+q$. Consider the maps $C : \mathcal{B}^{\bullet}_{tot} \to \mathcal{B}^{\bullet+1}_{tot}$, $\sigma_{tot} : \mathcal{B}^{\bullet+1} \to \mathcal{B}^{\bullet}_{tot}$ introduced in 7.3.3 and 7.3.7, and let $C = \mathsf{id} \otimes C$, $\sigma_{tot} = \mathsf{id} \otimes \sigma_{tot} : L^{p+q} \otimes \mathcal{B}_{tot} \to L^{p+q} \otimes \mathcal{B}_{tot}$ be the associated endomorphisms of the free module $L^{p+q} \otimes \mathcal{B}^{\bullet}_{tot}$.

The maps C and σ_{tot} preserve the Hodge bundle structure. The commutator $h = \{C, \sigma_{tot}\} : \mathcal{B}^{\bullet}_{tot} \to \mathcal{B}^{\bullet}_{tot}$ is invertible on $\mathcal{I}^{\bullet}_{tot} \subset \mathcal{B}^{\bullet}_{tot}$ by Corollary 8.3.4 and acts as $k\mathsf{id}$ on $\mathcal{B}^0_k \subset \mathcal{B}^{\bullet}_{tot}$. Therefore the endomorphism

$$h = \mathsf{id} \otimes h = \{C, \sigma_{tot}\} : L^{p+q} \otimes \mathcal{B}^{\bullet}_{tot} \to L^{p+q} \otimes \mathcal{B}^{\bullet}_{tot}$$

is invertible on $L^{p+q} \otimes \mathcal{I}^{\bullet}_{tot}$ and acts as $k\mathsf{id}$ on $L^{p+q} \otimes \mathcal{B}^0_k$.

Since the derivation $D : L^{p+q} \otimes \mathcal{B}^{\bullet} \to L^{p+q} \otimes \mathcal{B}^{\bullet+1}$ is weakly Hodge, it induces a map $D^{tot} : L^{p+q} \otimes \mathcal{B}^{\bullet}_{tot} \to L^{p+q} \otimes \mathcal{B}^{\bullet+1}_{tot}$, and $D\alpha \in L^{p+q} \otimes \mathcal{I}^1_{\geq k}$ if and only if the same holds for $D^{tot}\alpha$. To prove uniqueness, note that $D_0 = C$ is injective on $L^{p+q} \otimes \mathcal{B}^0_k$. If there are two sections β, β' satisfying the conditions of the proposition, then $D_0(\beta - \beta') = 0$, hence $\beta = \beta'$.

To prove existence, let $\gamma = (D^{tot}\alpha)_k$ be the component of the section $D^{tot}\alpha$ of augmentation degree k. Since $D^{tot} \circ D^{tot} = 0$, we have $C\gamma = D^{tot}_0\gamma = 0$ and $C \circ \sigma_{tot}\gamma = h\gamma$. Let $\beta = -\frac{1}{k} \circ \sigma_{tot}(\gamma)$. The section β is of Hodge type (p, q) and of augmentation degree k. Moreover, it satisfies

$$D^{tot}_0\beta = C\beta = -Ch^{-1}\sigma_{tot}(\gamma) = -h^{-1} \circ C \circ \sigma_{tot}\gamma = -h^{-1} \circ h\gamma = -\gamma.$$

Therefore $D_{tot}(\alpha + \beta)$ is indeed a section of $L^{p+q} \otimes (I^1_{tot})_{\geq k+1}$, which proves the proposition and finishes the proof of Theorem 9.1. □

9.5. The cotangent bundle

9.5.1. For every Kähler manifold M Theorem 9.1 provides a canonical formal hyperkähler structure on the total space $\overline{T}M$ of the complex-conjugate to the tangent bundle to M. In particular, we have a closed holomorphic 2-form Ω_I for the preferred complex structure $\overline{T}M_I$ on M.

Let T^*M be the total space to the cotangent bundle to M equipped with the canonical holomorphic symplectic form Ω. To obtain a hyperkähler metric of the formal neighborhood of the zero section $M \subset T^*M$, one can apply an appropriate version of the Darboux-Weinstein Theorem, which gives a local symplectic isomorphism $\kappa : \overline{T}M \to T^*M$ in a neighborhood of the zero section. However, this theorem is not quite standard in the holomorphic and formal situations. For the sake of completeness, we finish this section with a sketch of a construction of such an isomorphism $\kappa : \overline{T}M \to T^*M$ which can be used to obtain a hyperkähler metric on T^*M rather than on $\overline{T}M$.

9.5.2. We begin with the following preliminary fact on the holomorphic de Rham complex of the manifold $\overline{T}M_I$.

Lemma. *Assume given either a formal Hodge manifold structure on the $U(1)$-manifold $\overline{T}M$ along the zero section $M \subset \overline{T}M$, or an actual Hodge manifold structure on an open neighborhood $U \subset \overline{T}M$ of the zero section.*

(i) *For every point $m \in M$ there exists an open neighborhood $U \subset \overline{T}M_I$ such that the spaces $\Omega^\bullet(U)$ of holomorphic forms on the complex manifold $\overline{T}M_I$ (formally completed along $M \subset \overline{T}M$ if necessary) equipped with the holomorphic de Rham differential $\partial_I : \Omega^\bullet(U) \to \Omega^{\bullet+1}(U)$ form an exact complex.*

(ii) *If the subset $U \subset \overline{T}M$ is invariant under the $U(1)$-action on the manifold $\overline{T}M$, then the same is true for the subspaces $\Omega^\bullet_k(U) \subset \Omega^\bullet(U)$ of forms of weight k with respect to the $U(1)$-action.*

(iii) *Assume further that the canonical projection $\rho : \overline{T}M_I \to M$ is holomorphic for the preferred complex structure $\overline{T}M_I$ on $\overline{T}M$. Then both these claims also hold for the spaces $\Omega^\bullet(U/M)$ of relative holomorphic forms on U over M.*

Proof. The claim (i) is standard. To prove (ii), note that, both in the formal and in the analytic situation, the spaces $\Omega^\bullet(U)$ are equipped with a natural topology. Both this topology and the $U(1)$-action are preserved by the holomorphic Dolbeult differential ∂_I.

The subspaces $\Omega^\bullet_{fin}(U) \subset \Omega^\bullet(U)$ of $U(1)$-finite vectors are dense in the natural topology. Therefore the complex $\langle \Omega^\bullet_{fin}(U), \partial_I \rangle$ is also exact. Since the group $U(1)$ is compact, we have

$$\Omega^\bullet_{fin}(U) = \bigoplus_k \Omega^\bullet_k(U),$$

which proves (ii). The claim (iii) is, again, standard. □

9.5.3. We can now formulate and prove the main result of this subsection.

Proposition 9.2 *Assume given a formal polarized Hodge manifold structure on the manifold $\overline{T}M$ along the zero section $M \subset \overline{T}M$ such that the canonical projection $\rho : \overline{T}M_I \to M$ is holomorphic for the preferred complex structure $\overline{T}M_I$ on $\overline{T}M$. Let Ω_I be the associated formal holomorphic 2-form on $\overline{T}M_I$. Let T^*M be the total space of the cotangent bundle to the manifold M equipped with a canonical holomorphic symplectic form Ω. There exists a unique $U(1)$-equivariant holomorphic map $\kappa : \overline{T}M_I \to T^*M$, defined in a formal neighborhood of the zero section, which commutes with the canonical projections onto M and satisfies $\Omega_I = \kappa^*\Omega$. Moreover, if the polarized Hodge* *manifold structure on $\overline{T}M_I$ is defined in an open neighborhood $U \subset \overline{T}M$ of the zero section $M \subset \overline{T}M$, then the map κ is also defined in a (possibly smaller) open neighborhood of the zero section.*

Proof. By virtue of the uniqueness, the claim is local on M, so that we can assume that the whole M is contained in a $U(1)$-invariant neighborhood $U \subset \overline{T}M_I$ satisfying the conditions of Lemma 9.5.2. Holomorphic maps $\kappa : U \to T^*M$ which commute with the canonical projections onto M are in a natural one-to-one correspondence with holomorphic sections α of the bundle $\rho^*\Lambda^{1,0}(M)$ on $\overline{T}M_I$. Such a map κ is $U(1)$-equivariant if and only if the corresponding 1-form $\alpha \in \Omega^1(U)$ is of weight 1 with respect to the $U(1)$-action. Moreover, it satisfies $\kappa^*\Omega = \Omega_I$ if and only if $\partial_I \alpha = \Omega_I$. Therefore to prove the formal resp. analytic parts of the proposition it suffices to prove that there exists a unique holomorphic formal resp. analytic section $\alpha \in C^\infty(U, \rho^*\Lambda^{1,0}(M)$ which is of weight 1 with respect to the $U(1)$-action and satisfies $\partial_I \alpha = \Omega_I$.

The proof of this fact is the same in the formal and in the analytic situations. By definition of the polarized Hodge manifold the 2-form $\Omega_I \in \Omega^2(U)$ is of weight 1 with respect to the $U(1)$-action. Therefore by Lemma 9.5.2 (ii) there exists a holomorphic 1-form $\alpha \in \Omega^1(U)$ of weight 1 with respect to the $U(1)$-action and such that $\partial_I \alpha = \Omega_I$. Moreover, the image of the form Ω_I under the canonical projection $\Omega^2(U) \to \Omega^2(U/M)$ is zero. Therefore by Lemma 9.5.2 (iii) we can arrange so that the image of the form α under the projection $\Omega^1(U) \to \Omega^1(U/M)$ is also zero, so that α is in fact a section of the bundle $\rho^*\Lambda^{1,0}(M)$. This proves the existence part. To prove uniqueness, note that every two such 1-forms must differ by a form of the type $\partial_i f$ for a certain holomorphic function $f \in \Omega^0(U)$. Moreover, by Lemma 9.5.2 (ii) we can assume that the function f is of weight 1 with respect to the $U(1)$-action. On the other hand, by Lemma 9.5.2 (iii) we can assume that the function f is constant along the fibers of the canonical projection $\rho : \overline{T}M_I \to M$. Therefore we have $f = 0$ identically on the whole U. □

10. Convergence

10.1. Preliminaries

10.1.1. Let M be a complex manifold. By Theorem 8.1 every Kählerian connection $\nabla : \Lambda^1(M, \mathbb{C}) \to \Lambda^1(M, \mathbb{C}) \otimes \Lambda^1(M, \mathbb{C})$ on the cotangent bundle $\Lambda^1(M, \mathbb{C})$ to the manifold M defines a flat linear extended connection $D : S^1(M, \mathbb{C}) \to B^1(M, \mathbb{C})$ on M and therefore a formal Hodge connection

D on the total space $\overline{T}M$ of the complex-conjugate to the tangent bundle to M. By Proposition 5.2, this formal Hodge connection defines, in turn, a formal Hodge manifold structure on $\overline{T}M$ in the formal neighborhood of the zero section $M \subset \overline{T}M$.

In this section we show that if the Kählerian connection ∇ is real-analytic, then the corresponding formal Hodge manifold structure on $\overline{T}M$ is the completion of an actual Hodge manifold structure on an open neighborhood $U \subset \overline{T}M$ of the zero section $M \subset \overline{T}M$. We also show that if the connection ∇ comes from a Kähler metric ω on M, then the corresponding polarization Ω of the formal Hodge manifold $\overline{T}M$ defined in Theorem 9.1 converges in a neighborhood $U' \subset U \subset \overline{T}M$ of the zero section $M \subset \overline{T}M$ to a polarization of the Hodge manifold structure on U'. Here is the precise formulation of these results.

Theorem 10.1 *Let M be a complex manifold equipped with a real-analytic Kählerian connection $\nabla : \Lambda^1(M, \mathbb{C}) \to \Lambda^1(M, \mathbb{C}) \otimes \Lambda^1(M, \mathbb{C})$ on its cotangent bundle $\Lambda^1(M, \mathbb{C})$. There exists an open neighborhood $U \subset \overline{T}M$ of the zero section $M \subset \overline{T}M$ in the total space $\overline{T}M$ of the complex-conjugate to the tangent bundle to M and a Hodge manifold structure on $U \subset \overline{T}M$ such that its completion along the zero section $M \subset \overline{T}M$ defines a linear flat extended connection D on M with reduction ∇.*

Moreover, assume that M is equipped with a Kähler metric ω such that $\nabla\omega = 0$, and let $\Omega \in C^\infty_M(\overline{T}M, \Lambda^2(\overline{T}M, \mathbb{C}))$ be the formal polarization of the Hodge manifold structure on $U \subset \overline{T}M$ along $M \subset \overline{T}M$. Then there exists an open neighborhood $U' \subset U$ of $M \subset U$ such that $\Omega \in C^\infty(U', \Lambda^2(\overline{T}M, \mathbb{C})) \subset C^\infty_M(\overline{T}M, \Lambda^2(\overline{T}M, \mathbb{C}))$.

10.1.2. We begin with some preliminary observations. First of all, the question is local on M, therefore we may assume that M is an open neighborhood of 0 in the complex vector space $V = \mathbb{C}^n$. Fix once and for all a real structure and an Hermitian metrics on the vector space V, so that it is isomorphic to its dual $V \cong V^*$.

The subspace $\mathfrak{J} \subset C^\infty(M, \mathbb{C})$ of functions vanishing at $0 \in M$ is an ideal in the algebra $C^\infty(M, \mathbb{C})$, and $\mathfrak{J}$-adic topology on $C^\infty(M, \mathbb{C})$ extends canonically to the de Rham algebra $\Lambda^\bullet(M, \mathbb{C})$ of the manifold M and, further, to the Weil algebra $\mathcal{B}^\bullet(M, \mathbb{C})$ of M introduced in 6.3.5. Instead of working with bundle algebra $\mathcal{B}^\bullet(M, \mathbb{C})$ on M, it will be convenient for us to consider the vector space

$$\mathcal{B}^\bullet = C^\infty_{\mathfrak{J}}(M, B^\bullet(M, \mathbb{C})),$$

which is by definition the $\mathfrak{J}$-adic completion of the space $C^\infty(M, \mathcal{B}^\bullet(M, \mathbb{C}))$ of global sections of the Weil algebra. This vector space is canonically a (pro-)algebra over $\mathbb{C}$. Moreover, the Hodge bundle structure on $\mathcal{B}^\bullet(M, \mathbb{C})$ induces an $\mathbb{R}$-Hodge structure on the algebra $\mathcal{B}^\bullet$.

10.1.3. The $\mathfrak{J}$-adic completion $C^\infty_{\mathfrak{J}}(M, \mathbb{C})$ of the space of smooth functions on M is canonically isomorphic to the completion $\widehat{S}^\bullet(V)$ of the symmetric algebra of the vector space $V \cong V^*$. The cotangent bundle $\Lambda^1(M, \mathbb{C})$ is isomorphic to the trivial bundle $\mathcal{V}$ with fiber V over M, and the completed de Rham algebra $C^\infty_{\mathfrak{J}}(M, \Lambda^\bullet(M, \mathbb{C}))$ is isomorphic to the product

$$C^\infty_{\mathfrak{J}}(M, \Lambda^\bullet(M, \mathbb{C})) \cong S^\bullet(V) \otimes \Lambda^\bullet(V).$$

This is a free graded-commutative algebra generated by two copies of the vector space V, which we denote by $V_1 = V \subset \Lambda^1(V)$ and by $V_2 = V \subset S^1(V)$. It is convenient to choose the trivialization $\Lambda^1(M, \mathbb{C}) \cong \mathcal{V}$ in such a way that the de Rham differential $d_M : \Lambda^\bullet(M, \mathbb{C}) \to \Lambda^{\bullet+1}(M, \mathbb{C})$ induces an identity map $d_M : V_2 \to V_1 \subset C^\infty_{\mathfrak{J}}(M, \Lambda^1(M, \mathbb{C}))$.

10.1.4. The complex vector bundle $S^1(M, \mathbb{C})$ on M is also isomorphic to the trivial bundle $\mathcal{V}$. Choose a trivialization $S^1(M, \mathbb{C}) \cong \mathcal{V}$ in such a way that the canonical map $C : S^1(M, \mathbb{C}) \to \Lambda^1(M, \mathbb{C})$ is the identity map. Denote by

$$V_3 = V \subset C^\infty(M, S^1(M, \mathbb{C})) \subset \mathcal{B}^0$$

the subset of constant sections in $S^1(M, \mathbb{C}) \cong \mathcal{V}$. Then the Weil algebra $\mathcal{B}^\bullet$ becomes isomorphic to the product

$$\mathcal{B}^i \cong \widehat{S}^\bullet(V_2 \oplus V_3) \otimes \Lambda^i(V_1)$$

of the completed symmetric algebra $\widehat{S}^\bullet(V_2 \oplus V_3)$ of the sum $V_2 \oplus V_3$ of two copies of the vector space V and the exterior algebra $\Lambda^\bullet(V_1)$ of the third copy of the vector space V.

10.1.5. Recall that we have introduced in 7.2.2 a grading on the Weil algebra $\mathcal{B}^\bullet(M, \mathbb{C})$ which we call *the augmentation grading.* It induces a grading on the the Weil algebra $\mathcal{B}^\bullet$. The augmentation grading on $\mathcal{B}^\bullet$ is multiplicative, and it is obtained by assigning degree 1 to the generator subspaces $V_1, V_3 \subset \mathcal{B}^\bullet$ and degree 0 to the generator subspace $V_2 \subset \mathcal{B}^\bullet$. As in 7.2.2, we will denote the augmentation grading on $\mathcal{B}^\bullet$ by lower indices.

We will now introduce yet another grading on the algebra $\mathcal{B}^\bullet$ which we will call *the total grading.* It is by definition the multiplicative grading obtained by assigning degree 1 to *all* the generators $V_1, V_2, V_3 \subset \mathcal{B}^\bullet$ of the Weil algebra $\mathcal{B}^\bullet$ We will denote by $\mathcal{B}^\bullet_{k,n} \subset \mathcal{B}^\bullet$ the component of augmentation degree k and total degree n. Note that by definition $n, k \geq 0$ and, moreover, $n \geq k$.

Remark. In 7.2.2 we have also defined a finer *augmentation bigrading* on the Weil algebra $\mathcal{B}^\bullet(M, \mathbb{C})$ and it this bigrading that was denoted by double lower indices throughout Section 8. We will now longer need the augmentation bigrading, so there is no danger of confusion.

10.1.6. The trivialization of the cotangent bundle to M defines an isomorphism $\overline{T}M \cong M \times V$ and a constant Hodge connection on the pair $\langle \overline{T}M, M \rangle$. The corresponding extended connection $D^{const} : S^1(M, \mathbb{C}) \to \mathcal{B}^1(M, \mathbb{C})$ is the sum of the trivial connection

$$\nabla_1^{const} : S^1(M, \mathbb{C}) \to S^1(M, \mathbb{C}) \otimes \Lambda^1(M, \mathbb{C}) \subset \mathcal{B}^1(M, \mathbb{C})$$

on $S^1(M, \mathbb{C}) \cong \mathcal{V}$ and the canonical isomorphism

$$C = \mathsf{id} : S^1(M, \mathbb{C}) \to \Lambda^1(M, \mathbb{C}) \subset \mathcal{B}^1(M, \mathbb{C}).$$

The derivation $D^{const} : \mathcal{B}^\bullet \to \mathcal{B}^{\bullet+1}$ of the Weil algebra associated to the extended connection D^{const} by 6.4.5 is equal to

$$\begin{aligned} D^{const} &= C = \mathsf{id} : V_3 \to V_1 \\ D^{const} &= d_M = \mathsf{id} : V_2 \to V_1 \\ D^{const} &= d_M = 0 \text{ on } V_1 \end{aligned}$$

on the generator spaces $V_1, V_2, V_3 \subset \mathcal{B}^\bullet$. In particular, the derivation D^{const} preserves the total degree.

10.1.7. Let now $D : S^1(M, \mathbb{C}) \to \mathcal{B}^1(M, \mathbb{C})$ be the an arbitrary linear extended connection on the manifold M, and let

$$D = \sum_{k \geq 0} D_k : \mathcal{B}^\bullet \to \mathcal{B}^{\bullet+1}$$

be the derivation of the Weil algebra $\mathcal{B}^\bullet$ associated to the extended connection D by 6.4.5. The derivation D admits a finer decomposition

$$D = \sum_{k,n \geq 0} D_{k,n} : \mathcal{B}^\bullet \to \mathcal{B}^{\bullet+1}$$

according to both the augmentation and the total degree on $\mathcal{B}^\bullet$. The summand $D_{k,n}$ by definition raises the augmentation degree by k and the total degree by n.

10.1.8. Since the extended connection $D : S^1(M,\mathbb{C}) \to \mathcal{B}^1(M,\mathbb{C})$ is linear, its component $D_0 : S^1(M,\mathbb{C}) \to \Lambda^1(M,\mathbb{C})$ of augmentation degree 0 coincides with the canonical isomorphism $C : S^1(M,\mathbb{C}) \to \Lambda^1(M,\mathbb{C})$. Therefore the restriction of the derivation $D : \mathcal{B}^{\bullet} \to \mathcal{B}^{\bullet+1}$ to the generator subspace $V_3 \subset S^1(M,\mathbb{C}) \subset \mathcal{B}^0$ satisfies

$$D_0 = C = D_0^{const} = D_{0,0}^{const} : V_3 \to V_1 \subset \mathcal{B}^1.$$

In particular, all the components $D_{0,n}$ except for $D_{0,0}$ vanish on the subspace $V_3 \subset \mathcal{B}^{\bullet}$.

The restriction of the derivation D to the subspace $\Lambda^{\bullet}(M,\mathbb{C}) \subset \mathcal{B}^{\bullet}(M,\mathbb{C})$ by definition coincides with the de Rham differential $d_M : \Lambda^{\bullet}(M,\mathbb{C}) \to \Lambda^{\bullet+1}(M,\mathbb{C})$. Therefore on the generator subspaces $V_1, V_2 \subset \mathcal{B}^{\bullet}$ we have $D = d_M = D^{const}$. In particular, all the components $D_{k,n}$ except for $D_{1,0}$ vanish on the subspaces $V_1, V_2 \subset \mathcal{B}^{\bullet}$.

10.1.9. The fixed Hermitian metric on the generator spaces $V_1 = V_2 = V_3 = V$ extends uniquely to a metric on the whole Weil algebra such that the multiplication map $\mathcal{B}^{\bullet} \otimes \mathcal{B}^{\bullet} \to \mathcal{B}^{\bullet}$ is an isometry. We call this metric *the standard metric* on $\mathcal{B}^{\bullet}$. We finish our preliminary observations with the following fact which we will use to deduce Theorem 10.1 from estimates on the components $D_{n,k}$ of the derivation $D : \mathcal{B}^{\bullet} \to \mathcal{B}^{\bullet+1}$.

Lemma. *Let $D = \sum_{n,k} D_{n,k} : \mathcal{B}^{\bullet} \to \mathcal{B}^{\bullet+1}$ be a derivation associated to an extended connection D on the manifold M. Consider the norms $\|D_{k,n}\|$ of the restrictions $D_{k,n} : V_3 \to \mathcal{B}^1_{n+1,k+1}$ of the derivations $D_{k,n} : \mathcal{B}^{\bullet} \to \mathcal{B}^{\bullet+1}$ to the generator subspace $V_3 \subset \mathcal{B}^0$ taken with respect to the standard metric on the Weil algebra $\mathcal{B}^{\bullet}$. If for certain constants $C, \varepsilon > 0$ and for every natural $n \geq k \geq 0$ we have*

$$\|D_{k,n}\| < C\varepsilon^n, \tag{10.1}$$

then the formal Hodge connection on $\overline{T}M$ along M associated to D converges to an actual real-analytic Hodge connection on the open ball of radius ε in $\overline{T}M$ with center at $0 \in M \subset \overline{T}M$. Conversely, if the extended connection D comes from a real-analytic Hodge connection on an open neighborhood $U \subset \overline{T}M$, and if the Taylor series for this Hodge connection converge in the closed ball of radius ε with center at $0 \in M \subset \overline{T}M$, then there exists a constant $C > 0$ such that (10.1) *holds for every $n, k \geq 0$.*

Proof. The constant Hodge connection D^{const} is obviously defined on the whole $\overline{T}M$, and every other formal Hodge connection on $\overline{T}M$ is of the form

$$D = D^{const} + d^r \circ \Theta : \Lambda^0(\overline{T}M,\mathbb{C}) \to \rho^*\Lambda^1(M,\mathbb{C}),$$

where $d^r : \Lambda^0(\overline{T}M, \mathbb{C}) \to \Lambda^1(\overline{T}M/M, \mathbb{C})$ is the relative de Rham differential, and $\Theta \in C^\infty_M(\overline{T}M, \Lambda^1(\overline{T}M/M, \mathbb{C}) \otimes \rho^*\Lambda^1(M, \mathbb{C}))$ is a certain relative 1-form on the formal neighborhood of $M \subset \overline{T}M$ with values in the bundle $\rho^*\Lambda^1(M, \mathbb{C})$. Both bundles $\Lambda^1(\overline{T}M/M, \mathbb{C})$ and $\rho^*\Lambda^1(M, \mathbb{C})$ are canonically isomorphic to the trivial bundle $\mathcal{V}$ with fiber V on $\overline{T}M$. Therefore we can treat the 1-form Θ as a formal germ of a $\operatorname{End}(V)$-valued function on $\overline{T}M$ along M. The Hodge connection D converges on a subset $U \subset \overline{T}M$ if and only if this formal germ comes from a real-analytic $\operatorname{End}(V)$-valued function Θ on U.

The space of all formal Taylor series for $\operatorname{End}(V)$-valued functions on $\overline{T}M$ at $0 \in M \subset \overline{T}M$ is by definition equal to $\operatorname{End}(V) \otimes \mathcal{B}^0$. Moreover, for every $n \geq 0$ the component $\Theta_n \in \mathcal{B}^0_n \otimes \operatorname{End}(V) = \operatorname{Hom}(V, \mathcal{B}^0_n \otimes V)$ of total degree n of the formal power series for the function Θ at $0 \in \overline{T}M$ is equal to the derivation

$$\sum_{0 \leq k \leq n} D_{k,n} : V = V_3 \to V = V_1 \otimes \bigoplus_{0 \leq k \leq n} \mathcal{B}^0_{k,n}.$$

Every point $x \in \overline{T}M$ defines the "evaluation at x" map

$$\operatorname{ev}_x : C^\infty(\overline{T}M, \operatorname{End}(V)) \to \mathbb{C},$$

and the formal Taylor series for $\Theta \in \operatorname{End}(V) \otimes \mathcal{B}^0$ converges at the point $x \in \overline{T}M$ if and only if the series

$$\Theta(x) = \sum_{n \geq 0} \operatorname{ev}_x(\Theta_n) \in \operatorname{End}(V)$$

converges. But we have

$$\| \operatorname{ev}_x(\Theta_n) \| = \left\| \sum_{0 \leq k \leq n} D_{k,n} \right\| |x|^n,$$

where $|x|$ is the distance from the point x to $0 \in \overline{T}M$. Now the application of standard criteria of convergence finishes the proof of the lemma. □

10.2. Combinatorics

10.2.1. We now derive some purely combinatorial facts needed to obtain estimates for the components $D_{k,n}$ of the extended connection D. First, let a_n be the Catalan numbers, that is, the numbers defined by the recurrence relation

$$a_n = \sum_{1 \leq k \leq n-1} a_k a_{n-k}$$

and the initial conditions $a_1 = 1$, $a_n = 0$ for $n \leq 0$. As is well-known, the generating function $f(z) = \sum_{k \geq 0} a_k z^k$ for the Catalan numbers satisfies the equation $f(z) = f(z)^2 + z$ and equals therefore

$$f(z) = \frac{1}{2} - \sqrt{\frac{1}{4} - z}.$$

The Taylor series for this function at $z = 0$ converges for $4|z| < 1$, which implies that

$$a_k < C(4 + \varepsilon)^k$$

for some positive constant $C > 0$ and every $\varepsilon > 0$.

10.2.2. We will need a more complicated sequence of integers, numbered by two natural indices, which we denote by $b_{k,n}$. The sequence $b_{k,n}$ is defined by the recurrence relation

$$b_{k,n} = \sum_{p,q; 1 \leq p \leq k-1} \frac{q+1}{k} b_{p,q} b_{k-p,n-q}$$

and the initial conditions

$$\begin{cases} b_{k,n} & = 0 \quad \text{for} \quad k \leq 0, \\ b_{k,n} & = 0 \quad \text{for} \quad k = 1, n < 0, \\ b_{k,n} & = 1 \quad \text{for} \quad k = 1, n \geq 0, \end{cases}$$

which imply, in particular, that if $n < 0$, then $b_{k,n} = 0$ for every k. For every $k \geq 1$ let $g_k(z) = \sum_{n \geq 0} b_{k,n} z^n$ be the generating function for the numbers $b_{k,n}$. The recurrence relations on $b_{k,n}$ give

$$\begin{aligned} g_k(z) &= \frac{1}{k} \sum_{1 \leq p \leq k-1} g_{k-p}(z) \left(1 + z\frac{\partial}{\partial z}\right) (g_p(z)) \\ &= \frac{1}{2k} \sum_{1 \leq p \leq k-1} \left(2 + z\frac{\partial}{\partial z}\right) (g_p(z) g_{k-p}(z)), \end{aligned}$$

and the initial conditions give

$$g_1(z) = \frac{1}{1-z}.$$

10.2.3. Say that a formal series $f(z)$ in the variable z is *non-negative* if all the terms in the series are non-negative real numbers. The sum and product of two non-negative series and the derivative of a non-negative series is also obviously non-negative. For two formal series $s(z)$, $t(z)$ write $s(z) \ll t(z)$ if the difference $t(z) - s(z)$ is a non-negative power series.

Our main estimate for the generating functions $g_k(z)$ is the following.

Lemma. *For every $k \geq 1$ we have*

$$g_k(z) \ll a_k \frac{1}{(1-z)^{2k-1}},$$

where a_k are the Catalan numbers.

Proof. Use induction on k. For $k = 1$ we have $g_1(z) = \frac{1}{1-z}$ and $a_1 = 1$, which gives the base for induction. Assume that the claim is proved for all $p < k$. Since all the $g_n(z)$ are non-negative power series, this implies that for every p, $1 \leq p \leq k-1$ we have

$$g_p(z)g_{k-p}(z) \ll a_p a_{k-p} \frac{1}{(1-z)^{2p-1}} \frac{1}{(1-z)^{2k-2p-1}} = a_p a_{k-p} \frac{1}{(1-z)^{2k-2}}.$$

Therefore

$$\begin{aligned}
\left(2 + z\frac{\partial}{\partial z}\right)(g_p(z)g_{k-p}(z)) &\ll a_p a_{k-p}\left(2 + z\frac{\partial}{\partial z}\right)\frac{1}{(1-z)^{2k-2}} \\
&= a_p a_{k-p}\left(\frac{2}{(1-z)^{2k-2}} + \frac{(2k-2)z}{(1-z)^{2k-1}}\right) \\
&= a_p a_{k-p}\left(\frac{2k-2}{(1-z)^{2k-1}} - \frac{2k-4}{(1-z)^{2k-2}}\right) \\
&\ll (2k-2)a_p a_{k-p}\frac{1}{(1-z)^{2k-1}}.
\end{aligned}$$

Hence

$$\begin{aligned}
g_k(z) &= \frac{1}{2k}\sum_{1 \leq p \leq k-1}\left(2 + z\frac{\partial}{\partial z}\right)(g_p(z)g_{k-p}(z)) \\
&\ll \frac{2k-2}{2k}\sum_{1 \leq p \leq k-1} a_p a_{k-p}\frac{1}{(1-z)^{2k-1}} \\
&\ll \frac{1}{(1-z)^{2k-1}}\sum_{1 \leq p \leq k-1} a_p a_{k-p} = a_k\frac{1}{(1-z)^{2k-1}},
\end{aligned}$$

which proves the lemma. □

10.2.4. This estimate yields the following estimate for the numbers $b_{k,n}$.

Corollary. *The power series*

$$g(z) = \sum_{k,n} b_{k,n} z^{n+k} = \sum_{k \geq 1} g_k(z) z^k$$

converges for $z < 3-\sqrt{8}$. *Consequently, for every* C_2 *such that* $(3-\sqrt{8})C_2 > 1$ *there exists a positive constant* $C > 0$ *such that*

$$b_{n,k} < CC_2^{n+k}$$

for every n *and* k. *(One can take, for example,* $C_2 = 6$.*)*

Proof. Indeed, we have

$$g(z) \ll \sum_{k\geq 1} a_k z^k \frac{1}{(1-z)^{2k-1}} = (1-z)f\left(\frac{z}{(1-z)^2}\right), \tag{10.2}$$

where $f(z) = \frac{1}{2} - \sqrt{\frac{1}{4} - z}$ is the generating function for the Catalan numbers. Therefore

$$g(z) \ll (1-z)\left(\frac{1}{2} - \sqrt{\frac{1}{4} - \frac{z}{(1-z)^2}}\right),$$

and the right hand side converges absolutely when

$$\frac{|z|}{(1-z)^2} < \frac{1}{4}.$$

Since $3 - \sqrt{8}$ is the root of the quadratic equation $(1-z)^2 = 4z$, this inequality holds for every z such that $|z| < 3 - \sqrt{8}$. □

10.2.5. To study polarizations of Hodge manifold structures on $\overline{T}M$, we will need yet another recursive sequence of integers, which we denote by $b^m_{k,n}$. This sequence is defined by the recurrence relation

$$b^m_{k,n} = \sum_{p,q;1\leq p\leq k-1} \frac{q+m(k-p)}{k} b^m_{p,q} b_{k-p,n-q}$$

and the initial conditions

$$\begin{cases} b^m_{k,n} & = 0 \quad \text{unless} \quad k, m \leq 0, \\ b^m_{k,n} & = 0 \quad \text{for} \quad k = 1, n < 0, \\ b^m_{k,n} & = 1 \quad \text{for} \quad k = 1, n \geq 0. \end{cases}$$

10.2.6. To estimate the numbers $b^m_{k,n}$, consider the auxiliary sequence $c_{k,n}$ defined by setting

$$c_{k,n} = \sum_{p,q;1\leq p\leq k-1} c_{p,q} b_{k-p,n-q} \qquad k \geq 2,$$

and $c_{k,n} = b_{k,n}$ for $k \leq 1$. The generating series $c(z) = \sum_{k,n\geq 0} c_{k,n} z^{n+k}$ satisfies

$$c(z) = c(z)g(z) + \frac{z}{1-z},$$

so that we have $c(z) = \frac{z}{(1-z)(1-g(z))}$, which is non-singular when $|z|, |g(z)| < 1$ and $g(z)$ is non-singular. By (10.2) the latter inequality holds if

$$\left|(1-z)\left(\frac{1}{2} - \sqrt{\frac{1}{4} - \frac{z}{(1-z)^2}}\right)\right| < 1,$$

which holds in the whole disc where $g(z)$ converges, that is, for $|z| < 3 - \sqrt{8}$. Therefore, as in Corollary 10.2.4, we have

$$c_{k,n} < C6^{n+k} \tag{10.3}$$

for some positive constant C.

10.2.7. We can now estimate the numbers $b^m_{k,n}$.

Lemma. *For every m, k, n we have*

$$b^m_{k,n} \leq (2m)^{k-1} c_{k,n} b_{k,n}, \tag{10.4}$$

where $c_{k,n}$ are as in 10.2.6 *and $b_{k,n}$ are the numbers introduced in* 10.2.2. *Consequently, we have*

$$b^m_{k,n} < C(72m)^{n+k+m}$$

for some positive constant $C > 0$.

Proof. Use induction on k. The case $k = 1$ follows from the initial conditions. Assume the estimate (10.4) proved for all $b^m_{p,q}$ with $p < k$. Note that by the recurrence relations we have $b_{p,q} \leq b_{k,n}$ and $c_{p,q} \leq c_{k,n}$ whenever $p < k$. Therefore

$$\begin{aligned}
b^m_{k,n} &= \sum_{p,q;1\leq p\leq k-1} \frac{q + m(k-p)}{k} b^m_{p,q} b_{k-p,n-q} \\
&\leq \sum_{p,q;1\leq p\leq k-1} \frac{q}{k} b^m_{p,q} b_{k-p,n-q} + \sum_{p,q;1\leq p\leq k-1} b^m_{p,q} b_{k-p,n-q} \\
&\leq \sum_{p,q;1\leq p\leq k-1} 2^{p-1} c_{p,q} \frac{q}{k} b_{p,q} b_{k-p,n-q} + \sum_{p,q;1\leq p\leq k-1} 2^{p-1} b_{p,q} c_{p,q} b_{k-p,n-q} \\
&\leq 2^{k-2} c_{k,n} \sum_{p,q;1\leq p\leq k-1} \frac{q+1}{k} b_{p,q} b_{k-p,n-q} \\
&\quad + 2^{k-2} b_{k,n} \sum_{p,q;1\leq p\leq k-1} c_{p,q} b_{k-p,n-q} \\
&= 2^{k-2} c_{k,n} b_{k,n} + 2^{k-2} c_{k,n} b_{k,n} = 2^{k-1} c_{k,n} b_{k,n};
\end{aligned}$$

which proves (10.4) for $b^m_{k,n}$. The second estimate of the lemma now follows from (10.3) and Corollary 10.2.4. □

10.3. The main estimate

10.3.1. Let now $D = \sum_{k,n} D_{n,k} : \mathcal{B}^{\bullet} \to \mathcal{B}^{\bullet+1}$ be a derivation of the Weil algebra $\mathcal{B}^{\bullet}$ associated to a flat linear extended connection on M. Consider the restriction $D_{k,n} : \mathcal{B}^{\bullet}_{p,q} \to \mathcal{B}^{\bullet}p+k, q+n$ of the derivation $D_{k,n}$ to the component $\mathcal{B}^{\bullet}_{p,q} \subset \mathcal{B}^{\bullet}$ of augmentation degree p and total degree q. Since both $\mathcal{B}^{\bullet}_{p,q}$ and $\mathcal{B}^{\bullet}p+k, q+n$ are finite-dimensional vector spaces, the norm of this restriction with respect to the standard metric on $\mathcal{B}^{\bullet}$ is well-defined. Denote this norm by $\|D_{k,n}\|_{p,q}$.

By Lemma 10.1.9 the convergence of the Hodge manifold structure on $\overline{T}M$ corresponding to the extended connection D is related to the growth of the norms $\|D_{k,n}\|_{1,1}$. Our main estimate on the norms $\|D_{k,n}\|_{1,1}$ is the following.

Proposition 10.1 *Assume that there exist a positive constant C_0 such that for every n the norms $\|D_{1,n}\|_{1,1}$ and $\|D_{1,n}\|_{0,1}$ satisfy*

$$\|D_{1,n}\|_{1,1}, \|D_{1,n}\|_{1,0} < C_0^n.$$

Then there exists a positive constant C_1 such that for every n, k the norm $\|D_{k,n}\|_{1,1}$ satisfies

$$\|D_{k,n}\|_{1,1} < C_1^n.$$

10.3.2. In order to prove Proposition 10.1, we need some preliminary facts. Recall that we have introduced in 7.2.4 the total Weil algebra $\mathcal{B}^{\bullet}_{tot}(M, \mathbb{C})$ of the manifold M, and let

$$\mathcal{B}^{\bullet}_{tot} = C^{\infty}_{\mathfrak{I}}(M, \mathcal{B}^{\bullet}_{tot}(M, \mathbb{C}))$$

be the algebra of its smooth sections completed at $0 \subset M$. By definition for every $k \geq 0$ we have $\mathcal{B}^k_{tot} = \mathcal{B}^k \otimes \mathcal{W}^*_k$, where $\mathcal{W}_k$ is the $\mathbb{R}$-Hodge structure of weight k universal for weakly Hodge maps, as in 1.4.5. There exists a unique Hermitian metric on $\mathcal{W}_k$ such that all the Hodge components $\mathcal{W}^{p,q} \subset \mathcal{W}_k$ are orthogonal and all the Hodge degree components $w^{p,q}_k$ of the universal weakly Hodge map $w_k : \mathbb{R}(0) \to \mathcal{W}_k$ are isometries. This metric defines a canonical Hermitian metric on $\mathcal{W}^*_k$.

Definition. The *standard metric* on the total Weil algebra $\mathcal{B}^{\bullet}_{tot}$ is the product of the canonical metric and the standard metric on $\mathcal{B}^{\bullet}$.

10.3.3. By Lemma 7.1.4 the total Weil algebra $\mathcal{B}^{\bullet}_{tot}$ is generated by the subspaces $V_2, V_3 \subset \mathcal{B}^0 = \mathcal{B}^0_{tot}$ and the subspace $V_1 \otimes \mathcal{W}_1^* \subset \mathcal{B}^1_{tot}$, which we denote by V_1^{tot}. The ideal of relations for the algebra $\mathcal{B}^{\bullet}_{tot}$ is the ideal in $S^{\bullet}(V_2 \oplus V_3) \otimes \Lambda^{\bullet}(V_1^{tot})$ generated by $S^2(V_1) \otimes \Lambda^2(\mathcal{W}_1^*) \subset \Lambda^2(V_1^{tot})$.

The direct sum decomposition (7.3) induces a direct sum decomposition

$$V_1^{tot} = V_1^{ll} \oplus V_1^{o} \oplus V_1^{rr}$$

of the generator subspace $V_1^{tot} \subset \mathcal{B}^1_{tot}$. The subspaces $V_1^o \subset V_1^{tot}$ and $V_1^{ll} \oplus V_1^{rr} \subset V_1^{tot}$ are both isomorphic to the vector space V_1. More precisely. the universal weakly Hodge map $w_1 : \mathbb{R}(0) \to \mathcal{W}_1$ defines a projection

$$P : V_1^{tot} = V_1 \otimes \mathcal{W}_1^* \to V_1,$$

and the restriction of the projection P to either of the subspaces $V_1^o, V_1^{ll} \oplus V_1^{rr} \subset V_1^{tot}$ is an isomorphism. Moreover, either of these restrictions is an isometry with respect to the standard metrics.

10.3.4. The multiplication in $\mathcal{B}^{\bullet}_{tot}$ is not an isometry with respect to this metric. However, for every $b_1, b_2 \subset \mathcal{B}^{\bullet}_{tot}$ we have the inequality

$$\|b_1 b_2\| \leq \|b_1\| \cdot \|b_2\|.$$

Moreover, this inequality becomes an equality when $b_1 \subset \mathcal{B}^0_{tot}$. In particular, if we extend the map $P : V_1^{tot} \to V_1$ to a $\mathcal{B}^0$-module map

$$P : \mathcal{B}^1_{tot} \to \mathcal{B}^1,$$

then the restriction of the map P to either of the subspaces $\mathcal{B}^1_o, \mathcal{B}^1_{ll} \oplus \mathcal{B}^1_{rr} \subset \mathcal{B}^1_{tot}$ is an isometry with respect to the standard metric. Therefore the norm of the projection $P : \mathcal{B}^1_{tot} \to \mathcal{B}^1$ is at most 2.

10.3.5. The total and augmentation gradings on the Weil algebra $\mathcal{B}^{\bullet}$ extend to gradings on the total Weil algebra $\mathcal{B}^{\bullet}_{tot}$, also denoted by lower indices. The extended connection D on M induces a derivation $D^{tot} = \sum_{n,k} D^{tot}_{k,n} : \mathcal{B}^{\bullet}_{tot} \to \mathcal{B}^{\bullet+1}_{tot}$. As in 10.3.1, denote by $\|D^{tot}_{k,n}\|_{p,q}$ the norm of the map $D^{tot}_{k,n} : (\mathcal{B}^{\bullet}_{tot})_{p,q} \to (\mathcal{B}^{\bullet}_{tot})_{p+k,q+n}$ with respect to the standard metric on $\mathcal{B}^{\bullet}_{tot}$. The derivations $D^{tot}_{k,n}$ are related to $D_{k,n}$ by

$$D_{k,n} = P \circ D^{tot}_{k,n} : \mathcal{B}^0 \to \mathcal{B}^1,$$

and we have the following.

Lemma. *For every k, n and $p = 0, 1$ we have*

$$\|D_{k,n}\|_{p,1} = \|D^{tot}_{k,n}\|_{p,1}.$$

Proof. By definition we have $\mathcal{B}^{\bullet}_{0,1} \oplus \mathcal{B}^{\bullet}_{1,1} = V_1 \oplus V_2 \oplus V_3$. Moreover, the derivation $D_{k,n}$ vanishes on V_1, hence $D^{tot}_{k,n}$ vanishes on V^{tot}_1. Therefore it suffices to compare their norms on $V_2 \oplus V_3 \subset \mathcal{B}^0 = \mathcal{B}^0_{tot}$.

Since on $S^{\bullet}(V_2) \subset \mathcal{B}^0$ the derivation $D_{k,n}$ coincides with the de Rham differential, the derivation $D^{tot}_{k,n}$ maps the subspace V_2 into $\mathcal{B}^1_{ll} \oplus \mathcal{B}^1_{rr}$. Moreover, by Lemma 7.2.6 the derivation $D^{tot}_{k,n}$ maps V_3 either into $\mathcal{B}^1_o$ or into $\mathcal{B}^1_{ll} \oplus \mathcal{B}^1_{rr}$, depending on the parity of the number k. Since $D_{k,n} = P \circ D^{tot}_{k,n}$ and the map $P : \mathcal{B}^1_{tot} \to \mathcal{B}^1$ is an isometry on both $\mathcal{B}^1_o \subset \mathcal{B}^1_{tot}$ and $\mathcal{B}^1_{ll} \oplus \mathcal{B}^1_{rr} \subset \mathcal{B}^1_{tot}$, we have $\|D_{k,n}\| = \|D^{tot}_{k,n}\|$ on both $V_2 \subset \mathcal{B}^0$ and $V_3 \subset \mathcal{B}^0$, which proves the lemma. □

This lemma allows to replace the derivations $D_{k,n}$ in Proposition 10.1 with associated derivations $D^{tot}_{k,n}$ of the total Weil algebra $\mathcal{B}^{\bullet}_{tot}$.

10.3.6. Since the extended connection D is linear and flat, the construction used in the proof of Lemma 8.3.6 shows that

$$D^{tot}_k = h^{-1} \circ \sigma_{tot} \circ \sum_{1 \leq p \leq k-1} D^{tot}_p \circ D^{tot}_{k-p} : V_3 \to \left(\mathcal{B}^1_{tot}\right)_{k+1}. \tag{10.5}$$

where $\sigma_{tot} : \mathcal{B}^{\bullet+1}_{tot} \to \mathcal{B}^{\bullet}_{tot}$ is the canonical map constructed in 7.3.7, and $h : \mathcal{B}^{\bullet}_{tot} \to \mathcal{B}^{\bullet}_{tot}$ is as in Lemma 7.3.8. Both σ_{tot} and h preserve the augmentation degree. In order to obtain estimates on $\|D_{k,n}\|_{1,1} = \|D^{tot}_{k,n}\|_{1,1}$, we have to estimate the norms $\|h^{-1}\|$ and $\|\sigma_{tot}\|$ of the restrictions of maps h^{-1} and σ_{tot} on the subspace $(\mathcal{B}^{\bullet}_{tot})_{k+1} \subset \mathcal{B}^{\bullet}_{tot}$.

By Lemma 7.3.9 the map $h : (\mathcal{B}^{\bullet})_{k+1} \to (\mathcal{B}^{\bullet})_{k+1}$ is diagonalizable, with eigenvalues $k+1$ if k is even and $(k+1)/2, (k-1)/2$ if k is odd. Since $m \geq 2$, in any case on $(\mathcal{B}^{\bullet})_{k+1}$ we have

$$\|h^{-1}\| < \frac{3}{k}. \tag{10.6}$$

10.3.7. To estimate $\sigma_{tot} : \mathcal{B}^2_{tot} \to \mathcal{B}^2_{tot}$, recall that, as noted in 7.3.7, the map σ_{tot} is a map of $\mathcal{B}^0$-modules. The space $\mathcal{B}^2_{tot} = \mathcal{B}^0 \otimes \left(\mathcal{B}^2_{tot}\right)_2$ is a free $\mathcal{B}^0$-module generated by a finite-dimensional vector space $\left(\mathcal{B}^2_{tot}\right)_2$. The map σ_{tot} preserves the augmentation degree, hence it maps $\left(\mathcal{B}^2_{tot}\right)_2$ into the finite-dimensional vector space $\left(\mathcal{B}^1_{tot}\right)_2$. Therefore there exists a constant K such that

$$\|\sigma_{tot}\| \leq K \tag{10.7}$$

on $\left(\mathcal{B}^2_{tot}\right)_2$. Since we have $\|b_1 b_2\| = \|b_1\| \cdot \|b_2\|$ for every $b_1 \in \mathcal{B}^0$, $b_2 \in \mathcal{B}^2_{tot}$, and the map σ_{tot} is $\mathcal{B}^0$-linear, the estimate (10.7) holds on the whole $\mathcal{B}^2_{tot} = \mathcal{B}^0 \otimes \left(\mathcal{B}^2_{tot}\right)_2$. We can assume, in addition, that $K \geq 1$.

Remark. In fact $K = 2$, but we will not need this.

10.3.8. Let now $b_{k,n}$ be the numbers defined recursively in 10.2.2. Our estimate for $\|D^{tot}_{k,n}\|_{p,q}$ is the following.

Lemma. *In the assumptions and notations of Proposition* 10.1, *we have*

$$\|D^{tot}_{k,n}\|_{p,q} < q(3K)^{k-1}C_0^n b_{k,n}$$

for every k, n.

Proof. Use induction on k. The base of induction is the case $k = 1$, when the inequality holds by assumption. Assume that for some k we have proved the inequality for all $\|D^{tot}_{m,n}\|_{p,q}$ with $m < k$, and fix a number $n \geq k$.

Consider first the restriction of $D^{tot}_{k,n}$ onto the generator subspace $V_3 \subset \mathcal{B}^0$. Taking into account the total degree, we can rewrite (10.5) as

$$D^{tot}_{k,n} = h^{-1} \circ \sigma_{tot} \circ \sum_{1 \leq p \leq k-1} \sum_q D^{tot}_{k-p,n-q} \circ D^{tot}_{p,q} : V_3 \to \left(\mathcal{B}^1_{tot}\right)_{k+1}.$$

Therefore the norm of the map $D^{tot}_{k,n} : V_3 \to \mathcal{B}^1_{tot}$ satisfies

$$\|D^{tot}_{k,n}\} \leq \|h^{-1}\|\cdot\|\sigma_{tot}\|\cdot \sum_{1 \leq p \leq k-1} \sum_q \|D^{tot}_{k-p,n-q}\|_{p+1,q+1}\cdot\|D^{tot}_{p,q}\|_{1,1}.$$

Substituting into this the estimates (10.6), (10.7) and the inductive assumption, we get

$$\|D^{tot}_{k,n}\| < \frac{3}{k}K \sum_{1 \leq p \leq k-1} \sum_q (q+1)(3K)^{k-2}C_0^n b_{k-p,n-q}b_{p,q} = (3K)^{k-1}C_0^n b_{k,n}.$$

Since by definition $D^{tot}_{k,n}$ vanishes on $V_2 \subset \mathcal{B}^0$ and on $V_1^{tot} \subset \mathcal{B}^1_{tot}$, this proves that

$$\|D^{tot}_{k,n}\|_{p,1} < (3K)^{k-1}C_0^n b_{k,n}$$

when $p = 0, 1$. Since the map $D^{tot}_{k,n} : \mathcal{B}^{\bullet}_{tot} \to \mathcal{B}^{\bullet+1}_{tot}$ is a derivation, the Leibnitz rule and the triangle inequality show that for every p, q

$$\|D^{tot}_{k,n}\|_{p,q} < q(3K)^{k-1}C_0^n b_{k,n},$$

which proves the lemma. □

10.3.9. *Proof of Proposition* 10.1. By Lemma 10.3.5 we have $\|D_{k,n}\|_{p,1} = \|D^{tot}_{k,n}\|_{p,1}$, and by Lemma 10.3.8 we have

$$\|D_{k,n}\|_{p,1} = \|D^{tot}_{k,n}\|_{p,1} < (3K)^{k-1}C_0^n b_{n,k}.$$

Since $k \leq n$ and $K \geq 1$, this estimate together with Corollary 10.2.4 implies that

$$\|D_{k,n}\|_{p,1} < C(3K)^{k-1}C_0^n 6^{2n} < C(108KC_0)^n$$

for some positive constant $C > 0$, which proves the proposition. □

10.3.10. Proposition 10.1 gives estimates for the derivation $D : \mathcal{B}^{\bullet} \to \mathcal{B}^{\bullet+1}$ or, equivalently, for the Dolbeault differential

$$\bar{\partial}_J : \Lambda^{0,\bullet}(\overline{T}M_J) \to \Lambda^{0,\bullet+1}(\overline{T}M_J)$$

for the complementary complex structure $\overline{T}M_J$ on $\overline{T}M$ associated to the extended connection D on M. To prove the second part of Theorem 10.1, we will need to obtain estimates on the Dolbeult differential $\bar{\partial}_J : \Lambda^{p,0}(\overline{T}M_J) \to \Lambda^{p,1}(\overline{T}M_J)$ with $p > 0$. To do this, we use the model for the de Rham complex $\Lambda^{\bullet,\bullet}(\overline{T}M_J)$ constructed in Subsection 9.2.

Recall that in 9.2.3 we have identified the direct image $\rho_*\Lambda^{\bullet,\bullet}(\overline{T}M_J)$ of the de Rham algebra of the manifold $\overline{T}M$ with the free module over the Weil algebra $\mathcal{B}^{\bullet}(M,\mathbb{C})$ generated by a graded algebra bundle $L^{\bullet}(M,\mathbb{C})$ on M. The Dolbeult differential $\bar{\partial}_J$ for the complementary complex structure $\overline{T}M_J$ induces an algebra derivation $D : L^{\bullet}(M,\mathbb{C}) \otimes \mathcal{B}^{\bullet}(M,\mathbb{C}) \to L^{\bullet}(M,\mathbb{C}) \otimes \mathcal{B}^{\bullet+1}(M,\mathbb{C})$, so that the free module $\rho_*\Lambda^{\bullet,\bullet}(\overline{T}M_J) \cong L^{\bullet}(M,\mathbb{C}) \otimes \mathcal{B}^{\bullet}(M,\mathbb{C})$ becomes a differential graded module over the Weil algebra.

10.3.11. The algebra bundle $L^{\bullet}(M,\mathbb{C})$ is isomorphic to the de Rham algebra $\Lambda^{\bullet}(M,\mathbb{C})$. In particular, the bundle $L^1(M,\mathbb{C})$ is isomorphic to the trivial bundle $\mathcal{V}$ with fiber V over M. By 9.3.2 the relative de Rham differential

$$d^r : \Lambda^{\bullet}(\overline{T}M/M,\mathbb{C}) \to \Lambda^{\bullet+1}(\overline{T}M/M,\mathbb{C})$$

induces a derivation

$$d^r : L^{\bullet}(M,\mathbb{C}) \otimes \mathcal{B}^{\bullet}(M,\mathbb{C}) \to L^{\bullet+1}(M,\mathbb{C}) \otimes \mathcal{B}^{\bullet}(M,\mathbb{C}),$$

and we can choose the trivialization $L^1(M,\mathbb{C}) \cong \mathcal{V}$ in such a way that d^r identifies the generator subspace $V_3 \subset \mathcal{B}^0$ with the subspace of constant sections of $\mathcal{V} \cong L^1(M,\mathbb{C}) \subset L^1(M,\mathbb{C}) \otimes \mathcal{B}^0(M,\mathbb{C})$.

10.3.12. Denote by

$$\mathcal{L}^{\bullet,\bullet} = C^{\infty}_{\mathfrak{J}}(L^{\bullet}(M,\mathbb{C}) \otimes \mathcal{B}^{\bullet}_{tot}(M,\mathbb{C}))$$

the $\mathfrak{J}$-adic completion of the space of smooth sections of the algebra bundle $L^{\bullet}(M,\mathbb{C}) \otimes \mathcal{B}^{\bullet}_{tot}(M,\mathbb{C})$ on M. The space $\mathcal{L}^{\bullet,\bullet}$ is a bigraded algebra equipped with the derivations $d^r : \mathcal{L}^{\bullet,\bullet} \to \mathcal{L}^{\bullet+1,\bullet}$, $D^{tot} : \mathcal{L}^{\bullet,\bullet} \to \mathcal{L}^{\bullet,\bullet+1}$, which commute by Lemma 9.3.2. The algebra $\mathcal{L}^{\bullet,\bullet}$ is the free graded-commutative algebra generated by the subspaces $V_1, V_2, V_3 \subset \mathcal{L}^{0,\bullet} = \mathcal{B}^\bullet$ and the subspace $V = d^r(V_3) \subset \mathcal{L}^{1,0}$, which we denote by V_4.

10.3.13. As in 10.3.2, the given metric on the generator subspaces $V_1 = V_2 = V_3 = V_4 = V$ extends uniquely to a multiplicative metric on the algebra $\mathcal{L}^{\bullet,\bullet}$, which we call *the standard metric.* For every $k > 0$, introduce the total and augmentation gradings on the free $\mathcal{B}^\bullet$-module $\mathcal{L}^{k,\bullet} = \Lambda^k(V_4) \otimes \mathcal{B}^\bullet$ by setting $\deg \Lambda^k(V_4) = (0, 0)$. Let $D = D_{k,n}$ be the decomposition of the derivation $D : \mathcal{L}^{\bullet,\bullet} \to \mathcal{L}^{\bullet,\bullet+1}$ with respect to the total and the augmentation degrees. Denote by $\|D_{k,n}\|^p_q$ the norm with respect to the standard metric of the restriction of the derivation $D_{k,n} : \mathcal{L}^{p,0} \to \mathcal{L}^{p,1}$ to the component in $\mathcal{L}^{p,0}$ of total degree q.

10.3.14. Let now $\mathcal{L}^{\bullet,\bullet}_{tot} = \Lambda^\bullet(V_4) \otimes \mathcal{B}^\bullet_{tot}$ be the product of the exterior algebra $\Lambda^\bullet(V_4)$ with the total Weil algebra $\mathcal{B}^\bullet_{tot}$. We have the canonical identification $\mathcal{L}^{p,q}_{tot} = \mathcal{L}^{p,q} \otimes W^*_q$, and the canonical projection $P : \mathcal{L}^{\bullet,\bullet}_{tot} \to \mathcal{L}^{\bullet,\bullet}$, identical on $\mathcal{L}^{\bullet,0}_{tot} = \mathcal{L}^{\bullet,0}$. As in 10.3.4, the norm of the projection P on $\mathcal{L}^{\bullet,1}_{tot}$ is at most 2.

The derivation $D : \mathcal{L}^{\bullet,0} \to \mathcal{L}^{\bullet,1}$ induces a derivation $D^{tot} : \mathcal{L}^{\bullet,0}_{tot} \to \mathcal{L}^{\bullet,1}_{tot}$, related to D by $D = P \circ D^{tot}$. The gradings and the metric on $\mathcal{L}^{\bullet,\bullet}$ extend to $\mathcal{L}^{\bullet,\bullet}_{tot}$, in particular, we have the decomposition $D^{tot} = \sum_{k,n} D^{tot}_{k,n}$ with respect to the total and the augmentation degrees. Denote by $\|D^{tot}_{k,n}\|^p_q$ the norm with respect to the standard metric of the restriction of the derivation $D^{tot}_{k,n} : \mathcal{L}^{p,0}_{tot} \to \mathcal{L}^{p,1}_{C}$ to the component in $\mathcal{L}^{p,0}$ of total degree q. Since $\|P\| \leq 2$ on $\mathcal{L}^{\bullet,1}_{tot}$, we have

$$\|D^{tot}_{k,n}\|^p_q \leq 2 \cdot \|D_{k,n}\|^p_q. \tag{10.8}$$

10.3.15. The estimate on the norms $\|D^{tot}_{k,n}\|^{\mathcal{L}}_{p,q}$ that we will need is the following.

Lemma. *In the notation of Lemma* 10.3.8, *we have*

$$\|D^{tot}_{k,n}\|^p_q < 2(q + pk)(3K)^{k-1} C^n_0 b_{k,n},$$

for every p, q, k, n.

Proof. Since $D_{k,n}$ satisfies the Leibnitz rule, it suffices to prove the estimate for the restriction of the derivation $D^{tot}_{k,n}$ to the generator subspaces $V_2, V_3, V_4 \subset \mathcal{L}^{\bullet,0}$. By (10.8) it suffices to prove that on V_2, V_3, V_4 we have

$$\|D_{k,n}\|^p_q < (q + pk)(3K)^{k-1} C^n_0 b_{k,n}.$$

On the generator subspaces $V_2, V_3 \subset \mathcal{B}^0$ we have $p = 0$, and this equality is the claim of Lemma 10.3.8. Therefore it suffices to consider the restriction of the derivation $D_{k,n}$ to the subspace $V_4 = d^r(V_3) \subset \mathcal{L}^{1,0}$. We have $d^r \circ D_{k,n} = D_{k,n} \circ d^r : V_3 \to \mathcal{L}^{1,1}$. Moreover, $d^r = \mathrm{id} : V_3 \to V_4$ is an isometry. Since the operator $d^r : \mathcal{B}^{\bullet} \to \mathcal{L}^{1,\bullet}$ satisfies the Leibnitz rule and vanishes on the generators $V_1, V_2 \subset \mathcal{B}^{\bullet}$, the norm $\|d^r\|_k$ of its restriction to the subspace $\mathcal{B}^{\bullet}_k$ of augmentation degree k does not exceed k. Therefore

$$\begin{aligned}\|D_{k,n}\|_0^1 = \|D_{k,n}|_{V_4}\| &= \|D_{k,n} \circ d^r|_{V_3}\| = \|d^r \circ D_{k,n}|_{V_3}\| \\ &\leq \|D_{k,n}|_{V_3}| \cdot \left\|d^r_{\mathcal{B}^1_k}\right\| < k(3K)^{k-1}C_0^n b_{k,n},\end{aligned}$$

which proves the lemma. □

10.4. The proof of Theorem 10.1

10.4.1. We can now prove Theorem 10.1. Let $D : S^1(M, \mathbb{C}) \to \mathcal{B}^1(M, \mathbb{C})$ be a flat linear extended connection on M. Assume that its reduction $D_1 = \nabla : S^1(M, \mathbb{C}) \to S^1(M, \mathbb{C}) \otimes \Lambda^1(M, \mathbb{C})$ is a real-analytic connection on the bundle $S^1(M, \mathbb{C})$.

The operator $D_1 : S^1(M, \mathbb{C}) \to S^1(M, \mathbb{C}) \otimes \Lambda^1(M, \mathbb{C}) \subset \mathcal{B}^1(M, \mathbb{C})$ considered as an extended connection on M defines a Hodge connection $D_1 : \Lambda^0(\overline{T}M, \mathbb{C}) \to \rho^*\Lambda^1(M, \mathbb{C})$ on the pair $\langle \overline{T}M, M\rangle$, and this Hodge connection is also real-analytic. Assume further that the Taylor series at $0 \subset M \subset \overline{T}M$ for the Hodge connection D_1 converge in the closed ball of radius $\varepsilon > 0$.

10.4.2. Let $D = \sum_{n,k} D_{k,n} : \mathcal{B}^{\bullet} \to \mathcal{B}^{\bullet+1}$ be the derivation of the Weil algebra $\mathcal{B}^{\bullet}$ associated to the extended connection D. Applying Lemma 10.1.9 to the Hodge connection D_1 proves that there exists a constant $C > 0$ such that for every $n \geq 0$ the norm $\|D_{1,n}\|_{1,1}$ of the restriction of the derivation $D_{1,n}$ to the generator subspace $V_3 = \mathcal{B}^0_{1,1} \subset \mathcal{B}^0$ satisfies

$$\|D_{1,n}\|_{1,1} < CC_0^n,$$

where $C_0 = 1/\varepsilon$.

By definition the derivation $D_{1,n}$ vanishes on the generator subspace $V_1 \subset \mathcal{B}^1$. If $n > 0$, then it also vanishes on the generator subspace $V_2 = \mathcal{B}^0_{0,1} \subset \mathcal{B}^0$. If $n = 0$, then its restriction to $V_2 = \mathcal{B}^0_{0,1} \subset \mathcal{B}^0$ is the identity isomorphism $D_{1,0} = \mathrm{id} : V_2 \to V_1$. In any case, we have $\|D_{1,n}\|_{0,1} \leq 1$. Increasing if necessary the constant C_0, we can assume that for any n and for $p = 0, 1$ we have

$$\|D_{1,n}\|_{p,1} < C_0^n.$$

10.4.3. We can now apply our main estimate, Proposition 10.1. It shows that there exists a constant $C_1 > 0$ such that for every k, n we have

$$\|D_{k,n}\|_{1,1} < C_1^n.$$

Together with Lemma 10.1.9 this estimate implies that the formal Hodge connection D on $\overline{T}M$ along $M \subset \overline{T}M$ corresponding to the extended connection D converges to a real-analytic Hodge connection on an open neighborhood $U \subset \overline{T}M$ of the zero section $M \subset \overline{T}M$. This in turn implies the first claim of Theorem 10.1.

10.4.4. To prove the second claim of Theorem 10.1, assume that the manifold M is equipped with a Kähler form ω compatible with the Kählerian connection ∇, so that $\nabla\omega = 0$. The differential operator $\nabla : \Lambda^{1,1}(M) \to \Lambda^{1,1}(M) \otimes \Lambda^1(M, \mathbb{C})$ is elliptic and real-analytic. Since $\nabla\omega = 0$, the Kähler form ω is also real-analytic.

For every $p, q \geq 0$ we have introduced in 10.3.12 the space $\mathcal{L}^{p,q}$, which coincides with the space of formal germs at $0 \in M \subset \overline{T}M$ of smooth forms on $\overline{T}M$ of type (p, q) with respect to the complementary complex structure $\overline{T}M_J$. The spaces $\mathcal{L}^{\bullet,\bullet}$ carry the total and the augmentation gradings. Consider ω as an element of the vector space $\mathcal{L}_0^{2,0}$, and let $\omega = \sum_n \omega_n$ be the total degree decomposition. The decomposition $\omega = \sum_n \omega_n$ is the Taylor series decomposition for the form ω at $0 \in M$. Since the form ω is real-analytic, there exists a constant C_2 such that

$$\|\omega_n\| < C_2^n \tag{10.9}$$

for every n.

10.4.5. Let $\Omega = \sum_k \Omega_k = \sum_{k,n} \Omega_{k,n} \subset \mathcal{L}^{2,0}$ be the formal polarization of the Hodge manifold $\overline{T}M$ at $M \subset \overline{T}M$ corresponding to the Kähler form ω by Theorem 9.1. By definition we have $\omega = \Omega_0$. Moreover, by construction used in the proof of Proposition 9.1 we have

$$\Omega_k = -\frac{1}{k} \sum_{1 \leq p \leq k-1} \sigma_{tot}(D_{k-p}^{tot}\Omega_p), \tag{10.10}$$

where $D^{tot} : \mathcal{L}^{2,0} \to \mathcal{L}^{2,1}$ is the derivation associated to the extended connection D on M and $\sigma_{tot} : \mathcal{L}^{2,\bullet+1} \to \mathcal{L}^{2,\bullet}$ is the extension to $\mathcal{L}^{2,\bullet} = L_0^2 \otimes \mathcal{B}_{tot}^{\bullet}$ of the canonical endomorphism of the total Weil algebra $\mathcal{B}_{tot}^{\bullet}$ constructed in the proof of Proposition 9.1.

10.4.6. The map $\sigma_{tot} : \mathcal{L}^{2,1} \to \mathcal{L}^{2,0}$ is a map of $\mathcal{B}_{tot}^0$-modules. Therefore, as in 10.3.7, there exists a constant $K_1 > 0$ such that

$$\|\sigma_{tot}\| < K_1 \tag{10.11}$$

on $\mathcal{L}^{2,1}$. We can assume that $K_1 > 3K$, where K is as in (10.7). Together with the recursive formula (10.10), this estimate implies the following estimate on the norms $\|\Omega_{k,n}\|$ of the components $\Omega_{k,n}$ of the formal polarization Ω taken with respect to the standard metric.

Lemma. *For every k, n we have*

$$\|\Omega_{k,n}\| < (2K_1)^{k-1}C^n b_{k,n}^2,$$

where $C = \max(C_0, C_2)$ is the bigger of the constants C_0, C_2, and $b_{k,n}^2$ are the numbers defined recursively in 10.2.5.

Proof. Use induction on k. The base of the induction is the case $k = 1$, where the estimate holds by (10.9). Assume the estimate proved for all $\Omega_{p,n}$ with $p < k$, and fix a number n. By (10.10) we have

$$\Omega_{k,n} = -\frac{1}{k}\sum_{1\leq p\leq k-1}\sum_{q}\sigma_{tot}(D_{k-p,n-q}^{tot}\Omega_{p,q}).$$

Substituting the estimate (10.11) together with the inductive assumption and the estimate on $\|D_{k-p,n-q}^{tot}\|_q^2$ obtained in Lemma 10.3.15, we get

$$\begin{aligned}
\|\Omega_{k,n}\| &< \frac{1}{k}\sum_{1\leq p\leq k-1}\sum_{q}\|\sigma_{tot}\|\cdot\|D_{k-p,n-q}^{tot}\|_q^2\cdot\|\Omega_{p,q}\| \\
&< \frac{1}{k}\sum_{1\leq p\leq k-1}\sum_{q}K_1\cdot 2(q+2(k-p))(3K)^{k-p-1}C_0^{n-q}b_{k-p,n-q} \\
&\qquad\cdot(2K_1)^{p-1}C^q b_{p,q} \\
&< \frac{1}{k}(2K_1)^{k-1}C^n\sum_{1\leq p\leq k-1}\sum_{q}(q+2(k-p))b_{k-p,n-q}b_{p,q}^2 \\
&= (2K_1)^{k-1}C^n b_{k,n}^2,
\end{aligned}$$

which proves the lemma. □

10.4.7. This estimate immediately implies the last claim of Theorem 10.1. Indeed, together with Lemma 10.2.7 it implies that for every k, n

$$\|\Omega_{k,n}\| < (C_3)^n$$

for some constant $C_3 > 0$. But $\Omega = \sum_n \sum_{0\leq k\leq n}\Omega_{k,n}$ is the Taylor series decomposition for the formal polarization Ω at $0 \subset M$. The standard convergence criterion shows that this series converges in an open ball of radius $1/C_3 > 0$. Therefore the polarization Ω is indeed real-analytic in a neighborhood of $0 \subset M \subset \overline{T}M$. □

Appendix

A.1.1. In this appendix we describe a well-known Borel-Weyl type localization construction for quaternionic vector spaces (see, e.g. [HKLR]) which provides a different and somewhat more geometric approach to many facts in the theory of Hodge manifolds. In particular, we establish, following Deligne and Simpson ([D2], [S1]), a relation between Hodge manifolds and the theory of mixed $\mathbb{R}$-Hodge structures. For the sake of simplicity, we consider only Hodge manifold structures on the formal neighborhood of $0 \in \mathbb{R}^{4n}$ instead of actual Hodge manifolds, as in Section 10. To save the space all proofs are either omitted or only sketched.

A.1.2. Let SB be the Severi-Brauer variety associated to the algebra $\mathbb{H}$, that is, the real algebraic variety of minimal right ideals in $\mathbb{H}$. The variety SB is a twisted $\mathbb{R}$-form of the complex projective line $\mathbb{C}P^1$.

For every algebra map $I : \mathbb{C} \to \mathbb{H}$ let the algebra $\mathbb{H} \otimes_{\mathbb{R}} \mathbb{C}$ act on the 2-dimensional complex vector space $\mathbb{H}_I$ by left multiplication, and let $\widehat{I} \subset \mathbb{H} \otimes_{\mathbb{R}} \mathbb{C}$ be the annihilator of the subspace $I(\mathbb{C}) \subset \mathbb{H}_I$ with respect to this action. The subspace $\widehat{I}$ is a minimal right ideal in $\mathbb{H} \otimes_{\mathbb{R}} \mathbb{C}$. Therefore it defines a $\mathbb{C}$-valued point $\widehat{I} \subset \mathrm{SB}(\mathbb{C})$ of the real algebraic variety SB. This establishes a bijection between the set $\mathrm{SB}(\mathbb{C})$ and the set of algebra maps from $\mathbb{C}$ to $\mathbb{H}$.

A.1.3. Let Shv(SB) be the category of flat coherent sheaves on SB. Say that a sheaf $\mathcal{E} \in \mathrm{Ob}\,\mathrm{Shv(SB)}$ is *of weight p* if the sheaf $\mathcal{E} \otimes \mathbb{C}$ on $\mathbb{C}P^1 = \mathrm{SB} \otimes \mathbb{C}$ is a sum of several copies of the sheaf $\mathcal{O}(p)$.

Consider a quaternionic vector space V. Let $\mathcal{I} \in \mathbb{H} \otimes \mathcal{O}_{\mathrm{SB}}$ be the tautological minimal left ideal in the algebra sheaf $\mathbb{H} \otimes \mathcal{O}_{\mathrm{SB}}$, and let $V_{loc} \in \mathrm{Ob}\,\mathrm{Shv(SB)}$ be the sheaf defined by

$$V_{loc} = V \otimes \mathcal{O}_{\mathrm{SB}} / \mathcal{I} \cdot V \otimes \mathcal{O}_{\mathrm{SB}}.$$

The correspondence $V \mapsto V_{loc}$ defines a functor from quaternionic vector spaces to Shv(SB). It is easy to check that this functor is a full embedding, and its essential image is the subcategory of sheaves of weight 1. Call V_{loc} *the localization* of the quaternionic vector space V. For every algebra map $\mathcal{I} : \mathbb{C} \to \mathbb{H}$ the fiber $V_{loc}|_{\widehat{I}}$ of the localization V_{loc} over the point $\widehat{I} \subset \mathrm{SB}(\mathbb{C})$ corresponding to the map $i : \mathbb{C} \to \mathbb{H}$ is canonically isomorphic to the real vector space V with the complex structure V_I.

A.1.4. The compact Lie group $U(1)$ carries a canonical structure of a real algebraic group. Fix an algebra embedding $I : \mathbb{C} \to \mathbb{H}$ and let the group $U(1)$ act on the algebra $\mathbb{H}$ as in 1.1.2. This action is algebraic and induces therefore an algebraic action of the group $U(1)$ on the Severi-Brauer variety

SB. The point $\widehat{I} : \operatorname{Spec} \mathbb{C} \subset \mathrm{SB}$ is preserved by the $U(1)$-action. The action of the group $U(1)$ on the complement $\mathrm{SB} \setminus \widehat{I}(\operatorname{Spec} \mathbb{C}) \subset \mathrm{SB}$ is free, so that the variety SB consists of two $U(1)$-orbits. The corresponding orbits of the complexified group $\mathbb{C}^* = U(1) \times \operatorname{Spec} \mathbb{C}$ on the complexification $\mathrm{SB} \times \operatorname{Spec} \mathbb{C} \cong \mathbb{C}P$ are the pair of points $0, \infty \subset \mathbb{C}P$ and the open complement $\mathbb{C}P \setminus \{0, \infty\} \cong \mathbb{C}^* \subset \mathbb{C}P$.

Let $\operatorname{Shv}^{U(1)}(\mathrm{SB})$ be the category of $U(1)$-equivariant flat coherent sheaves on the variety SB. The localization construction immediately extends to give the equivalence $V \mapsto V_{loc}$ between the category of equivariant quaternionic vector spaces and the full subcategory in $\operatorname{Shv}^{U(1)}(\mathrm{SB})$ consisting of sheaves of weight 1. For an equivariant quaternionic vector space V, the fibers of the sheaf V_{loc} over the point $\widehat{I} \subset \mathrm{SB}(\mathbb{C})$ and over the complement $\mathrm{SB} \setminus \widehat{I}(\operatorname{Spec} \mathbb{C})$ are isomorphic to the space V equipped, respectively, with the preferred and the complementary complex structures V_I and V_J.

A.1.5. The category of $U(1)$-equivariant flat coherent sheaves on the variety SB admits the following beautiful description, due to Deligne.

Lemma ([D2],[S1]). (i) *For every integer n the full subcategory*

$$\operatorname{Shv}^{U(1)}_n(\mathrm{SB}) \subset \operatorname{Shv}^{U(1)}(\mathrm{SB})$$

of sheaves of weight n is equivalent to the category of pure $\mathbb{R}$-Hodge structures of weight n.

(ii) *The category of pairs $\langle \mathcal{E}, W_\bullet \rangle$ of a flat $U(1)$-equivariant sheaf*

$$\mathcal{E} \in \operatorname{Ob} \operatorname{Shv}^{U(1)}(\mathrm{SB})$$

and an increasing filtration $W_\bullet$ on $\mathcal{E}$ such that for every integer n

$$W_n\mathcal{E}/W_{n-1}\mathcal{E} \quad \text{is a sheaf of weight} \quad n \quad \text{on} \quad \mathrm{SB} \tag{A.1}$$

is equivalent to the category of mixed $\mathbb{R}$-Hodge structures. (In particular, it is abelian.)

A.1.6. For every pure $\mathbb{R}$-Hodge structure V call the corresponding $U(1)$-equivariant flat coherent sheaf on the variety SB *the localization* of V and denote it by V_{loc}. For the trivial $\mathbb{R}$-Hodge structure $\mathbb{R}(0)$ of weight 0 the sheaf $\mathbb{R}(0)_{loc}$ coincides with the structure sheaf $\mathcal{O}$ on SB. If V, W are two pure $\mathbb{R}$-Hodge structures, then the space $\operatorname{Hom}(V_{loc}, W_{loc})$ of $U(1)$-equivariant maps between the corresponding sheaves coincides with the space of weakly Hodge maps from V to W in the sense of Subsection 1.4.

For every pure $\mathbb{R}$-Hodge structure V the space $\Gamma(\mathrm{SB}, V_{loc})$ of the global sections of the sheaf V_{loc} is equipped with an action of the group $U(1)$ and carries therefore a canonical $\mathbb{R}$-Hodge structure of weight 0. This $\mathbb{R}$-Hodge structure is the same as the universal $\mathbb{R}$-Hodge structure $\Gamma(V)$ of weight 0 constructed in Lemma 1.4.6. This explains our notation for the functor $\Gamma : \mathcal{WH}odge_{\geq 0} \to \mathcal{WH}odge_0$.

A.1.7. Assume given a complex vector space V and let M be the formal neighborhood of $0 \in V$. Let $\mathcal{B}^{\bullet}$ be the Weil algebra of the manifold M, as in 10.1.2. For every $n \geq 0$ the vector space $\mathcal{B}^n$ is equipped with an $\mathbb{R}$-Hodge structure of weight n, so that we can consider the localization $\mathcal{B}^n_{loc}$. The sheaf $\oplus\mathcal{B}^{\bullet}_{loc}$ is a commutative algebra in the tensor category $\mathrm{Shv}^{U(1)}(\mathrm{SB})$. We will call *the localized Weil algebra.*

The augmentation grading on $\mathcal{B}^{\bullet}$ defined in 7.2.2 is compatible with the $\mathbb{R}$-Hodge structures. Therefore it defines an augmentation grading on the localized Weil algebra $\mathcal{B}^{\bullet}_{loc}$. The finer augmentation bigrading on $\mathcal{B}^{\bullet}$ does not define a bigrading on $\mathcal{B}^{\bullet}_{loc}$. However, it does define a bigrading on the complexified algebra $\mathcal{B}^{\bullet}_{loc} \otimes \mathbb{C}$ of $\mathbb{C}^*$-equivariant sheaves on the manifold $\mathrm{SB} \otimes \mathbb{C} \cong \mathbb{C}P$.

A.1.8. Assume now given a flat extended connection $D : \mathcal{B}^{\bullet} \to \mathcal{B}^{\bullet+1}$ on M. Since the derivation D is weakly Hodge, it corresponds to a derivation $D : \mathcal{B}^{\bullet}_{loc} \to \mathcal{B}^{\bullet+1}_{loc}$ of the localized Weil algebra $\mathcal{B}^{\bullet}_{loc}$. It is easy to check that the complex $\langle\mathcal{B}^{\bullet}_{loc}, D\rangle$ is acyclic in all degrees but 0. Denote by $\mathcal{H}^0$ the 0-th cohomology sheaf $H^0(\mathcal{B}^{\bullet}_{loc})$. The sheaf $\mathcal{H}^0$ carries a canonical algebra structure. Moreover, while the derivation D does not preserve the augmentation grading on $\mathcal{B}^{\bullet}_{loc}$, it preserves the decreasing *augmentation filtration* $(\mathcal{B}^{\bullet}_{loc})_{\geq \bullet}$. Therefore we have a canonical decreasing filtration on the algebra $\mathcal{H}^0$, which we also call the augmentation filtration.

A.1.9. It turns out that the associated graded quotient algebra $\operatorname{gr}\mathcal{H}^0$ with respect to the augmentation filtration does not depend on the extended connection D. To describe it, introduce the $\mathbb{R}$-Hodge structure W of weight -1 by setting $W = V$ as a real vector space and

$$\begin{aligned} W^{-1,0} &= V \subset V \otimes_{\mathbb{R}} \mathbb{C}, \\ W^{0,-1} &= \overline{V} \subset V \otimes_{\mathbb{R}} \mathbb{C}. \end{aligned} \tag{A.2}$$

The k-th graded piece $\operatorname{gr}_k \mathcal{H}^0$ with respect to the augmentation filtration is then isomorphic to the symmetric power $S^k(W_{loc})$ of the localization W_{loc} of the $\mathbb{R}$-Hodge structure W. In particular, it is a sheaf of weight $-n$, so that up to a change of numbering the augmentation filtration on $\mathcal{H}^0$ satisfies the condition (A.1). The extension data between these graded pieces depend on the extended connection D. The whole associated graded algebra $\operatorname{gr}\mathcal{H}^0$ is isomorphic to the completed symmetric algebra $\widehat{S}^{\bullet}(W_{loc})$.

A.1.10. Using standard deformation theory, one can show that the algebra map $\mathcal{H}^0 \to \mathcal{B}^0_{loc}$ is the universal map from the algebra $\mathcal{H}^0$ to a complete commutative pro-algebra in the tensor category of $U(1)$-equivariant flat coherent sheaves of weight 0 on SB. Moreover, the localized Weil algebra $\mathcal{B}^{\bullet}_{loc}$ coincides with the relative de Rham complex of $\mathcal{B}^0_{loc}$ over $\mathcal{H}^0$. Therefore one can recover, up to an isomorphism, the whole algebra $\mathcal{B}^{\bullet}_{loc}$ and, consequently, the extended connection D, solely from the algebra $\mathcal{H}^0$ in $\operatorname{Shv}^{U(1)}(\mathrm{SB})$. Together with Lemma A.1.5 (ii) this gives the following, due also to Deligne (in a different form).

Proposition A.1 *The correspondence* $\langle \mathcal{B}^{\bullet}, D\rangle \mapsto \mathcal{H}^0$ *is a bijection between the set of all isomorphism classes of flat extended connections on* M *and the set of all algebras* $\mathcal{H}^0$ *in the tensor category of mixed* $\mathbb{R}$*-Hodge structures equipped with an isomorphism* $\operatorname{gr}^W_{-1}\mathcal{H}^0 \cong W$ *between the* -1*-th associated graded piece of the weight filtration on* $\mathcal{H}^0$ *and the pure* $\mathbb{R}$*-Hodge structure* W *defined in* (A.2) *which induces for every* $n \geq 0$ *an isomorphism* $\operatorname{gr}^W_{-n}\mathcal{H}^0 \cong S^n W$.

Remark. The scheme $\operatorname{Spec}\mathcal{H}^0$ over SB coincides with the so-called twistor space of the manifold $\overline{T}M$ with the hypercomplex structure given by the extended connection D (see [HKLR] for the definition). Deligne's and Simpson's ([D2], [S1]) approach differs from ours in that they use the language of twistor spaces to describe the relation between $U(1)$-equivariant hypercomplex manifolds and mixed $\mathbb{R}$-Hodge structures. Since this requires some additional machinery, we have avoided introducing twistor spaces in this paper.

A.1.11. We will now try to use the localization construction to eludicate some of the complicated linear algebra used in Section 8 to prove our main theorem. As we have already noted, the category $\mathcal{WH}odge$ of pure $\mathbb{R}$-Hodge structures with weakly Hodge maps as morphisms is identified by localization with the category $\operatorname{Shv}^{U(1)}(\mathrm{SB})$ of $U(1)$-equivariant flat coherent sheaves on SB. Moreover, the functor $\Gamma : \mathcal{WH}odge_{\geq 0} \to \mathcal{WH}odge_0$ introduced in Lemma 1.4.6 is simply the functor of global sections $\Gamma(\mathrm{SB}, \cdot)$.

A.1.12. Consider the localized Weil algebra $\mathcal{B}^{\bullet}_{loc}$ with the derivation $C : \mathcal{B}^{\bullet}_{loc} \to \mathcal{B}^{\bullet+1}_{loc}$ associated to the canonical weakly Hodge derivation introduced in 7.3.2. The differential graded algebra $\langle \mathcal{B}^{\bullet}_{loc}, C\rangle$ is canonically an algebra over the completed symmetric algebra $\widehat{S}^{\bullet}(V)$ generated by the constant sheaf on SB with the fiber V. Moreover, it is a free commutative algebra generated by the complex

$$V \longrightarrow V(1) \tag{A.3}$$

placed in degrees 0 and 1, where $V(1)$ is the $U(1)$-equivariant sheaf of weight 1 on SB corresponding to the $\mathbb{R}$-Hodge structure given by $V(1)^{1,0} = V$ and $V(1)^{0,1} = \overline{V}$.

A.1.13. The homology sheaves of the complex (A.3) are non-trivial only in degree 1. This non-trivial homology sheaf is a skyscraper sheaf concentrated in the point $\widehat{I}(\operatorname{Spec}\mathbb{C}) \subset \mathrm{SB}$ with fiber V. The associated sheaf on the complexification $\mathrm{SB} \otimes \mathbb{C} \cong \mathbb{C}P$ splits into the sum of skyscraper sheaf with fiber V concentrated at $0 \in \mathbb{C}P$ and the skyscraper sheaf with fiber $\overline{V}$ concentrated at $\infty \in \mathbb{C}P$. This splitting corresponds to the splitting of the complex (A.3) itself into the components of augmentation bidegrees $(1,0)$ and $(0,1)$.

A.1.14. Let now $\mathcal{I}^{\bullet} = \mathcal{B}^{\bullet}_{\geq 0, \geq 0}$ be the sum of the components in the Weil algebra $\mathcal{B}^{\bullet}$ of augmentation bidegree greater or equal than $(1,1)$. The subspace $\mathcal{I}^{\bullet} \subset \mathcal{B}^{\bullet}$ is compatible with the $\mathbb{R}$-Hodge structure. The crucial point in the proof of Theorem 8.1 is Proposition 7.1, which claim the acyclycity of the complex $\langle \Gamma(\mathcal{I}^{\bullet}), C\rangle$. It is this fact that becomes almost obvious from the point of view of the localization construction. To show it, we first prove the following.

Lemma. *The complex*

$$\langle \mathcal{I}^{\bullet}_{loc}, C\rangle$$

of $U(1)$-equivariant sheaves on SB *is acyclic.*

Proof. It suffices to prove that the complex $I^{\bullet}_{loc} \otimes \mathbb{C}$ of sheaves on $\mathrm{SB} \otimes \mathbb{C} \cong \mathbb{C}P$ is acyclic. To prove it, let $p, q \geq 1$ be arbitrary integer, and consider the component $(B^{\bullet}_{loc})_{p,q}$ of augmentation bidegree (p,q) in the localized Weil algebra $\mathcal{B}^{\bullet}_{loc}$. By definition we have

$$(B^{\bullet}_{loc})_{p,q} = (B^{\bullet}_{loc})_{p,0} \otimes (B^{\bullet}_{loc})_{0,q}$$

Since the complex $(B^{\bullet}_{loc})_{p,0} = S^p\,(B^{\bullet}_{loc})_{1,0}$ has homology concentrated at $0 \in \mathbb{C}P$, while the complex $(B^{\bullet}_{loc})_{0,q} = S^q\,(B^{\bullet}_{loc})_{0,1}$ has homology concentrated at $\infty \in \mathbb{C}P$, their product is indeed acyclic. □

Now, we have $\Gamma(\mathcal{I}^{\bullet}) = \Gamma(\mathrm{SB}, \mathcal{I}^{\bullet}_{loc})$, and the functor $\Gamma(\mathrm{SB}, \cdot)$ is exact on the full subcategory in $\mathrm{Shv}^{U(1)}(\mathrm{SB})$ consisting of sheaves of positive weight .

Therefore the complex $\langle \Gamma(\mathcal{I}^\bullet), C\rangle$ is also acyclic, which gives an alternative proof of Proposition 7.1.

A.1.15. We would like to finish the paper with the following observation. Proposition A.1 can be extended to the following claim.

Proposition A.2 *Let M be a complex manifold. There exists a natural bijection between the set of isomorphism classes of germs of Hodge manifold structures on $\overline{T}M$ in the neighborhood of the zero section $M \subset \overline{T}M$ and the set of multiplicative filtrations $F^\bullet$ on the sheaf $\mathcal{O}_{\mathbb{R}}(M) \otimes \mathbb{C}$ of $\mathbb{C}$-valued real-analytic functions on M satisfying the following condition:*

- *For every point $m \in M$ let $\widehat{\mathcal{O}}_m$ be the formal completion of the local ring $\mathcal{O}_m$ of germs of real-analytic functions on M in a neighborhood of m with respect to the maximal ideal. Consider the filtration $F^\bullet$ on $\widehat{\mathcal{O}}_m \otimes \mathbb{C}$ induced by the filtration $f^\bullet$ on the sheaf $\mathcal{O}_{\mathbb{R}}(M) \otimes \mathbb{C}$, and for every $k \geq 0$ let $W_{-k}\widehat{\mathcal{O}}_m \subset \widehat{\mathcal{O}}_m$ be the k-th power of the maximal ideal in $\widehat{\mathcal{O}}_m$. Then the triple $\langle \widehat{\mathcal{O}}_m, F^\bullet, W^\bullet\rangle$ is a mixed $\mathbb{R}$-Hodge structure. (In particular, $F^k = 0$ for $k > 0$.)*

If the Hodge manifold structure on $\overline{T}M$ is such that the projection $\rho : \overline{T}M_I \to M$ is holomorphic for the preferred complex structure $\overline{T}M_I$ on $\overline{T}M$, then it is easy to see that the first non-trivial piece $F^0\mathcal{O}_{\mathbb{R}}(M) \otimes \mathbb{C}$ of the filtration $F^\bullet$ on the sheaf $\mathcal{O}_{\mathbb{R}}(M) \otimes \mathbb{C}$ coincides with the subsheaf $\mathcal{O}(M) \subset \mathcal{O}_{\mathbb{R}}(M) \otimes \mathbb{C}$ of holomorphic functions on M. Moreover, since the filtration $F^\bullet$ is multiplicative, it is completely defined by the subsheaf $F^{-1}\mathcal{O}_{\mathbb{R}}(M) \otimes \mathbb{C} \subset \mathcal{O}_{\mathbb{R}}(M) \otimes \mathbb{C}$. It would be very interesting to find an explicit description of this subsheaf in terms of the Kählerian connection ∇ on M which corresponds to the Hodge manifold structure on $\overline{T}M$.

Bibliography

[B] A. Besse, *Einstein manifolds*, Springer-Verlag, 1987.

[Beau] A. Beauville, *Variétés Kähleriennes dont la première classe de Chern est nulle*, J. Diff. Geom. **19** (1983), 755–782.

[BG] O. Biquard and P. Gauduchon, *Hyperkähler metrics on cotangent bundles of Hermitian symmetric spaces*, in "Proc. of the Special Session an Geometry and Physics held at Aarhus University and of the Summer School at Odense University, Denmark, 1995", Lecture Notes in Pure and Applied Mathematics, volume 184, Marcel Dekker Inc. 1996, 768 pages.

[B] Charles P. Boyer, *A note on hyper-Hermitian four-manifolds*, Proc. Amer. Math. Soc. **102** (1988), 157–164.

[C] E. Calabi, *Métriques kähleriennes et fibrés holomorphes*, Ann. Ecol. Norm. Sup. **12** (1979), 269–294.

[DS] A. Dancer and A. Swann, *The structure of quaternionic kähler quotients*, in "Proc. of the Special Session an Geometry and Physics held at Aarhus University and of the Summer School at Odense University, Denmark, 1995", Lecture Notes in Pure and Applied Mathematics, volume 184, Marcel Dekker Inc. 1996, 768 pages.

[D1] P. Deligne, *Cohomologie a support propre et construction du foncteur $f^!$*, Appendix in: Robin Hartshorne, Residues and Duality, Lecture Notes in Math. vol. 20, Springer - Verlag, 1966.

[D2] P. Deligne, a letter to Simpson, quoted in [S1].

[H1] N.J. Hitchin, *The self-duality equations on a Riemann surface*, Proc. London Math. Soc. **55** (1987), 59–126.

[H2] N.J. Hitchin, *Metrics on moduli spaces*, Proc. Lefschetz Centennial Conference (Mexico City 1984), Contemp. Math. **58**, Part I, AMS, Providence, RI, 1986.

[HKLR] N.J. Hitchin, A. Karlhede, U. Lindström, M. Roček, *Hyperkähler metrics and supersymmetry*, Comm. Math. Phys (1987).

[K] D. Kaledin, *Integrability of the twistor space for a hypercomplex manifold*, alg-geom preprint 9612016.

[Kr1] P.B. Kronheimer, *A hyper-Kählerian structure on coadjoint orbits of a semisimple complex group*, J. of LMS, **42** (1990), 193–208.

[Kr2] P.B. Kronheimer, *Instantons and the nilpotent variety*, J. Diff. Geom. **32** (1990), 473–490.

[N] H. Nakajima, *Instantons on ALE spaces, quiver varieties, and Kac-Moody algebras*, Duke Math. J. **76** (1994), 365–416.

[S1] C. Simpson, *Mixed twistor structures*, alg-geom preprint, 1997.

[S2] C. Simpson, *Moduli of representations of the fundamental group of a smooth projective variety*, I: Publ. Math. I.H.E.S. **79** (1994), 47–129; II: Publ. Math. I.H.E.S. **80** (1994), 5–79.

[S3] C. Simpson, *Higgs bundles and local systems*, Publ. Math. I.H.E.S. **75** (1992), 5–95.

INDEPENDENT UNIVERSITY OF MOSCOW
B. VLASSIEVSKY, 11
MOSCOW, RUSSIA, 121002
E-mail: KALEDIN@BALTHI.DNTTM.RSSI.RU

Index of terms and notation

Authors index

Made in the USA
Las Vegas, NV
31 January 2025

17310225R00144